CONTROL AND DYNAMIC SYSTEMS

Advances in Theory and Applications

Volume 38

CONTRIBUTORS TO THIS VOLUME

J. A. DE ABREU-GARCIA
ITZHACK Y. BAR-ITZHACK
KENNETH R. BOFF
WILLIAM J. CODY
DOUGLAS G. DEWOLF
PAUL R. FREY
T. T. HARTLEY
GEORGE MEYER
WILLIAM B. ROUSE
JOSEF SHINAR
P. F. SINGER
G. ALLAN SMITH
D. D. SWORDER
HENDRIKUS G. VISSER
STEPHEN A. WHITMORE

CONTROL AND DYNAMIC SYSTEMS

ADVANCES IN THEORY AND APPLICATIONS

Edited by

C. T. LEONDES

School of Engineering and Applied Science
University of California, Los Angeles
Los Angeles, California
and
College of Engineering
University of Washington
Seattle, Washington

VOLUME 38: ADVANCES IN AERONAUTICAL SYSTEMS

ACADEMIC PRESS, INC.
Harcourt Brace Jovanovich, Publishers
San Diego New York Boston
London Sydney Tokyo Toronto

This book is printed on acid-free paper. ∞

Academic Press, Inc.
San Diego, California 92101

United Kingdom Edition published by
Academic Press Limited
24-28 Oval Road, London NW1 7DX

Library of Congress Catalog Card Number: 64-8027

ISBN 0-12-012738-5 (alk. paper)

Printed in the United States of America
90 91 92 93 9 8 7 6 5 4 3 2 1

CONTENTS

CONTRIBUTORS vii
PREFACE ix

Aircraft Automatic Flight Control System With Model Inversion 1

G. Allan Smith and George Meyer

Information Systems for Supporting Design of Complex Human-Machine Systems 41

William B. Rouse, William J. Cody, Kenneth R. Boff, and Paul R. Frey

Formulation of a Minimum Variance Deconvolution Technique for Compensation of Pneumatic Distortion in Pressure Sensing Devices 101

Stephen A. Whitmore

Synthesis and Validation of Feedback Guidance Laws for Air-To-Air Interceptions 153

Josef Shinar and Hendrikus G. Visser

Multistep Matrix Integrators for Real-Time Simulation 211

J. A. De Abreu-Garcia and T. T. Hartley

The Role of Image Interpretation in Tracking and Guidance 273

D. D. Sworder and P. F. Singer

Continuous Time Parameter Estimation: Analysis via a Limiting Ordinary Differential Equation 307

Douglas G. DeWolf

In-Flight Alignment of Inertial Navigation Systems 369

Itzhack Y. Bar-Itzhack

INDEX 397

CONTRIBUTORS

Numbers in parentheses indicate the pages on which the authors' contributions begin.

J. A. De Abreu-Garcia (211), *Department of Electrical Engineering, The University of Akron, Akron, Ohio 44325*

Itzhack Y. Bar-Itzhack (369), *Department of Aerospace Engineering, Technion-Israel Institute of Technology, Haifa, 32000 Israel*

Kenneth R. Boff (41), *United States Air Force, Armstrong Aerospace Medical Research Laboratory, Wright Patterson Air Force Base, Ohio 43455*

William J. Cody (41), *Search Technology, Inc., Norcross, Georgia 30092*

Douglas G. DeWolf (307), *Hughes Aircraft Company, Los Angeles, California 90009*

Paul R. Frey (41), *Search Technology, Inc., Norcross, Georgia 30092*

T. T. Hartley (211), *Department of Electrical Engineering, The University of Akron, Akron, Ohio 44325*

George Meyer (1), *NASA Ames Research Center, Moffett Field, California 94035*

William B. Rouse (41), *Search Technology, Inc., Norcross, Georgia 30092*

Josef Shinar (153), *Faculty of Aerospace Engineering, Technion-Israel Institute of Technology, Haifa, 32000 Israel*

P. F. Singer (273), *Hughes Aircraft Company, El Segundo, California 90245*

G. Allan Smith (1), *Department of Electrical Engineering, Yale University, New Haven, Connecticut 06520*

D. D. Sworder (273), *Department of AMES, University of California, San Diego, La Jolla, California 92093*

Hendrikus G. Visser (153), *Faculty of Aerospace Engineering, Delft University of Technology, Delft, The Netherlands*

Stephen A. Whitmore (101), *NASA Ames Research Center, Moffett Field, California 94035*

PREFACE

In the 1940s military and civil aircraft were relatively simple and uncomplicated, as were their avionics systems. The intervening period and the last decade in particular have witnessed a tremendous growth in the capability and performance of aircraft and their avionics systems. In military aircraft the trend is toward future aircraft designs which will include significant nonlinear features in their aerodynamic and propulsion characteristics; in addition, they may be required to operate over extreme flight envelopes with implications for advanced automatic or semiautomatic flight control systems. Parallel advances are occurring in military aircraft avionics systems. Such aircraft as the ATF (Advanced Tactical Fighter) will have the computing power of three Cray supercomputers on board, and very necessarily so. Military aircraft sensor systems are becoming increasingly more powerful and their sensor data is being "fused." Current advanced trends include "smart aircraft skins," wherein the sensor systems might actually become an integral part of the aircraft skin. Other advanced trends of very impressive proportions are also occurring. Separately, in commercial aircraft, triply redundant inertial guidance systems have now been standard for almost two decades. Integrated GPS (Global Positioning Satellite) and AHARS (Attitude and Heading Reference Systems) seem to be an eventual certainty. The "glass cockpit," wherein display devices are multifunctional, is now a very significant reality. The MLS (Microwave Landing System), a very high precision capability landing system which is capable of category 3, zero-zero (visibility) landings and which has been aborning for over two decades, starting with the deliberations of RTCA/SC-117 (Radio Technical Commission for Aeronautics/Special Committee —117), seems to be headed for eventual, albeit delayed, introduction into commercial aviation. All these trends and many more make it most appropriate to devote a theme volume to "Advances in Aeronautical Systems Dynamics and Control," the theme for this volume in this Academic Press series.

The first contribution, "Aircraft Automatic Flight Control System with Model

Inversion," by G. Allan Smith and George Meyer, lays a foundation for a rather powerfully capable approach to aircraft trajectory control by the technique of inverse-model follower control systems. The basic idea behind the inverse-model follower is to embed an inverse model of the aircraft force and moment generating processes into the control system. The series combination of this inverse model provides the controls—aircraft control surface deflections, throttle, and thrust angle for aircraft, where applicable—necessary to produce the commanded acceleration input in response to a commanded trajectory acceleration input. A feedback control loop around this open-loop feed-forward control then results in an overall closely linear system which can compensate for model uncertainty and external disturbances. Of course, many questions arise as a result of all this, and this first contribution is devoted to answering such questions. Because of the fundamental importance of advanced aircraft trajectory capabilities, particularly in military aircraft, this is a most appropriate contribution with which to begin this volume.

The next contribution, "Information Systems for Supporting Design of Complex Human–Machine Systems," by W. B. Rouse, W. J. Cody, K. R. Boff, and P. R. Frey, is concerned with the design of complex systems and how information systems can support the design process. With particular reference to the introductory comments for this preface, it is eminently clear that the design process for modern aeronautical systems is becoming an increasingly elaborate and challenging process. Indeed, the design decision process is a distributed process, typically across temporal, organizational, geographic, and disciplinary attributes. Further, for complex aeronautical systems, design decisions emerge both from individuals' activities as well as from group processes when individuals collaborate. Further, in the military aircraft area, the design process involves a continual iteration with respect to operational requirements, which it seems are in a constant state of flux. In addition, national politics have a very heavy influence on the design process as a number of specific aircraft system designs have clearly manifested. In the area of commercial aircraft, continual iterations with the customer community on the international scene have a heavy influence on the design process. In any event, this contribution which treats comprehensively the many issues involved in the design of complex human–machine systems, namely aircraft, is an essential element of this volume.

Recent advances in aircraft performance and maneuver capability have dramatically complicated the problem of flight control augmentation. With increasing regularity, aircraft system designs require that aerodynamic parameters derived from pneumatic measurements be used as control system feedbacks. These requirements necessitate that pneumatic data be measured with accuracy and fidelity. To date this has been a difficult task. The primary difficulty in obtaining high frequency pressure measurements is pressure distortion due to frictional attenuation and pneumatic resonance within the sensing system. Typically, most of the distortion occurs within the pneumatic tubing used to transmit pressure impulses from the surface of the aircraft to the measurement transducer. The next contribution, "Formulation of a Minimum Variance Deconvolution Technique for Compensation of Pneumatic Distortion in Pressure Sensing Devices," by Stephen A. Whitmore, presents techniques for achieving essential accuracy and fidelity in these pneumatic sensor systems and, because of the fundamental importance of this issue in modern aircraft, constitutes an essential element of this volume.

One of the fundamental tasks in modern military aircraft is the interception of adversary aircraft with the essential objective of disrupting a hostile mission. The next contribution, "Synthesis and Validation of Feedback Guidance Laws for Air-to-Air Interceptions," by Josef Shinar and Hendricks G. Visser, provides an in-depth analysis of these issues and powerful and practically implementable techniques for guidance laws for air-to-air interceptions. Because of the rather comprehensive nature of this contribution, it will be an important source reference for workers in this field for years to come, and, as such, it constitutes an essential element of this volume.

Real-time simulation of aeronautical systems is fundamental in the analysis, design, and testing of today's increasingly conplex aeronautical systems. Perhaps more important is the fact that simulation, including 3-D vision and motion simulation techniques, is an essential element in pilot training for both commercial and military aircraft. For instance, imagine, if you will, the first time the Boeing 747 airplane was flown, the huge significance of this event, and the absolute requirement that it be a completely successful flight. This is a classic example of the enormous significance of adequately valid real-time simulation techniques. The next contribution, "Multistep Matrix Integrators for Real-Time Simulation," by J. A. De Abreu-Garcia and T. T. Hartley, is a comprehensive treatment of issues which are of fundamental significance in achieving faithful real-time simulation results for both linear and nonlinear systems. Training simulators have achieved such a high level of capability that experienced pilots have felt the stresses and strains of flying under actual conditions. They have played a key role in the design of every single aircraft in modern times, particularly their cockpit instrumentation. Clearly, this contribution is also an essential element of this volume.

An essential characteristic of all modern aeronautical systems is their avionics system, which is composed of many elements, in particular sensor systems. The matter of sensor systems signal processing and, in particular, electro-optical (EO) sensor systems signal processing is of particular significance and a continual challenge for military aircraft. There are many aspects of this problem including target dynamics, stochastic processes, system modeling, and others. In a rather remarkably comprehensive treatment of the basic and applied aspects of this problem, the next contribution, "The Role of Image Interpretation in Tracking and Guidance," by D. D. Sworder and P. F. Singer, provides an excellent articulation of the many issues in this broad problem area. Because of the increasing pervasiveness of the issues treated in this contribution, in particular, with the increasing trend to the requisite implementation of "Smart" EO systems (trackers), this contribution plays a rather fundamental role in the overall contents of this volume.

Parameter estimation in the presence of a stochastic environment has been a problem of theoretical interest for about three decades. Over the past decade, however, it has been of increasing applied interest in various advanced aeronautical systems. Examples have included parameter estimation in VSTOL aircraft, guidance system parameter estimates, and other issues. More generally, in aerospace systems the requirement for parameter estimation has, in some cases, resulted in extremely challenging and significant modeling issues. For example, the MX guidance system requires approximately a 90 element state vector, and the Trident vehicle requires approximately a 130 element state vector in the overall vehicle and guidance system parameter estimation. The next contribution, "Continuous Time Parameter Estimation: Analysis Via a Limiting Ordinary Differential Equation," by Douglas G. Dewolf, presents an analysis of many of the fundamental analytical issues in this area and a number of new results of fundamental significance.

Inertial navigation systems are now an integral part of the avionics suite of all commercial and military aircraft. Their operation on both aircraft have many common features. However, there are also distinctions. One of these, in the case of military aircraft and, in some cases, the air-to-ground missiles they might carry, is the requirement for in-flight alignment. In-flight alignment is required for military aircraft in some operational instances because of the requirement for instant takeoff or "scrambling" as it is sometimes called. The next contribution, "In-Flight Alignment of Inertial Navigation Systems," by Itzhack Y. Bar-Itzhack, concludes this volume with an in-depth treatment of the issues and techniques in this area.

The authors are all to be commended for their superb contributions which will provide, in this volume, a unique and significant reference source for many years to come for practicing professionals as well as those involved with advancing the state of the art.

AIRCRAFT AUTOMATIC FLIGHT CONTROL SYSTEM WITH MODEL INVERSION

G. ALLAN SMITH
Electrical Engineering Department
Yale University
New Haven, Connecticut

GEORGE MEYER
NASA Ames Research Center
Moffett Field, California

NOMENCLATURE 2
I. INTRODUCTION 5
II. HISTORICAL BACKGROUND 6
III. CONTROL SYSTEM CONCEPT 8
IV. CONTROL SYSTEM IMPLEMENTATION 11
A. COMMAND SECTION 11
B. CONTROL SECTION 11
V. MODEL INVERSION PROCESS 13
VI. SIMULATION RESULTS 20
A. VERTICAL-ATTITUDE MANEUVERING 21
B. TRANSITION RUN FROM FORWARD FLIGHT TO HOVER 24
C. OTHER TRAJECTORIES 27
VII. ONGOING RESEARCH 27
VIII. CONCLUSIONS 28
APPENDIX A. COMMAND SECTION AND REGULATOR SECTION DESIGN 30
APPENDIX B. MATHEMATICAL DETAILS 35
REFERENCES 39

NOMENCLATURE

A	aircraft acceleration vector
A_c	smooth commanded acceleration vector
A_i	rough commanded acceleration vector
EF	force equations error vector
EM	moment equations error vector
$E_2(\theta)$	elementary direction cosine matrix for rotation about the second axis through an angle θ, similar notation for other angles and axes, where

$$E_2(\theta) = \begin{bmatrix} \cos\theta & 0 & -\sin\theta \\ 0 & 1 & 0 \\ \sin\theta & 0 & \cos\theta \end{bmatrix}$$

F	force vector acting on aircraft in Earth reference axes
F_b	force vector acting on aircraft in body axes
G	gravity vector
J	aircraft moment of inertia matrix, where

$$J = \begin{bmatrix} I_{xx} & 0 & I_{xz} \\ 0 & I_{yy} & 0 \\ -I_{xz} & 0 & I_{zz} \end{bmatrix}$$

m	aircraft mass
M_b	moment vector acting on aircraft in body axes
R	aircraft position vector
R_c	smooth commanded position vector
R_i	rough commanded position vector
STOL	short takeoff and landing

$S(\omega)$	skew symmetric matrix function of the angular velocity vector, where $$S(\omega) = \begin{bmatrix} 0 & \omega(3) & -\omega(2) \\ -\omega(3) & 0 & \omega(1) \\ \omega(2) & -\omega(1) & 0 \end{bmatrix}$$
TBP	direction cosin transformation matrix from perpendicular reference axes to body axes; symbol implies P to B
TPB	inverse (transpose) of TBP; symbol implies B to P
TBR	direction cosine transformation matrix from Earth reference axes to body axes; symbol implies R to B
TRB	inverse (transpose) of TBR; symbol implies B to R
TPR	direction cosine transformation matrix from Earth reference to perpendicular reference axes; symbol implies R to P
U	generalized aircraft controls: roll, pitch, and yaw
V	aircraft velocity vector
VATOL	vertical attitude takeoff and landing
V_c	smooth commanded velocity vector
V_i	rough commanded velocity vector
V_w	wind velocity vector
V_{wb}	velocity of aircraft with respect to wind in body axes
α	aircraft angle of attack
α_c	commanded angle of attack, six-degree-of-freedom trim variable
β	aircraft sideslip angle
β_c	commanded side slip angle for six-degree-of-freedom trim
ΔA_c	translational regulator acceleration correction vector
$\Delta\dot{\omega}_c$	rotational regulator angular acceleration correction vector
Γ_c	commanded flight-path angle for six-degree-of-freedom trim

Γ_v	flight-path angle
ε	error signal
ζ	damping ratio of a second order system
θ_{pc}	commanded pitch angle, six-degree-of-freedom trim variable in perpendicular reference
ϕ_c	commanded roll angle, six-degree-of freedom trim variable
ϕ_E, θ_E, ψ_E	roll, pitch, and yaw Euler angles with respect to Earth reference axes
ϕ_p, θ_p, ψ_p	roll, pitch, and yaw Euler angles with respect to perpendicular axes
ϕ_{pc}	commanded roll angle for six-degree-of-freedom trim in perpendicular reference
ϕ_v	roll angle about velocity vector
ω	aircraft angular velocity vector, body axes
ω_i	rough commanded angular velocity vector
$\dot{\omega}$	aircraft angular acceleration vector, body axes
$\dot{\omega}_c$	smooth commanded angular acceleration vector
$\dot{\omega}_T$	total commanded angular acceleration vector
ψ_c	commanded heading angle for six-degree-of freedom trim
ψ_{pc}	commanded yaw angle, six-degree-of-freedom trim variable in perpendicular reference
ψ_v	heading angle

SUBSCRIPTS

b	body axes
c	commanded
E	Earth reference axes
i	rough input
p	perpendicular reference axes

REF	reference
T	total
v	derived from velocity vector
w	wind

SUPERSCRIPTS

$(\dot{\ })$	time derivative
-1	matrix inverse

I. INTRODUCTION

Powerful new airborne digital computers and the even more flexible and higher speed machines projected for the future will make it possible to design very complex and effective trajectory control systems for advanced aircraft. Some future aircraft designs will include significant nonlinear features in their aerodynamic and propulsion characteristics; in addition, they may be required to operate over extreme flight envelopes. This requirement may impose a high pilot work load and demand high levels of pilot skill and endurance. Thus, there may be requirements for automatic control over parts of the flight regime or for semiautomatic control in which a pilot would directly command trajectory accelerations, climb rates, or g-levels in coordinated turns by means of a sidestick controller. Examples include automatic night landing on a carrier at sea in foul weather, terrain following by a helicopter, or precise maneuvering in the vertical attitude by a vertical attitude takeoff and landing aircraft (VATOL).

The inverse-model-follower control system discussed herein can cope with a wide range of nonlinear characteristics and severe mission requirements. The basic idea behind the inverse-model-follower control system is to embed an inverse model of the aircraft force and moment generating processes into the control system. The series combination of this inverse model and the actual aircraft then approximates a unity response. A commanded trajectory acceleration input to the inverse model provides the controls—aircraft control surface deflections, throttle, (and thrust angle for aircraft where applicable)—necessary to produce the commanded acceleration input. When the actual aircraft controls are made to correspond to these calculated controls, the aircraft will produce the desired accelerations of its trajectory.

Under the circumstances of a perfect model, there might appear to be no need for conventional feedback control, for the open-loop feed-forward control could achieve the desired trajectory. Of course this is not strictly true because the aircraft model will contain inaccuracies and also because extraneous forces caused by wind and turbulence are not known. Therefore, a feedback control loop is used, but it can be easily designed by standard linear principles, since if the loop is closed around the series combination of inverse model and actual aircraft it includes only an element that responds linearly, at least to a very close approximation. Furthermore, the feedback is called upon only to compensate for model uncertainty and external disturbances. Thus, all the advantageous features of linear feedback control theory should be applicable to the closed-loop regulator design.

Of course, many questions arise from such simplified assertions as the foregoing, and the body of this publication will be devoted to answering such questions. It must, however, be acknowledged that both simulator and flight-test results are limited and that many improvements are desirable. In fact, the major claim of this work is only that a sound foundation has been established for a somewhat different approach to aircraft trajectory control and that the opportunities for extension, modification, and improvement are substantial.

II. HISTORICAL BACKGROUND

The inverse-model control system concept was formulated in the Flight Dynamics and Control Branch at NASA Ames Research Center. There was a need for an automatic trajectory control system for a new experimental aircraft: the augmentor wing jet short takeoff and landing (STOL) research aircraft.

An initial attempt to apply available modern control-theory techniques was not fruitful. The aircraft had very nonlinear aerodynamic effects owing to the expulsion of air bled from the jet engine compressor through slots in the aircraft wing flap system to augment lift for short takeoff and landing operations. The augmented lift was a nonlinear function of flap setting and engine throttle; moreover, the engine hot thrust was diverted through nozzles by up to 90° for landing.

Efforts to deal with the problems posed by this aircraft led to the formulation of an inverse-model control system [1]. This control system employed a nonlinear mathematical model of the aerodynamics and propulsion formulated in over three dozen tables of lift, drag, and moment data, some as a function of a single variable, such as angle of attack, and several two-dimensional tables as a function of angle of attack and engine thrust. These data were used to construct trim maps to provide plots of lift and drag coefficients (from total aerodynamics and thrust) as a function of angle of attack and engine throttle. These force and moment characteristics were inverted by an inverse table lookup scheme.

Theoretical and simulator studies of this concept led to flight tests in which a preprogrammed trajectory with climbing and turning flight was carried out successfully [2]. The concept was then applied to a simulator study of the automatic control of a Navy aircraft during carrier landing in conditions of low visibility and high sea states [3]. Satisfactory automatic landings under extreme conditions of carrier deck motion and severe wind turbulence were demonstrated in simulation. The automatic system was also extended to permit the human pilot to control the aircraft with the same basic control system through the use of a sidestick to command horizontal acceleration and vertical velocity in the vertical plane, as well as g-level in coordinated turns; air speed was commanded by a finger lever. This study used essentially the same model-inversion scheme as the original augmentor wing system.

The concept was next applied to control of a proposed vertical attitude takeoff and landing aircraft (VATOL). This simulation study used a very complete mathematical model of the aircraft aerodynamics and propulsion [4], and a new method of model inversion was used. Instead of attempting to construct inverse tables, the complete nonlinear model formulation with equations, tables, and all nonlinear coupling effects was incorporated directly into the control system. It was inverted by making six computer passes through the model in the usual forward direction with small perturbations of six control and attitude variables to construct a six-by-six Jacobian matrix of partial derivatives; the matrix was then inverted to determine the control variables corresponding to a commanded acceleration vector. This complete procedure was carried out at each computer cycle time, that is, every 0.05 sec. This (Newton-Raphson) approach gave very satisfactory performance.

The initial studies of the vertical-attitude maneuver were then extended to cover transition from vertical-attitude hover to conventional flight with 3-g turns in climbing trajectories up to 900 ft/sec. This simulation study was reported in detail in [5] and further refined as reported in [6] and [7].

Essentially the same system was installed in a Bell Aircraft UH-1H helicopter and successfully flight tested in 1986. The vehicle flew a racetrack pattern terminating in a descending helix to hover. Flight-test results were closely correlated to prior simulation results. This material has not yet been published.

A discussion of the inverse-model control system using the Newton-Raphson inversion scheme for control of the VATOL aircraft comprises most of this chapter. A line of theoretical investigations of the requirements for successful inversion of nonlinear systems has been carried out in parallel with the simulation and flight-test results mentioned above [8-10]. This theoretical approach still relies basically on an inversion of a complete nonlinear model of the aircraft force and moment generating processes. However, it has a somewhat different computational flavor as it is presently being investigated in simulation studies.

This work has not yet been flight tested nor have the simulation results been published. Nevertheless, it promises a significant extension of the prior work and a brief discussion of it is therefore included in the final section of this chapter (Ongoing Research).

III. CONTROL SYSTEM CONCEPT

As indicated above, the research investigation of the inverse-model control system concept is an ongoing process. Although the underlying approach remains essentially the same, there are significant differences in the details as modified, improved, and extended versions of the digital computer program are developed. In order to have a particular model for the following discussion, the design of the system as applied to a conceptual VATOL aircraft [11] will be used. This aircraft is designed to operate from small Navy ships from which it could take off and land in the vertical attitude by engaging a landing gear hook to a fixture on the side of the ship as illustrated in Fig. 1. The control system developed for this aircraft is, with only minor differences, the same as that used for the UH-1H helicopter, which, as already mentioned, has been successfully flight tested.

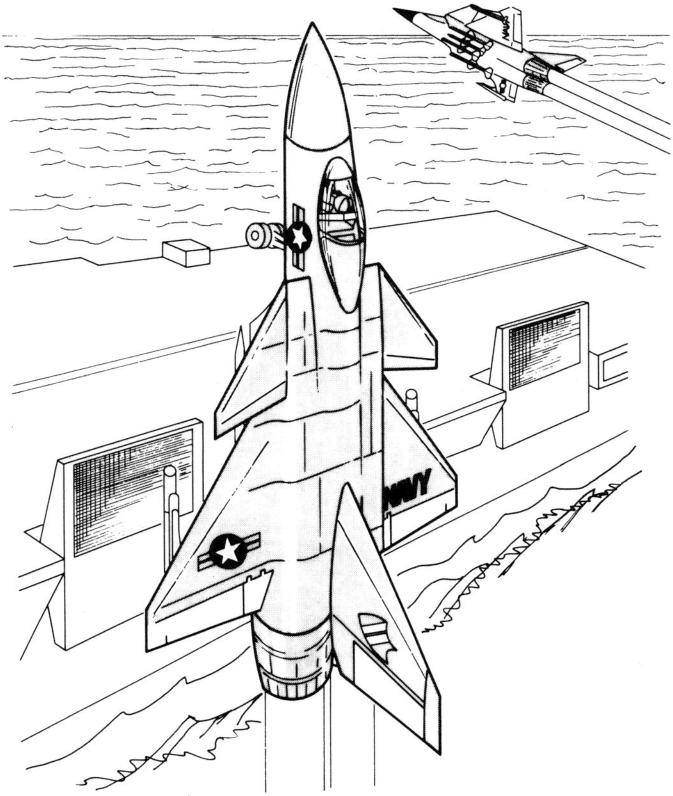

Fig. 1. VATOL aircraft used in simulation study: artist's rendition.

The control-system concept has several key features. Fundamentally, it combines a feed-forward and feedback control, with the forward path split into a command section and a control section (Fig. 2). The feed-forward controller is shown in solid lines, and the feedback control is shown in dashed lines. Most signals in this control system are three-dimensional column vectors with appropriate subscripts.

The command section generates smooth, executable, and consistent acceleration A_c, velocity V_c, and position R_c command vectors in the reference runway inertial axis system (north, east, and down) in response to corresponding rough trajectory command inputs A_i, V_i, and R_i. These rough inputs may be supplied

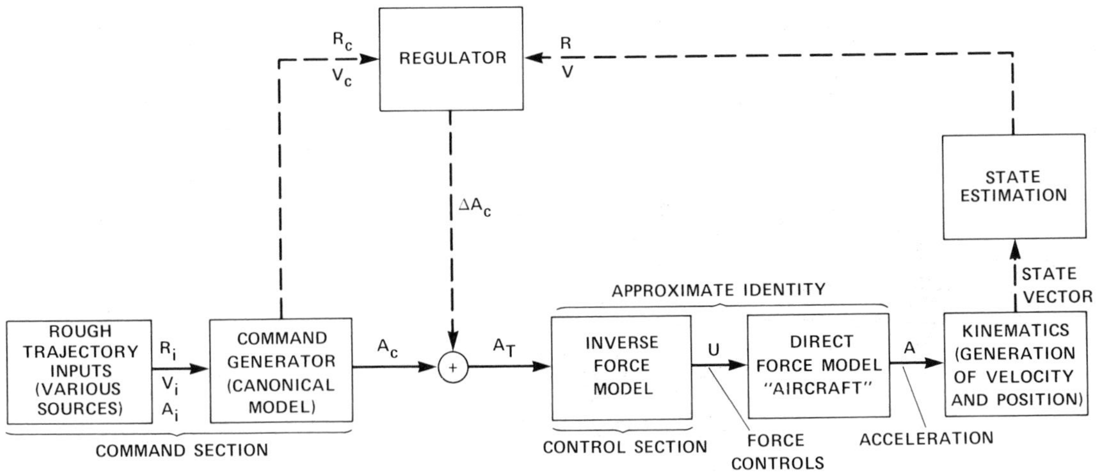

Fig. 2. Essential aspects of the Ames control concept.

from an air-traffic-control system, from ground vectoring, from a trajectory time-history stored in the computer, or by the pilot.

IV. CONTROL SYSTEM IMPLEMENTATION

A. COMMAND SECTION

In the command section, the command generator (Fig. 2) accepts the rough trajectory commands, which may not be executable because of discontinuities or deficiencies in some components of the vectors, and produces a complete and consistent set of smooth, executable acceleration, velocity, and position vectors. This smoothing is accomplished by using a canonical model of the aircraft. The model consists of strings of four integrators with appropriate feedback to give the desired dynamic response for each of the three channels, which correspond to the components (north, east, down) of the input command vectors. Adjustment of gains and limits provides smooth, executable vector commands. The smooth commanded acceleration (A_c) output of the command generator is added to the closed-loop feedback acceleration (ΔA_c) output of the regulator to give the total commanded acceleration vector (A_T).

There are several schemes, essentially filters, that can provide the smooth and consistent trajectory acceleration, velocity, and position vector commands in response to the rough, stepwise-discontinuous, or incomplete (having missing components) vector inputs. The mechanism whereby this is accomplished is not essential to understanding the overall action of the control system; however, a more detailed discussion of the command generator used for this application is given in Appendix A.

B. CONTROL SECTION

The control section includes a complete nonlinear inverse model so it is appropriate to consider first the aircraft mathematical model itself (Fig. 3). This model has been specialized to represent the VATOL aircraft used in this simulation. The four generalized controls (U) are normalized thrust and three angular acceleration controls for pitch, roll, and yaw which are converted to various aircraft control deflections. The VATOL engine thrust is capable of swiveling through ±15° in both pitch and yaw.

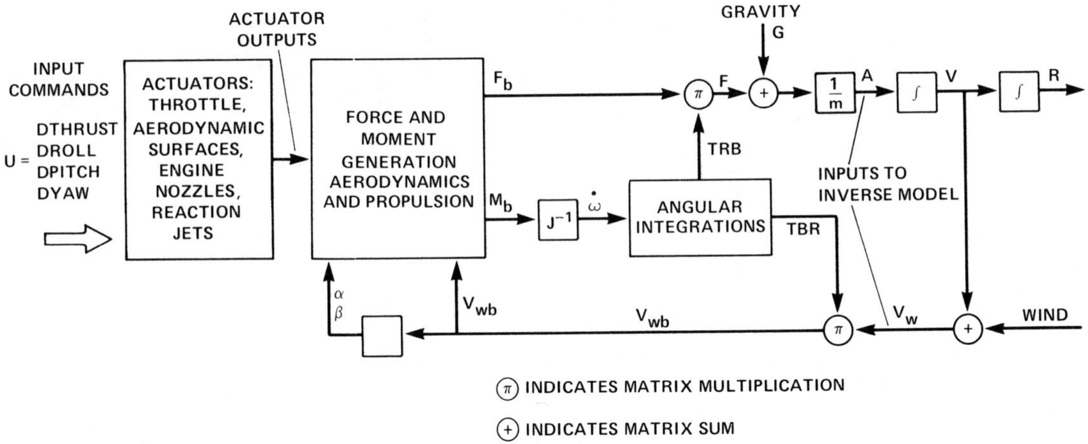

Fig. 3. Aircraft model.

The four generalized input control commands are divided appropriately between various aircraft controls, which include throttle, elevons, rudder, flaps, engine-nozzle angles, and engine bleed-air reaction jets at the wing tips for roll control in the hover mode. The three generalized angular acceleration controls include the elevons, rudder, and nozzle deflection angles in yaw and pitch, as well as the bleed air reaction jets.

The outputs of the actuators in Fig. 3 are input to the force and moment generating section where aerodynamic forces and moments in body axes are computed as a result of angle of attack α and the velocity with respect to wind in body axes V_{wb}. A detailed engine model with afterburner is used in calculations of thrust and ram drag, including such refinements as engine gyroscopic moments and thrust losses caused by nozzle turning angle, and the bleed air used for roll control. The body-axis moments M_b are used with the aircraft inertia matrix J and the Euler angular-rate equations to calculate angular acceleration $\dot{\omega}$, angular velocity, and aircraft attitude which is expressed as TBR, the direction cosine-matrix of rotation from reference axes to body axes (Appendix B). The inverse (transpose TRB) is used to transform forces F_b from body axes to reference axes. The addition of the gravity vector G and division by the aircraft mass m yields the acceleration vector A in reference axes. This acceleration vector is the essential output of the direct model, although integrations to give velocity V and position R are included in the figure.

The control section of the feed-forward controller (Fig. 2) is functionally the inverse of the aircraft force and moment model just described. The input is the total acceleration command vector A_T, and it is applied to the model inverse to determine the corresponding thrust and attitude control positions. The overall operation of the trajectory control system should now be clear (through reference to the figures), and attention will be directed to the details of the model-inversion process which differs substantially from the inverse table lookup scheme used for the initial applications of this concept.

V. MODEL INVERSION PROCESS

In the early implementations of this control system concept [1,2] the model inversion was carried out with the aid of extensive tables of nonlinear aerodynamic data that related lift and drag coefficients to angle of attack and thrust. Simulation and flight results were satisfactory for these aircraft for which the configuration allowed the force and moment equations to be treated separately. An exhaustive search technique was then used to solve two-dimensional tables.

The new model-inversion technique (Newton-Raphson) is suitable for more complex configurations with serious nonlinearities and with aerodynamic and thrust relationships that are functions of more than two variables. A symbolic explanation of this model-inversion scheme is presented in Fig. 4. The upper diagram indicates how the aircraft model shown in Fig. 3 is normally used in a conventional simulation study. For any particular set of linear and angular velocity vectors V_W and ω, knowledge of the control U and the aircraft attitude TBR allows the output accelerations to be calculated by using equations for the various aerodynamic and propulsion forces and moments.

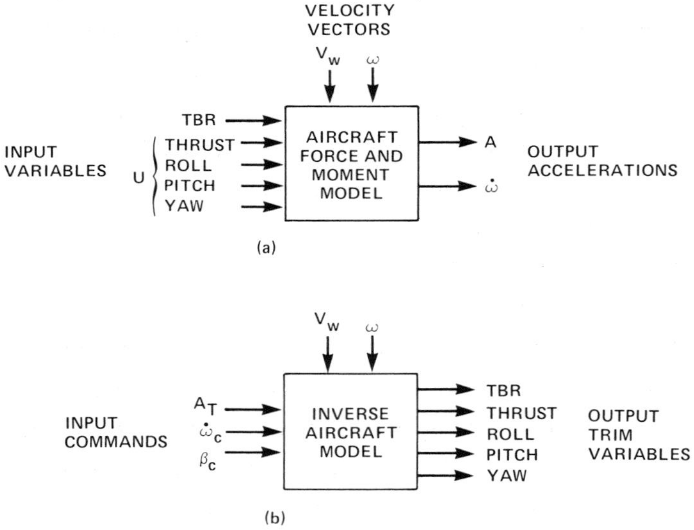

Fig. 4. Model inversion process. (a) Direct process. (b) Inverse process.

The diagram in Fig. 4b represents the model-inversion problem; it will be recognized by those engaged in aircraft simulation studies as the approach for computing the control positions required to trim the aircraft at a given flight condition.

The aircraft velocities and commanded accelerations and sideslip are specified as inputs, and it is desired to calculate as outputs the corresponding aircraft attitude and controls that would produce the commanded inputs. The inverse problem in Fig. 4b cannot be directly solved as illustrated in that diagram because we do not have analytical expressions for directly calculating the trim variables. Instead, the method applies an iterative Newton-Raphson trim procedure to the aircraft model of Fig. 4a.

The trim procedure is quite standard so only a brief discussion of particular features will be given here. The trim procedure employs the force and moment vector equations in body axes:

$$F_b + TBR_c(G - mA_T) = EF,$$

$$M_b + S(\omega_c)J\omega_c - J\dot{\omega}_c = EM.$$

These equations apply to the aircraft model embedded in the control system and have, therefore, commanded variables as indicated by the subscript c. These two vector equations represent six scalar equations. The first three are force-error equations and the last three are moment-error equations. The force represented in body axes is F_b. The gravity vector G and the commanded acceleration vector A_T are both represented in inertial reference axes. They are transformed by TBR_c so that all terms are in body axes.

If the aircraft model is in force trim, the three components of the error EF are zero. The M_b term is the torque vector represented in body axes, ω_c is the angular velocity, $\dot{\omega}_c$ is the commanded angular acceleration, J is the aircraft inertia matrix, and $S(\omega_c)$ (Appendix B) is the skew symmetric matrix function of the angular velocity. If the model is in moment trim, the three components of EM are zero.

For trim in conventional flight, the rotation matrix TBR_c (Appendix B) is implemented as the product of five elementary direction-cosine matrices that represent heading ψ, flight-path angle Γ, roll angle ϕ, commanded sideslip β (usually zero), and angle of attack α. The angles ψ, Γ, and β are fixed during a trim cycle, and roll and angle of attack are two of the trim variables. The other trim variables are the four controls U. Trim variables are the six quantities that are adjusted to obtain trim (zero or acceptably low errors for the trim equations).

The trim procedure first selects trial values of the six trim variables (input variables in Fig. 4a). These are rough estimates for the initial trim or values from the preceding trim cycle as the simulation proceeds. These trial trim variables are then used along with the commanded velocity vectors to calculate the resulting forces and moments that are substituted into the trim vector equations along with the commanded accelerations to find errors (EF and EM). If the errors are below specified tolerance values, the model is considered trimmed, and the four trim variables that constitute the aircraft controls U are sent to the actual aircraft actuators. Nominal tolerances were taken as 0.001 g for the force equations and 0.005 rad/sec^2 for the moment equations.

If any errors exceed their tolerance, a perturbation procedure is initiated. One trim variable is perturbed by a small amount from the trial value, and the force and moment calculations are repeated to determine new values of the six trim-equation errors. This is done six times, once for a perturbation of each of the six trim variables. A six-by-six Jacobian matrix of error derivatives of each force and moment with respect to each trim variable is formed where, for example, the third-row fifth-column term is the partial derivative of vertical force with respect to angle of attack. The six Jacobian matrix rows correspond to the the X-, Y- and Z-axis forces and to the roll, pitch, and yaw moments, respectively. The six Jacobian columns correspond to perturbations of the roll, pitch, yaw and throttle controls and to angle of attack and roll angle, respectively.

The Newton-Raphson procedure then inverts this matrix to give estimated changes in the trial values of each trim variable. These new trim variables are then used to compute forces, moments, and the angular transformations required in the trim equations; if the six equation errors are less than the tolerance test, the model is considered trimmed. If not, the entire process is repeated with seven more passes through the model, one pass for each trim variable perturbation and one to verify trim. A single set of seven passes is sufficient for most flight conditions, but two or more sets are needed for parts of the trajectory when commanded accelerations are changing rapidly or when external disturbances are encountered. Only a single pass is required when the trim variables from the preceding computer cycle give error values that are less than the tolerance.

The use of the error equations in the trim procedure should now be clear. There are, however, several interesting details involved in the actual system. An important consideration arises when one attempts to trim the aircraft in a vertical attitude. When the aircraft is in a vertical-attitude hover, all velocities may become zero, so that angle of attack is undefined. Therefore, for low velocities in the vertical attitude, the angle of attack is replaced as a trim variable by the pitch attitude θ. Angle of attack is, of course, still computed and used, though not as a trim variable, at all velocities for which it is defined. Furthermore, in hover, θ is nearly 90° with respect to reference axes, which is a singular point for the transformation TBR_C. Therefore, a change is made for low speed, and aircraft attitude is specified by only the three conventional Euler angles, ϕ_P, θ_P, and ψ_P; furthermore, these angles are measured with respect to an inertial system rotated 90° from the runway reference axis system.

This inertial system is called the perpendicular system, and its axes are positive up, east, and north, respectively. For this condition, the transformation matrix from the perpendicular system to body axes is designated TBP_C, and is implemented as the product of three elementary direction-cosine matrices representing the Euler angles. The pitch angle about the aircraft Y-axis and the yaw angle about the Z-axis are taken as trim variables. The other Euler angle is a commanded heading angle about the vertical X-axis; it is the angle between north and the aircraft landing gear in the vertical attitude; it remains constant during a trim cycle. It is assumed that suitable on-board instrumentation will be available to measure any angles required.

A constant transformation matrix TPR is used to rotate from the reference axes 90° about the east horizontal axis to the perpendicular system. These rotations are used in the force-error equations to get all terms in the perpendicular-axis system so that the force-error equation becomes

$$TPB_c F_b + TPR(G - mA_T) = EF.$$

In the simulation, a preselected pitch attitude of 60° from the horizontal was used to switch between the two different sets of force-error equations and to substitute pitch attitude for angle of attack as a trim variable. Operation was such that only very slight transients were observable in the system. This minimal influence was expected, because the transfer is merely a computational operation, and no switching of physical sensors would be involved in an actual aircraft installation.

Other details in the trim procedure may be seen in Fig. 5, which is a diagram of the overall actual system. Note that there are several differences between it and the simplified configuration shown in Fig. 2. The control section has been split into translational and rotational sections, each with a command generator and a regulator. We have been discussing the six-degree-of-freedom trim whose input signal is the total commanded-acceleration vector A_T and whose output is the rough commanded attitude matrix TBR_c, which contains two of the trim variables. Furthermore the angular velocity ω_c and angular acceleration $\dot{\omega}_C$ are taken as zero for the six-degree-of-freedom trim.

The matrix TBR_C is input to the rotational command generator, which is functionally similar to the translational command generator, and whose output, a smooth commanded angular acceleration vector $\dot{\omega}_C$, is combined with the output of the rotational regulator $\Delta\dot{\omega}_C$ to give a total commanded angular acceleration vector $\dot{\omega}_T$. The rotational regulator compares the actual aircraft attitude TBR and angular velocity ω with a smooth commanded attitude matrix TBR_{sc} and angular velocity ω_c to generate the closed-loop angular acceleration command increment $\Delta\omega_C$. The total angular acceleration $\dot{\omega}_T$ is input to a four-degree-of-freedom Newton-Raphson trim section, which is entirely analogous to the previous six-degree-of-freedom trim section except that only four trim equations are used: the three scalar moment equations and the first (aircraft longitudinal axis) of the three force scalar equations.

The four trim variables are roll, pitch, yaw, and thrust. The actual aircraft attitude matrix TBR and angular velocity vector ω are used in these trim equations. The outputs of the four-degree-of-freedom trim are the four trim variables, which become the controls U; they are input to the actual aircraft. This trim differs from the six-degree-of-freedom trim in that it has a commanded angular acceleration input and an angular velocity input in addition to the same commanded translational acceleration input A_T. The Newton-Raphson technique again requires one pass through the model to determine initial errors, and four additional passes for perturbations of the four trim variables.

The simple diagram of Fig. 2 does not distinguish between the two trim implementations. As shown in Fig. 5, the six-degree-of-freedom trim is used to

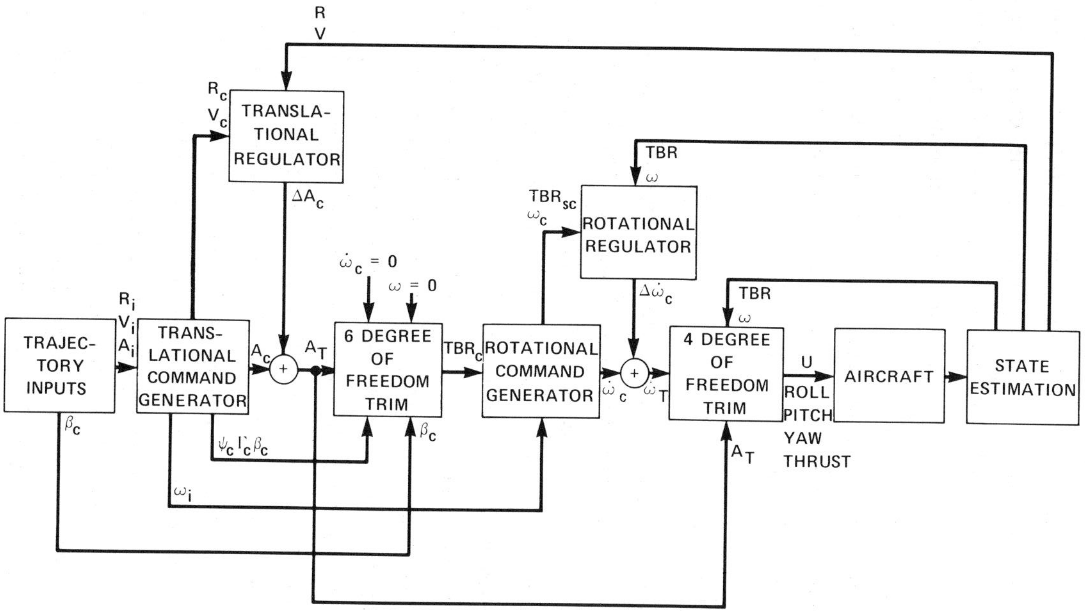

Fig. 5. Complete control system concept.

determine aircraft attitude required for the commanded translational acceleration, but assuming zero angular velocity and acceleration. Two of the trim variables, angle of attack and roll angle, are used along with the commanded heading, flight-path angle, and sideslip to determine the commanded attitude for conventional flight (two of the trim variables, pitch angle, and roll angle are used along with commanded heading to determine TBR for trim with the perpendicular reference system). The four-degree-of freedom trim uses the actual aircraft attitude and angular velocity and determines the four control variables necessary to produce the command angular acceleration vector. The four-degree-of-freedon trim switches from reference axes to perpendicular axes at the same time that the six-degree-of-freedom trim switches.

VI. SIMULATION RESULTS

A series of simulation studies was performed using the control system just described in the VATOL aircraft (previously discussed). The aircraft aerodynamic and propulsion characteristics were taken from a conceptual study of a proposed aircraft [12-14]. Maneuvers were first performed in the vertical-attitude/hover mode in all directions, both individually and simultaneously and with various aircraft roll rates about the vertical axes. Accelerations up to 0.3 g and velocities up to 50 ft/sec were investigated. The response to gusts and steady winds was also studied. Then, in the conventional flight mode, major attention was directed to the transition between horizontal flight and hover. Tests were also carried out at high subsonic speed, with turns up to 4 g and flight-path angles to 10°.

Eight trajectories were reported in [11]. The results of only two different flight trajectories are presented here: maneuvers in the vertical attitude at low speed and maneuvers during a transition run from conventional level flight to hover. Time-histories are shown for each trajectory; they include aircraft accelerations, velocities, and displacements, as well as attitude angles and control variables. The results shown are preliminary in that no particular effort was made to optimize transient performance by gain or limit adjustments, since stable and reasonably satisfactory performance resulted from the initial settings.

A. VERTICAL-ATTITUDE MANEUVERING

Performance during maneuvers in the vertical attitude is shown in Fig. 6. The aircraft is initialized in hover in a vertical attitude at zero velocity with the landing gear facing north and thrust equal to weight; the initial altitude was 1000 ft. A series of simple commands that would exercise the complete control system was used for the simulation. A step command of 3 ft/sec^2 horizontal acceleration was applied at 3 sec for a 15-sec period, followed by a command of 3 ft/sec^2 acceleration vertically at 30 sec for 10 sec. A roll command of 20°/sec about the vertical for 5 sec was applied at 50 sec. Thus, at 64 sec the aircraft was translating horizontally at 45 ft/sec, translating up at 30 ft/sec, and rolling at 20°/sec. Simultaneous commands were then given to bring the aircraft back to the hover condition, which was essentially achieved by 90 sec.

Curve 1 in Fig. 6 shows both the rough horizontal acceleration command $A_i(1)$ (the first component of the rough commanded-acceleration column vector in Fig. 2) and the smooth command $A_c(1)$ from the command generator. The total commanded acceleration in curve 2 in Fig. 6 is $A_T(1)$, and it includes the feedback corrective acceleration signal $\Delta A_c(1)$. The control system performance can be evaluated by noting how closely the actual acceleration $A(1)$, indicated by the solid line on the same plot, follows $A_T(1)$. The tracking is quite close except at the four points labeled a-d on curve 2 in Fig. 6, which show an initial aircraft acceleration in the direction opposite to the command. This reflects the nonminimum phase character (right half-plane zero) of the aircraft transfer function as the engine nozzle swivels forward to rotate the aircraft for translation, thus producing a momentary force opposing the intended translation. This is similar to the initial downward acceleration caused by elevator deflection when starting a climb in a conventional aircraft.

In curve 3 of Fig. 6, the velocity reaches a steady value of 45 ft/sec, and the horizontal translation in the curve 4 finally increases to almost 3000 ft. Curves 5 and 6 of Fig. 6 show the vertical-acceleration commands $A_i(3)$, $A_c(3)$ and responses $A_T(3)$, $A(3)$. There is no nonminimum phase effect in the vertical channel since the acceleration is due to engine thrust. The actual acceleration follows the total command very closely except for a transient effect (point e in curve 6, Fig. 6) during the roll-rate application. Curve 7 shows that vertical

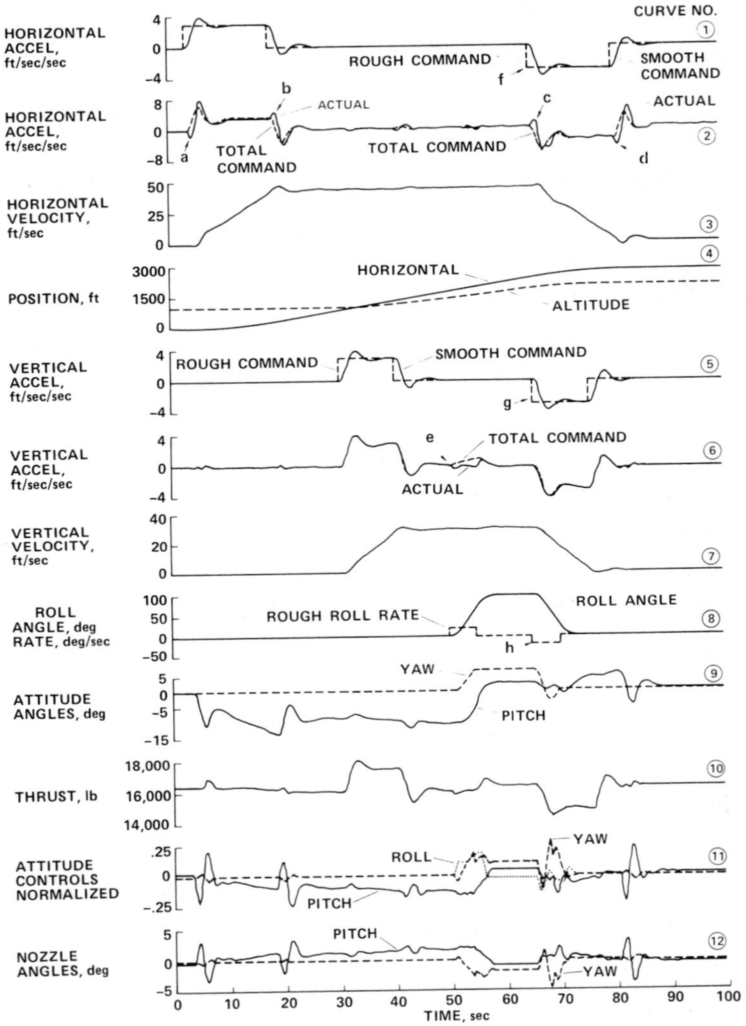

Fig. 6. Vertical-attitude maneuvers from hover.

velocity responds with essentially no overshoot and reaches a value of 30 ft/sec. The altitude response is shown along with horizontal distance on curve 4. Both the rough roll-rate command and the roll-angle response are shown on curve 8.

In the low speed, vertical-attitude regime, the aircraft attitude is measured by Euler angles in the perpendicular axis system. The pitch angle in curve 9 (Fig. 6) is not quite zero in hover, owing to an engine offset from the aircraft centerline. As shown in this plot, the aircraft pitches forward and the yaw angle remains zero until the aircraft starts to roll. Basically the aircraft thrust vector is deflected to provide the horizontal acceleration; that is, a deflection of about 0.1 rad to give a forward acceleration of 0.1 g. No deflection would be required to sustain a constant velocity in the absence of aerodynamic forces. However, at an airspeed of 54 ft/sec, an aircraft pitch of 12° is needed to overcome aerodynamic forces and engine ram drag effects. As the aircraft rotates about the vertical, the thrust vector is deflected in yaw and less in pitch until, after 100° of roll, there are about 6° of deflection in yaw and 2° in pitch. The size of the angles indicates differences in aerodynamic and propulsion forces in the two axes. During all of these commanded rotational maneuvers, the trajectory deviated less than 2 ft from the smooth command. The thrust curve in curve 10 (Fig. 6) shows the correlation with the vertical acceleration.

The normalized pitch channel signal in curve 11 of Fig. 6 reflects the pitching-moment requirements needed for angular acceleration and to counteract aerodynamic moments as airspeed increases. The yaw control (dashed line) indicates the shift from pitch to yaw control as the aircraft rolls. The roll control (shown by the dotted line) initiates the roll at 50 sec. Curve 12 shows the engine nozzle pitch and yaw deflections which remain well below their maximum limits of 15°.

At 63 sec, simultaneous commands (points f-h, Fig. 6) are applied to all three axes to reverse the original commands and restore the aircraft to a hover condition at an altitude of 2000 ft. The commands and responses are nearly the inverse of the original except for coupling effects caused by the initial roll angle. It will be noted that the axes are quite well decoupled during the separate commands of the first 50 sec and that the major coupling effects for simultaneous commands occur in the controls and angular response so that as a basic translational acceleration control system, the trajectory response in the different axes remains nearly decoupled for all commands.

The trim equations were successfully iterated in one Newton-Raphson cycle for over 95% of the run and required only two cycles during the transient adjustments to step commands. It should be recalled that at the end of any particular computer cycle time the trim equations must be balanced. This may require only one Newton-Raphson iteration if the trajectory change is moderate (95% of the time) or it may require two iteration cycles if acceleration changes are more significant. In extreme cases three or even four iteration cycles have been observed during one computer cycle time. On the other hand, during relatively undisturbed flight, no iteration may be required during a computer cycle time, for the prior controls still provide the commanded acceleration within the tolerance limits. Other vertical-attitude maneuvers showed similarly satisactory response to lateral translational commands.

B. TRANSITION RUN FROM FORWARD FLIGHT TO HOVER

Performance during a 60-sec transition run at constant altitude is shown in Fig. 7. The aircraft is initialized in conventional level flight at a velocity of 180 ft/sec at an altitude of 1000 ft. A deceleration command of 3 ft/sec^2 is given at 5 sec and is increased to 12 ft/sec^2 at 33 sec. The commanded acceleration is reduced to zero at 41 sec when the corresponding velocity command reaches zero. The vertical-acceleration command is zero throughout the run.

The point of particular interest in this run is the response during lift-curve slope reversal beginning at about 22 sec. The lift-curve slope reversal near an angle of attack of 32° presents a problem to the essentially linear Newton-Raphson technique. In order to achieve trim, it was necessary to reduce the predicted corrections to 80% to prevent hunting in the algorithm. The difficulty was compounded by the configuration with a large canard surface whose local angle of attack caused the slope of the lift curve for that surface to reverse at a different speed than for the main wing. Most of the same variables shown in Fig. 6 are also shown in Fig. 7 where they display many of the same correlations.

The step deceleration command $A_i(1)$ at 5 sec shown in curve 1 of Fig. 7 results in the smooth command $A_c(1)$ on the same plot and the total command $A_T(1)$ and the aircraft response $A(1)$ on curve 2. The velocity V on curve 3

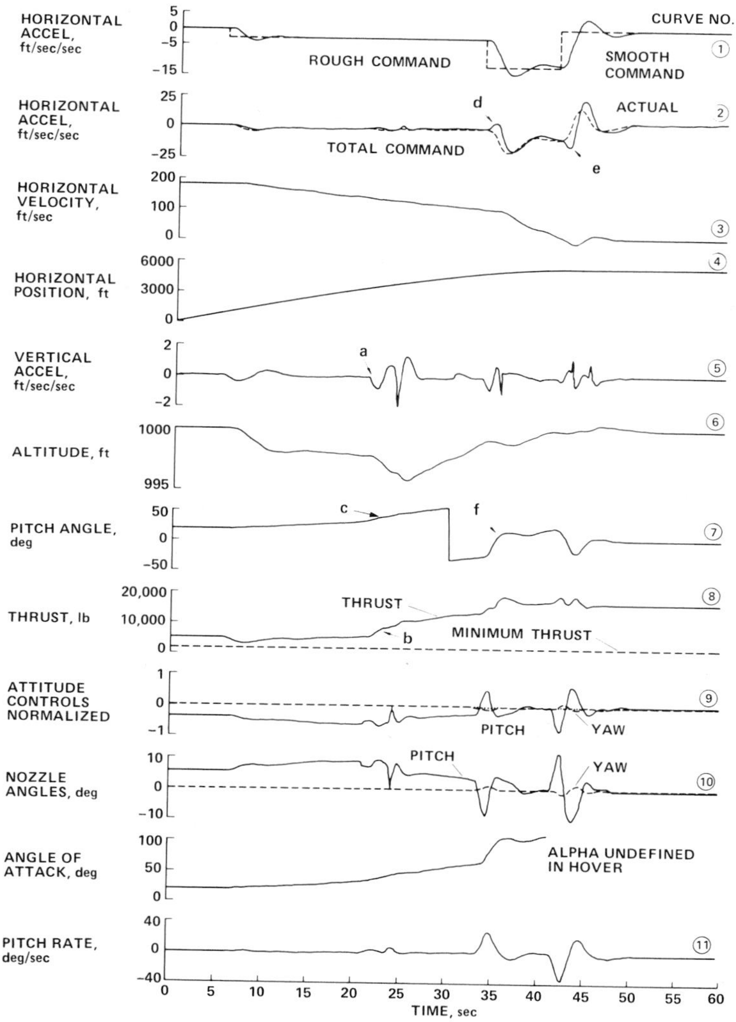

Fig. 7. Transition from conventional flight to hover.

begins to decrease, and the position R on curve 4 continues to increase but at a slower rate. A slight vertical-acceleration response occurs between 5 and 15 sec accompanied by a 2-ft loss of altitude. The pitch angle (curve 7) measured from the horizontal, and the angle of attack (curve 11) begin to increase from their initial values of 19° as the aircraft responds to the pitch control command in curve 9 and the corresponding pitch nozzle in curve 10. The thrust decreases almost to its minimum value during the initial deceleration.

As airspeed decreases, the angle of attack in curve 11 (Fig. 7) increases to maintain lift. The aerodynamic lift curve of the wing reaches its peak at an angle of attack of about 32° at which point it reverses, following wing stall. This effect is seen at 22 sec where a vertical acceleration transient (point a, curve 5) occurs as the trim process tries to obtain increased lift by increasing the angle of attack, At this point, the trim routine must maintain altitude by using additional engine thrust at a higher pitch attitude, shown by the thrust (point b, curve 8) and pitch angle (point c, curve 7)) as the aircraft transitions to the vertical attitude mode. The airspeed is about 130 ft/sec at this time. Only 5 ft of altitude were lost in the transition.

The aircraft continues to slow down and the attitude increases until at 30 sec a pitch attitude of 60° is reached, and the attitude direction-cosine matrix is switched from the runway reference-axis to the perpendicular axis system. This is seen in curve 7 of Fig. 7, where the pitch angle jumps from 60° to –30°. Only a slight response was observed in vertical acceleration. At 33 sec the deceleration command is increased to 12 ft/sec^2, and transients similar to those seen in Fig. 6 occur in all variables. The nonminimum phase behavior on curve 2 is denoted by points d and e. The rapid rise to a vertical attitude (i.e., zero pitch angle) is shown by f in curve 7 of Fig. 7. When the commanded velocity reaches zero, the deceleration command is removed and the aircraft comes to a hover position at zero velocity.

The transition starting at 22 sec required two trim cycles for about 5 sec and was equally troublesome when other transition runs were made going from hover to high speed while climbing at 30 ft/sec. There was less that 2 ft of sidewise deviation from the trajectory except when wind gusts were applied. This aircraft has such a clean aerodynamic profile that deceleration along the flight path is difficult to achieve. An initial deceleration command of much more than 0.1 g would have called for a thrust from the trim process of less than the engine idle setting. The maximum deceleration obtainable varies from about 0.1 g at low speed to 0.4 g at an air speed of 700 ft/sec in conventional flight. Of course, much higher deceleration can be achieved in the vertical-attitude mode.

C. OTHER TRAJECTORIES

Several other trajectories [11] for which data are not shown here were simulated. In particular, a turning, climbing, accelerating trajectory with a speed change of from 300 ft/sec to 900 ft/sec and a climb at a 10° flight-path angle showed very satisfactory performance, including the execution of turns of 2, 3, and 4 g. Some trajectories were subjected to wind gusts and to initial offsets; the transient results were quite satisfactory.

If too abrupt or overly severe commands were attempted, the trim process would fail. Excessive commands that caused control position or rate to saturate could also cause trim failure. Unrealistically sharp wind gust disturbances also caused trim difficulties. These problems are being addressed in ongoing research.

VII. ONGOING RESEARCH

The theoretical investigations of nonlinear systems reported in [8-10] have continued to explore characteristics required of nonlinear systems so that the nonlinearities can be minimized or eliminated by suitable transformations. Current simulations indicate that many practical but highly nonlinear aircraft configurations can be transformed to a linear uncoupled representation and that the effect of nonminimum phase zeroes can be compensated. New coordinates are selected in which the state equation has no zeros so that it can be inverted directly. The necessary and sufficient conditions for the existence of such coordinates are given in [8-10]. Such linearization requires derivatives of several orders of the force and moment functions.

An effective approach for obtaining the required derivatives is to model the functions by multidimensional polynomials and then take the derivatives analytically. The current research uses a polynomnial that includes the linear, squared, and binomial cross product terms of a 12-state variable representation for a total of 91 terms for each of the three force and three moment functions. Simulation results show that although many calls to the complete nonlinear model are required to determine the coefficients, the resulting approximation is accurate over a large enough region to require updating only once in 60 cycle times for the maneuvers tested.

VIII. CONCLUSIONS

A simulator study was carried out to confirm the advantages of using a Newton-Raphson model-inversion technique, developed to replace a prior inverse table lookup technique, as a design basis for an automatic trajectory control system for aircraft with highly nonlinear characteristics. The simulation used a detailed mathematical model of aerodynamic and propulsion system performance of a vertical-attitude takeoff and landing aircraft (VATOL or tail sitter). Tests were carried out over an extensive flight envelope from vertical-attitude hover to conventional flight at high subsonic speeds.

Some specific comments regarding this simulation and other related research are given below.

1. The results presented here and those of other simulation and flight tests show that the control system performs satisfactorily over a large proportion of the flight envelope.

2. The reversal of the lift curve slope near an angle of attack of 32° initially caused instability in the trim algorithm, but the instability was overcome by minor modifications of the technique.

3. The use of a perpendicular inertial-axis reference system for vertical-attitude hover maneuvers was quite satisfactory. No appreciable transients occurred, although the aircraft-attitude matrix was simultaneously switched at three places in the control circuits. The switching point is not critical and can be anywhere between 45° and 80°, so a suitable hysteresis loop can be designed to prevent any switching chatter when going through the switching region in either direction.

4. The system response to wind gusts (not shown here; see Ref. [11]) was satisfactory for various reasonable combinations of wind magnitude and direction. However, trim failure could be induced for sufficiently severe disturbances. Logic to cope with such conditions is under investigation.

5. The description of the control-system concept mentioned that the rough trajectory command inputs can be generated on line by a human pilot rather than from precomputed trajectory segments. For a previous simulation [3], runs were made for the conventional flight mode in which the pilot's fore-and-aft stick position commanded vertical velocity and the sideway stick displacement commanded g level in a turn; throttle position commanded airspeed. Under these conditions, the aircraft performance was quite satisfactory. It should be noted that only a few components of the rough trajectory command input vectors were thus generated and that some of these were in the body-axis system rather than in the reference-axis system. Nevertheless, suitable input transformations allowed the command generator to provide a full set of smooth commanded vectors in reference axes.

6. The credibility of these simulation results is supported by the flight-test history of the concept in three fixed-wing aircraft: a DHC-6 Twin Otter aircraft [2], the augmentor wing jet STOL research aircraft [1], and the Quiet Short Haul Research Aircraft (QSRA) [15]; all used multidimensional table lookup for model inversion, and showed good agreement between simulation and flight-test results. A flight-test program using the Newton-Raphson model inversion technique in a fourth aircraft, a UH-1H helicopter, has successfully controlled that aircraft over a complex maneuvering trajectory in spite of strong nonlinear coupling between axes.

7. Flight envelope and control limits are conservatively imposed on the model motion by means of position and rate limiters. A more efficient switching logic is being investigated.

8. The procedure allows the construction of the complete state and control motions consistent with the a priori model of the aircraft state equations, whether or not the open-loop dynamics are stable, provided that the effect of right half-plane zeros are negligible. Whereas the necessary and sufficient conditions for linearity are known in general, practical algorithms for transforming the state and control vectors when the system is not a strictly feedback one, are being developed, and significant progress has been made.

9. The extent to which the dynamics seen by the regulator are linearized and decoupled depends directly on the accuracy of the a priori model and state estimation. For the cases investigated, the accuracy was such that the first -order effects of any nonlinearities were indeed removed. The resulting uncertainty in regulator

dynamics was handled by standard robust regulator design techniques. However, the regulator design approach given here should not be used when large modeling errors are suspected.

APPENDIX A. COMMAND SECTION AND REGULATOR SECTION DESIGN

The command section of Fig. 2 supplies a smooth commanded acceleration vector A_C which is the primary open loop command to the system. It also provides a consistent set of smooth commanded velocity and position vectors V_C and R_C to the regulator. The regulator compares these commanded velocity and position vectors with the actual trajectory velocity and position vectors R and V, which come from an estimation section that is not considered here in detail. It could involve on-board inertial instrumentation, air velocity data from pressure sensors and vanes, radar altimeter, ground-based radar position data and estimated steady wind, and even satellite position data (GPS), all manipulated to give estimates of the aircraft position, velocity, and attitude.

The regulator provides an incremental acceleration vector ΔA_C based on errors between the commanded and actual trajectory. This increment is added to the open-loop commanded acceleration A_C to give A_T as input to the control section. The body of this chapter has been devoted to showing how well the actual aircraft acceleration A can be made to follow the commanded acceleration vector A_T. This appendix is concerned with the mechanism whereby this total acceleration command is generated.

TRAJECTORY INPUT SECTION

The rough trajectory input could be generated in several ways depending on the mission requirements. For these simulation studies, the primary inputs were commanded accelerations or acceleration rates in each of the three reference inertial system directions. Integrators then generated the corresponding velocities and positions, as shown in Fig. A.1. For the simulation data shown in Figs. 6 and 7, step inputs of acceleration were used as the inputs to the second integrators. For other tests it was convenient to use step inputs of acceleration rate as the input of

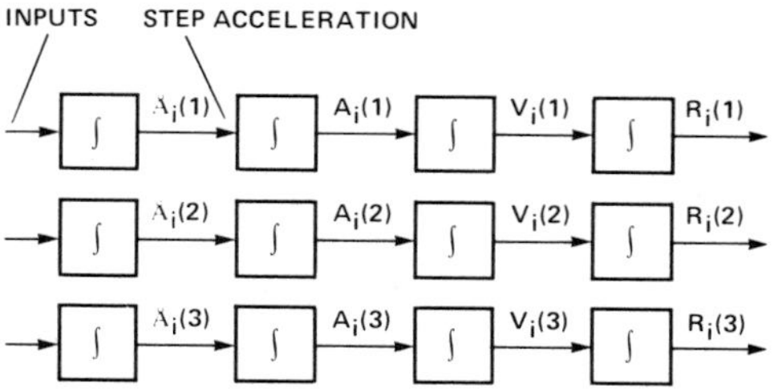

Fig. A.1. Rough trajectory input section.

the first integrator to get a ramp of acceleration at the second integrator input. inputs to the second integrator to give varying acceleration commands.

TRANSLATIONAL COMMAND GENERATOR SECTION

The function of the translational command generator in Fig. 5 is to accept the rough input trajectory command vectors of position R_i, velocity V_i, and acceleration A_i, which may not be executable by the aircraft because of missing vector components, excessive magnitudes, excessive rates of change, or discontinuities, and to provide smooth, executable, consistent trajectory command vectors R_c, V_c, and A_c. The command generator achieves these objectives by taking the form of a canonical model of the aircraft with a dynamic response that is the desired dynamic response of the aircraft. The canonical model represents the aircraft as three strings of four integrators each, with suitable feedback to give the desired dynamic response. Each string corresponds to one of the Earth reference axes (north, east, down). Figure A.2 shows a single string representing the first channel with the integrator outputs as smooth commanded position, velocity, acceleration, and acceleration rate. The input of primary interest for this discussion is the rough acceleration command A_i. The corresponding output is the smooth acceleration command A_c. Feedback loops within the command generator are formed with error signals between corresponding input and output quantities. The transient response of this system depends on the feedback gains which can be selected by various pole placement methods.

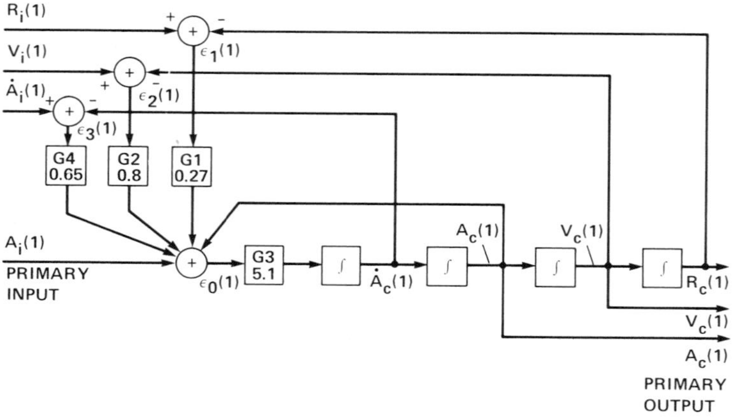

Fig. A.2. One of three parallel channels in the translational command generator.

For steady-state and slowly varying inputs, the feedback loops will tend to maintain low values of the primary error signal ε_0 and of the secondary error signals between input and output quantities ε_1, ε_2, and ε_3 whenever possible; but if an input is missing or inconsistent the other errors will adjust to keep the primary error low at the expense of standoff errors between the other variables. In any event, the output quantities R_c, V_c, and A_c will be smooth consistent sets of position, velocity, and acceleration since they are the outputs of consecutive integrators.

The actual command generator used for this study was designed as shown in Fig. A.3, which is a complete three-dimensional flow diagram with vector quantities and three channel integrators. A number of amplitude and rate limiters are included, as well as a differencing circuit to calculate the derivative of input acceleration A_i. The limiters are adjusted to limit the output commands to suitable levels as discussed more fully in [3].

The gains of the command generator were selected, as previously mentioned, to give the dynamic response that was desired for the actual aircraft. Since the channels are uncoupled, the single-axis diagram of Fig. A.2 can be referred to. For this study the gains were selected as follows. The force generation servo, which was essentially represented by the first two integrators, was specified to have a second-order response with natural frequency of $\omega_n = 1.2$ rad/sec and a damping

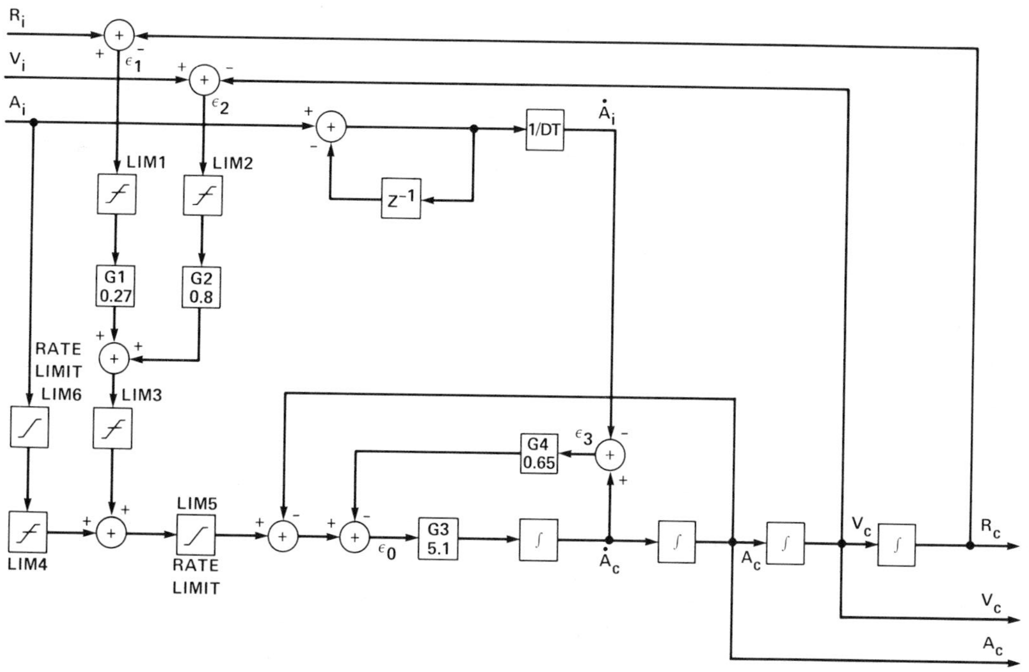

Fig. A.3. Complete translational command generator with three-channel integrators.

ratio of $\zeta = 0.6$. The translational response, associated with the last two integrators was specified to have a natural frequency of $\omega_n = 0.98$ rad/sec and a damping ratio of $\zeta = 0.96$. This leads to a desired transfer function that is a cascade of two second-order systems. The denominator of this transfer function then determines the characteristic equation and is expressed as a quartic equation in the desired natural frequencies and damping ratios. The characteristic equation of the same system as modeled in Fig. A.3 (assuming for the moment that it represents a single channel) is

$$s^4 + G_4G_3s^3 + G_3s^2 + G_2G_3s + G_1G_3 = 0.$$

Matching coefficients between the two characteristic equations leads to the values $G_1 = 0.27$, $G_2 = 0.8$, $G_3 = 5.1$, and $G_4 = 0.65$. The same values were used for all three channels, although a more sophisticated design could easily tailor each channel independently, since they are decoupled. The operation of the command generator can be observed in Fig. 6, where curve 1 shows the rough and smooth horizontal acceleration commands, which are the input and output of the command generator. Similar data for the vertical channel is shown in curve 5.

TRANSLATIONAL REGULATOR SECTION

The translational regulator was designed on similar principles. As shown in Fig. 5, it generates an acceleration correction signal ΔA_c. This signal is a sum of weighted errors between smooth-commanded and actual position and velocity vectors as shown in Fig. A.4. The position and velocity error gains were initially selected to be the same as the position and velocity error gains in the command generator. This is a reasonable approach to close the loops around the aircraft with the same gains that were used to close the loops around the canonical model. This gave generally satisfactory response. However, these gains were not particularly sensitive, and some improvement in the first two channels was seen when the gains in the first two channels were reduced to the values on Fig. A.4.

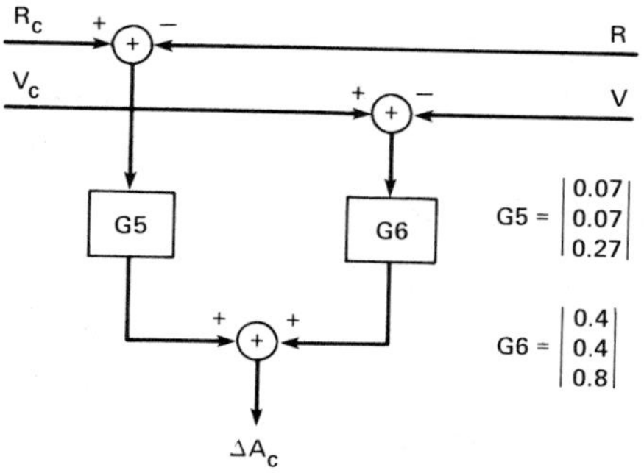

Fig. A.4. Translational regulator.

APPENDIX B. MATHEMATICAL DETAILS

COORDINATE SYSTEMS

Several coordinate systems are used in this material. All are right-handed, three-axis orthogonal systems. Vectors are expressed by their three components in the appropriate system. Figure B.1 illustrates the chain of coordinate systems that transforms quantities from inertial space runway coordinates to aircraft body axes. The inertial space runway axis system is centered at the runway threshold with its first axis pointing north, second axis east, and third axis down. The heading axis system is rotated through the heading angle ψ_V, about the third axis of the runway axis system. The velocity axis system is centered at the aircraft center of gravity with its first axis along the direction of the aircraft velocity vector with respect to the relative wind (airmass). The velocity axis system thus results from a rotation of the runway axis system about its third axis through the horizontal flight-path angle ψ_V, followed by a rotation about its second axis through the vertical flight-path angle Γ_V. The aircraft is assumed to be in an air mass that has a steady wind velocity vector with respect to the ground. The next rotation, about the No. 1 velocity axis through the roll angle ϕ, defines what has been called the wind-tunnel axis system. Rotation about the third axis of this system through the side slip angle (negative β) leads to the stability axis system, and a final rotation about its second axis through the angle of attack α defines the aircraft body axis system. It should be noted that this series of five rotations achieves the same aircraft body-axis orientation as the three conventional rotations through the Euler angles ϕ_E, θ_E, and ψ_E. Three of these axis system are used extensively. The runway axis system (r) is used for basic trajectory definition and for differentiation when inertial accelerations are involved. The velocity axis system (v) is particularly important because the aircraft center of gravity is fixed at the origin and the relative velocity vector is aligned along the No. 1 axis. The aircraft lift and drag are easily expressed in this system. The aircraft body axis system (b) is used for developing forces and torques due to the engine and control surfaces and for body angular rate calculations.

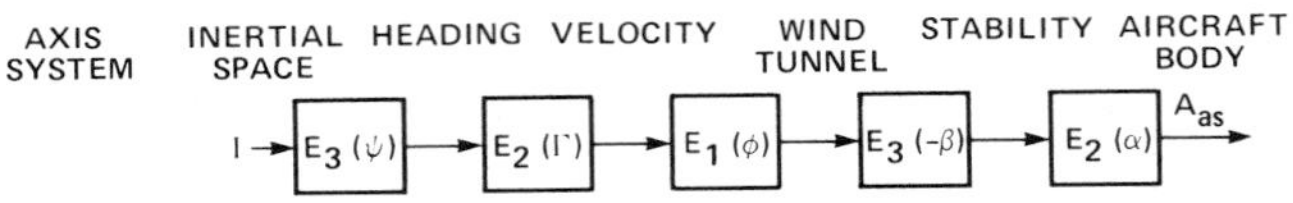

Fig. B.1. Coordinate axis systems.

The coordinates of a vector are transformed from one of these coordinate systems to another by multiplication by the direction cosine matrix between the systems. The general symbol for the transformation is T followed by the designation of the final system, that is, B or V, etc. and then by the designation of the initial system, that is, R,V, etc. For example TBR is the transformation matrix from runway axes to body axes. The rotation from runway axis to aircraft body axes is

$$TBR = E_3(\psi)E_2(\Gamma)E_1(\phi)E_3(-\beta)E_2(\alpha).$$

It should be noted that this does not represent the rotation of the vector.The vector remains fixed but is represented in various coordinate systems that are rotated with respect to each other.

SKEW SYMMETRIC MATRIX FUNCTION OF THE ANGULAR VELOCITY VECTOR

The skew symmetric matrix function of the angular velocity vector $S(\omega)$ is defined as:

$$S(\omega) = \begin{bmatrix} 0 & \omega_3 & -\omega_2 \\ -\omega_3 & 0 & \omega_1 \\ \omega_2 & -\omega_1 & 0 \end{bmatrix},$$

where the vector

$$\omega = \begin{bmatrix} \omega_1 \\ \omega_2 \\ \omega_3 \end{bmatrix}.$$

VECTOR CROSS PRODUCT

The vector cross product $\overline{\omega} \times \overline{V}$ where $\overline{\omega}$ and $\overline{V}$ are three-component vectors is equivalent to the matrix product $-S(\omega)V$.

DERIVATIVE OF A DIRECTION COSINE MATRIX

The derivative of the transformation matrix TBR is given by the matrix product $\dot{TBR} = S(\omega)TBR$.

MATRIX FORM OF THE ROTATIONAL DYNAMIC EQUATION

Let M_r and M_b be the total torque (aerodynamic and propulsive) expressed in the runway inertial frame and the body frame, respectively. Let h_r and h_b be the angular momentum vector in inertial and body coordinates. Let J_r and J_b be the inertia matrix in runway and body axes. Let ω_b be the angular velocity vector of the body-axes with respect to the inertial axes expressed in body-axis coordinates and let TBR be the rotation matrix from the runway axes to the body axes. Note that the vector quantities are all three-element column vectors; that is,

$$h_b = \begin{bmatrix} h_b(1) \\ h_b(2) \\ h_b(3) \end{bmatrix},$$

and the matrices are three-by-three arrays.

Now

$$M_b = TBRM_r$$

$$h_b = TBRh_r$$

$$T\dot{B}R = S(\omega_b)TBR.$$

Neglecting spinning rotors, we have

$$h_b = J_b\,\omega_b,$$

and because J_b is constant in the body frame,

$$\dot{h}_b = J_b\dot{w}_b;$$

furthermore, the basic dynamic relationship in inertial axes is

$$\dot{h}_r = m_r.$$

The expression for angular acceleration using the easily available quantities m_b and h_b may now be found by the following manipulations:

$$h_b = TBRh_r,$$

$$\dot{h}_b = T\dot{B}Rh_r + TBR\dot{h}_r = J_b\dot{\omega}_b.$$

Then,

$$\dot{\omega}_b = J_b^{-1}\left[S(\omega_b)TBR\ TBR^{-1}h_b + TBR\ \ M_r\right],$$

$$\dot{\omega}_b = J_b^{-1}\left[S(\omega_b)h_b + M_b\right],$$

$$\dot{\omega}_b = J_b^{-1}\left[M_b + S(\omega_b)J_b\omega_b\right].$$

REFERENCES

1. G. Meyer and L. Cicolani, "Application of Nonlinear System Inverses to Automatic Flight Control Design System Concepts and Flight Evaluation," AGARDograph No. 251, Theory and Applications of Optimal Control in Aerospace Systems, July 1981.
2. G. Meyer and W. Wherend, "DHC-6 Flight Tests of the Total Automatic Flight Control System (TAFCOS) Concept on a DHC-6 Twin-Otter Aircraft," NASA TP-1513, 1980.
3. G. Smith and G. Meyer, "Application of the Concept of Dynamic Trim Control to Automatic Landing of Carrier Aircraft," NASA TP-1512, 1980.
4. G. Smith and G. Meyer, "Total Aircraft Flight Control System—Balanced Open and Closed Loop Control with Dynamic Trim Maps," Proceedings of the 3rd Digital Avionics Systems Conference, Fort Worth, Texas, Nov. 1979.
5. G. Smith and G. Meyer, "Application of Dynamic Trim Control and Nonlinear System Inverses to Automatic Control of a Vertical Attitude Takeoff and Landing Aircraft," Proceedings of the 4th Digital Avionics Systems Conference, St. Louis, Missouri, Nov. 1981.
6. G. Smith and G. Meyer, "Aircraft Automatic Digital Flight Control System with Inversion of the Model in the Feed Forward Path," Proceedings of the 6th Digital Avionics Systems Conference, Baltimore, Maryland, Dec. 1984.
7. G. Smith and G. Meyer, "Aircraft Automatic Flight Control System with Model Inversion," *AIAA J. Guidance, Control and Dynamics*, May-June 1987, Vol. 10, No. 3 (1987).
8. G. Meyer, "The Design of Exact Nonlinear Model Followers, " Proceedings of the Joint Automatic Control Conference, University of Virginia, Charlottesville, Virginia, June 1981.

9. G. Meyer, "Nonlinear System Approach to Control System Design," First Annual Aircraft Controls Workshop, NASA Langley Research Center, Hampton, Virginia, NASA CP-2296, 1983.

10. G. Meyer, R. Hunt, and R. Su, "Design of a Helicopter Autopilot by Means of Linearizing Transformations," Guidance and Control Panel, 35th Symposium, Lisbon, Portugal (AGARD Conference Proceeding No. 321, 1983).

11. G. Smith, G. Meyer, and M. Nordstrom, "Aircraft Automatic-Flight-Control System with Inversion of the Model in the Feed-Forward Path Using a Newton-Raphson Technique for the Inversion," NASA TM-88209, 1986.

12. H. Driggers, "Study of Aerodynamic Technology for VSTOL Fighter/Attack Aircraft," NASA CR-152132, 1978.

13. "USAF Stability and Control DATCOM," Air Force Flight Dynamics Laboratory, Wright-Patterson AFB, Ohio, Oct. 1960 (April 1978, revision).

14. R. Fortenbaugh, "A Mathematical Model for Vertical Attitude Takeoff and Landing (VATOL) Aircraft Simulation," Prepared under Contract NAS2-10294 by Vought Corporation, Dallas, Texas, for NASA Ames Research Center, Dec. 1980.

15. J. Franklin, C. Hynes, G. Hardy, J. Martin, and R. Innis, "Flight Evaluation of Augmented Controls for Approach and Landing of Powered-Lift Aircraft," *AIAA J. Guidance, Control and Dynamics*, Sept-Oct 1986, Vol. 9, No. 5 (1986).

INFORMATION SYSTEMS FOR SUPPORTING DESIGN OF COMPLEX HUMAN-MACHINE SYSTEMS

William B. Rouse[1], William J. Cody[1], Kenneth R. Boff[2] and Paul R. Frey[1]

[1]Search Technology, Inc.
Norcross, Georgia 30092

[2]United States Air Force
Armstrong Aerospace Medical Research Laboratory
Wright Patterson Air Force Base, Ohio 43455

I. INTRODUCTION

This chapter is concerned with the design of complex systems, and how information systems can support the design process. In order to support design, it is necessary to understand the nature of the design process -- who is involved, when, how, and where? The answer to this question is much more subtle than might at first be imagined [39, 43, 45].

Considering who designs systems, we defined a designer to be anyone who intentionally influences the function and form of the resulting system. This definition quickly leads to the conclusion that designers are elusive and design decision making is pervasive. This is due to the distributed nature of the design process, typically across temporal, organizational, geographic and disciplinary attributes. Further, for complex systems, design decisions emerge both from individuals' activities as well as from group processes when individuals collaborate.

Within this broad perspective on design, there are a variety of views of the nature of decision making of individual designers. Curtis and his colleagues [13] emphasize the relationship between the individual and the design group, as well as the relationship of both with the organization. Schon [52] characterizes design decision making as a process of iteratively uncovering phenomena and seeking explanations.

Rasmussen [35, 36] describes the "space" within which design occurs in terms of levels of abstraction and aggregation of the design artifact -- we have found this construct to be quite useful and return to it in later discussion. Ullman and his colleagues [53] describe elemental activities associated with design in terms of maintenance and updating of the design "state." Goel and Pirolli [18] depict the human information processing underlying design activities.

Within this spectrum from the roles of individuals in organizations to basic human information processing, we have found it most useful to adopt a level of description consistent with that of Schon and Rasmussen. This choice reflects what we found was the most appropriate level of description for design support. In other words, the level of description chosen was dictated by the purpose for which design was being described, i.e., developing a design information system.

Our goal in this chapter is to establish the basis for and describe an information system that enhances designers' abilities to access and use the range of information that pertains in design decision making. As will be shown, useful information exists in many places, forms and states. The support system can be thought of as a computer-based "Designers' Associate" which knows about, can access, and can help the designer to apply this information, thereby improving his decisions. While we recognize the distributed nature of design decision making and envision the Designers' Associate as supporting the special demands of collaborative design, this chapter focuses primarily on the support it will provide to the individual.

A. Information Seeking and Utilization

Considering the range of disciplines, people, and organizations involved with designing complex systems, it is seldom feasible to identify particular decisions or problems pursued by a sufficiently large number of people to justify development of a design support system. Design activities are too discipline-specific and context-specific to allow for a homogeneous solution to supporting all activities. There is, however, a common denominator across all activities -- information seeking and utilization. All designers, regardless of context, discipline, and organization, seek and utilize a wide range of information in the process of design decision making and problem solving [39, 41, 43].

Information can be categorized into that which serves as input to the design process, that which is generated by the process, and the output information that results from the process. Input information includes system requirements, technology alternatives, scientific and technical data, and information on past designs, company practices, and industry standards. Information generated includes sketches, diagrams, and models; analyses and results; and design documentation. Output information includes engineering drawings and parts lists; manufacturing plans for hardware and software; and, of course, design artifacts, i.e., physical encodings of information.

B. Design as Questions and Answers

Information seeking and utilization is a rather amorphous concept. On one hand it seems to be a reasonably concrete activity, but on the other is seems to include almost all activities. Consequently, we felt a strong need to adopt a unifying metaphor to crystallize the meaning and intent of the concept of information seeking and utilization.

The predominant metaphor of design decision making is that of "economic man," analytically decomposing problems, generating all feasible alternatives, and using explicit multi-attribute criteria to choose the optimal solution. Our extensive interviews of designers [11, 43, 45] proved to us that this metaphor is not realistic. Leifer [25] has posed piloting through design space as a metaphor. For us, this metaphor implied too much directed control to be realistic.

Schon's [52] notion of design as a process of iteratively uncovering phenomena and seeking explanations is a much better characterization than top-down decomposition or vehicle control. We hypothesized that design is best characterized as a process of asking questions and pursuing answers as a designer develops and evaluates his or her ideas and the evolving design product [44]. By reviewing the results of a series of questionnaire, interview, and observational studies of designers, we derived a tentative list of 25 types of question within four classes: What? What if? How? Why?

Virtually, any design activity can be fit into the questions/answers metaphor. For example, the activity of sketching can be viewed as pursuing an answer to the question, "What will it look like?" While this may be stretching the metaphor somewhat, we are primarily concerned that it is useful, not that it is correct. The "proof of the pudding" is in the design support system that results from adopting this metaphor -- in other words, to the extent that information access and utilization, as well as the designed artifact, are enhanced, the metaphor has been "proven" to be useful.

II. A THEORY OF DESIGN

Using the questions/answers metaphor, we reviewed the results of several previous data collection efforts aimed at understanding the nature of design -- these efforts are summarized in [45]. The result was a theory of design decision making and problem solving [44].

Central to this theory is the three-dimensional design space depicted in Figs. 1 and 2. Fig. 1 depicts the abstraction and aggregation dimensions of the design space. This characterization is based on Rasmussen's constructs [35, 36]. The definition of aggregation is obvious from the figure. Abstraction is more subtle.

The concept of abstraction can be used to describe the types of representation relevant to design -- representation provides the context

LEVEL OF AGGREGATION	LEVEL OF ABSTRACTION		
	Purpose	Function	Form
System			
Subsystem			
Assembly			
Component			

Fig. 1. The abstraction and aggregation dimensions of the design space

PURPOSE	FUNCTION	FORM
Explore Problem/Need	**Conceptualize Solution Functionality**	**Compose Form of Solution**
• Study current requirements (e.g., Statement of Work) - read and analyze	• Review functionality of past designs - read and analyze	• Review forms of past designs - read and analyze
• Study scenarios of operational need - view and analyze	• Synthesize/derive input/output relationships - create and represent	• Synthesize form of solution - create, visualize, and "sketch"
• Review requirements for past designs - read and analyze	• Develop model of functionality - integrate, analyze, & test	• Prototype/mockup solution - integrate and fabricate
• Explicate performance attributes and criteria - integrate and decide	• Predict performance (exercise model) - calculate/simulate, analyze and interpret	• Measure performance (collect data) - observe, measure, analyze, and interpret

Fig 2. The task dimension of the design space

within which design questions are posed and answers pursued. The three levels shown in Fig. 1 can be defined as follows:

- Purpose: representation of design requirements, objectives to be met, problems to be solved, etc. via scenarios, simulations, requirements, documents, etc.
- Function: representation of relationships (i.e., physical, computational, etc.) via diagrams, equations, simulations, etc.
- Form: representation of appearance (i.e., geometry, assembly, etc.) via drawings, pictures, mockups, etc.

Fig. 2 depicts the task dimension of the design space. The tasks shown, and perhaps others that are similar in nature, can be viewed as designers' proximal intentions as they ask questions and pursue answers.

Based on the three-dimensional design space depicted in Figs. 1 and 2, the following theory of design decision making and problem solving emerged. Simply stated, design behaviors can be characterized in terms of asking questions and seeking answers, within a context that can be represented as trajectories or sequences of tasks at various levels of abstraction and aggregation.

The purpose of this theory is to provide a basis for developing a design information system. Our concern, therefore, was whether or not this theory would be useful in this manner. To assess the utility of the theory, we used trajectories in the design space as a basis for creating design scenarios which designers were asked to rate relative to the "reasonableness" of their depiction of design tasks [44]. Results of this study indicated that the theory's characterization of design behaviors was satisfactory to the sample of designers we queried.

Of course, these results do not directly relate to the "bottom line" for the theory. The real test of the theory is whether or not a design information system based on the theory turns out to be usefully supportive. To this end, we now consider the design support requirements implied by the design space characterization.

III. REQUIREMENTS FOR DESIGN SUPPORT

What should a design information system do to support designers whose behaviors can be described by the above theory of design? To answer this questions, we utilized an analysis methodology that we have found useful for designing information systems for applications in command and control [50], nuclear power [48], and manufacturing [40]. The steps of this methodology are depicted in Fig. 3. In this section, we describe this analysis methodology in the process of applying it to developing an information system for design support.

A. Step 1: Define Tasks

This step is concerned with defining the tasks to be supported. For this application, this step was completed by adopting the characterization of the design space in Figs. 1 and 2. Thus, there are twelve tasks to be supported, each of which can occur at various levels of aggregation.

B. Step 2: Map to General Tasks

Use of the analysis methodology results in identification of the support functionality necessary for the tasks specified in Step 1. Alternative support concepts are indexed in terms of one or more general tasks. This set of 13 general tasks and 17 associated support concepts were derived from an analysis of the design documentation of over 100 past support systems -- see [40] for the most recent exposition on these sets of tasks and support concepts.

Thus, the set of general tasks is the intervening mechanism for identifying support concepts. Therefore, in this step of the analysis process, one maps or links each of the tasks specified in Step 1 to one or more of the general tasks.

Two analysts independently performed the mapping from each of the 12 design tasks in Fig. 2 to the 13 general tasks. Results were then compared and a consolidated mapping produced which had 64 links, or an

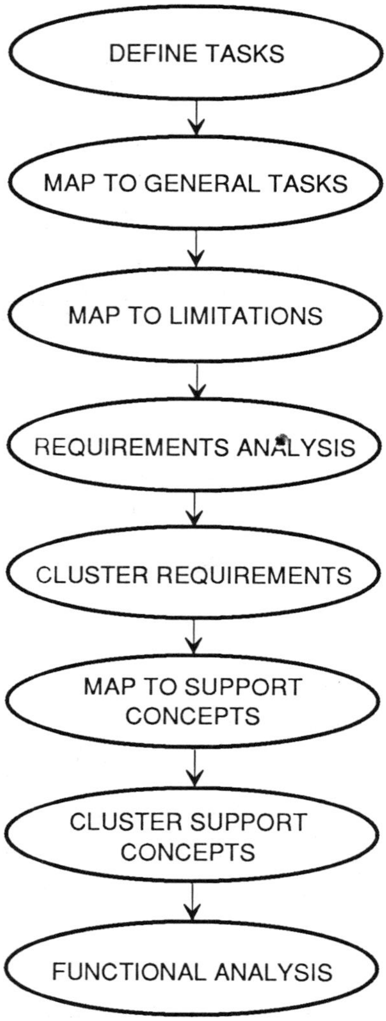

Fig. 3. Design methodology

average of somewhat over 5 links per design task. Each of these links was annotated with the reason for the link and a design-specific interpretation of the connection. This set of 64 design-specific instances of the general tasks served as the input to the next step.

C. Step 3: Map to Limitations

The next step involves considering the limitations that designers are likely to face in performing these 64 tasks. Limitations can be characterized in more than one way.

Sage [51] and Zachary [54] have considered the psychological limitations associated with decision making and problem solving in complex systems. The difficulty with tabulations of psychological limitations is that they are very general. With a little stretching, it is easy to imagine mapping each of the 64 tasks to all of the psychological limitations. Such a result is not useful.

We have found it more useful to think in terms of task limitations that are tailored to the context of interest. For information seeking and utilization within design, two types of task limitation are relevant. As shown in Fig. 4, the task limitations of interest relate to getting things and doing things.

Each of the 64 tasks was reviewed to determine whether or not each of the 10 limitations in Fig. 4 was likely to affect the task. We assumed a journeyman designer and concluded that the "why" and "when" limitations in Fig. 4 would not be relevant. Applying each of the remaining 7 limitations to the 64 tasks resulted in 213 task-limitation pairs that represented potential needs for support.

D. Step 4: Requirements Analysis

Each of the 213 requirements was then analyzed to determine in a very design-specific manner what needed to be done to overcome the limitation. This was accomplished by reviewing each of the 213 in terms of the nature of the limitation (Fig. 4), general task involved, and design task involved (Fig. 2). The results of this analysis formed a computer-

- Not knowing about objects
- Not knowing where to get objects
- Not knowing why to get objects
- Not knowing how to get objects
- Not being able to get objects

- Not knowing what to do
- Not knowing when to do it
- Not knowing why to do it
- Not knowing how to do it
- Not being able to do it

Fig. 4. Potential task limitations

readable database describing the requirements to overcome each relevant combination of limitations, general tasks, and design tasks.

E. Step 5: Clustering Requirements

The 213 requirements, including the associated context-specific interpretations, were sorted into clusters with common limitations and general task attributes. This type of sorting was needed in order to map to the aforementioned list of support concepts which is indexed by general tasks. Limitations were used as a second sorting attribute because the interpretation of support concepts is influenced by the nature of the limitation that the support is to help overcome. This process resulted in 43 clusters of requirements which served as input to the next step.

F. Step 6: Map to Support Concepts

Each of the 43 clusters was then mapped to one or more of the 17 general support concepts. The mapping was determined by the nature of the general task associated with the cluster. Each relevant support concept was then interpreted within the context of the limitation associated with the cluster, as well as the design tasks (Fig. 2) with links to this cluster.

It was necessary to develop and maintain a highly structured and controlled vocabulary for expressing the specific instances of support needed. This was dictated by the need to consistently and comprehensively manage the analysis process. After several iterations, 613 instances of support resulted.

G. Step 7: Cluster Support Concepts

The database of support concepts was then sorted using the terms of the aforementioned controlled vocabulary. The types of term included:

- o Support verb: Actions of the information system to support the designer
- o Designer verb: Actions of the designer that are enhanced by the support system's actions
- o Primary object: Object of the designer's actions
- o Modifying object: Noun plus preposition that modifies a primary object

Typical entries in the database are of the form (support verb) (designer verb) (modifying object) (primary object). An example entry is: (Explain procedure) (to locate) (sources of) (drawing tools/packages).

Support verbs are summarized and defined in Fig. 5. This information is the primary input to the next step, functional analysis. Designer verbs, modifying objects, and primary objects are listed in Fig. 6.

The meaning of the words and phrases on these lists are quite straightforward, with one exception. Both the singular and plural of drawings of form, model of functionality, and prototype appear as primary objects. The plural refers to retrieving relevant candidates, while the singular refers to creating a candidate.

It is interesting to note that with 6 support verbs, 11 designer verbs, 8 modifying objects, and 36 primary objects, there over 19,000 possible combinations. The 613 actual instances of support account for roughly 3% of these alternatives. Thus, the results of the analysis process

SUPPORT	NUMBER	DEFINITION
Search	109	Use attributes or labels to identify and/or locate objects.
Execute	78	Perform procedures to access, construct, evaluate, measure, etc.
Indicate	21	Display variables, relevant procedures, necessary activities, etc.
Transform	2	Modify, filter, and/or highlight observed or computed variables.
Explain	325	Interpret procedures, measures, variables, transforms, etc.
Tutor	78	Coach in use of procedures to access, construct, evaluate, etc.
Total	613	---------------------

Fig. 5. Summary of support verbs

represent a much more structured and focused conclusion than a purely combinatoric tour de force would indicate.

H. Step 8: Functional Analysis

This step is concerned with converting the results of Step 7 into a functional architecture for a design information system. The purpose of this architecture is to provide the basis for realizing the 613 instances of support that resulted from the requirements analysis. The question/answer metaphor played a strong role in the formulation of the requirements analysis. Similarly, this metaphor played a central role in conceptualizing a support system architecture.

DESIGNER VERBS	MODIFYING OBJECTS	PRIMARY OBJECTS
Access Construct Create Evaluate Identify Locate Measure Monitor Obtain Run Select	Availability of Between acceptance and rejection of Collection of Deviations from expectations of Execution of Explanations of Processing of Sources of	Data collection plan Deviations of measured performance Deviations of predicted performance Drawing tools/packages Drawings of form Drawings of forms Experimental variables Explanations of forms Explanations of functions Forms of past designs Functionality of past designs Information on forms Information on functions Information on operational needs Input/output representations Measured performance Model of functionality Models of functionality Model's predictions Model's variables Modeling tools/packages Off-the-shelf forms Off-the-shelf functions Off-the-shelf prototypes Past designs (forms) Past designs (functions) Past designs (requirements) Performance attributes and criteria Performance Performance data Prototype Prototypes/mockups Prototyping tools/packages Relevant input/output representations Requirements information for current design Requirements information for past designs

Fig. 6. Summary of designer verbs, modifying objects, and primary objects

IV. AN ARCHITECTURE FOR DESIGN SUPPORT

The information environment within which designers ask questions and pursue answers can be depicted as shown in Fig. 7. The central focus of the information environment is the designer's ideas and the emerging design artifact [44].

Designers ask many questions and pursue many answers in the process of formulating, representing, manipulating, and evaluating their ideas. This involves accessing many information sources and a variety of tools. However, the goal is not to use information or exercise tools. These things are only a means to the end of the designer pursuing his or her ideas, and generating information in terms of manifestations of these ideas.

From this perspective, for an information system to be perceived as valuable, it must increase the perceived benefits of utilizing the information and/or decrease the perceived costs of accessing the information [4, 38, 39]. The central issue, therefore, becomes one of determining how to configure the types of support summarized in Figs. 5 and 6 into an information environment with high perceived benefits and low perceived costs. The overall architecture for design support is shown in Fig. 8. Several aspects of this architecture are of particular importance.

First, any of the primary functions of Search, Execute, Indicate, and Transform can be accessed without having to access the other functions. Thus, the designer is not constrained to following a particular design trajectory. In fact, the designer's behaviors need not conform to the underlying questions/answers metaphor.

A second aspect of note is the ability to access the functions of Explain and Tutor from each primary function. In this way, explanations and tutoring can be tailored to the type of support being utilized and the current state of the design process. Note also that Tutor can be accessed either via a primary support function or via Explain. This provides the possibility of smoothly transitioning from an explanation that is not understood to more in-depth tutoring.

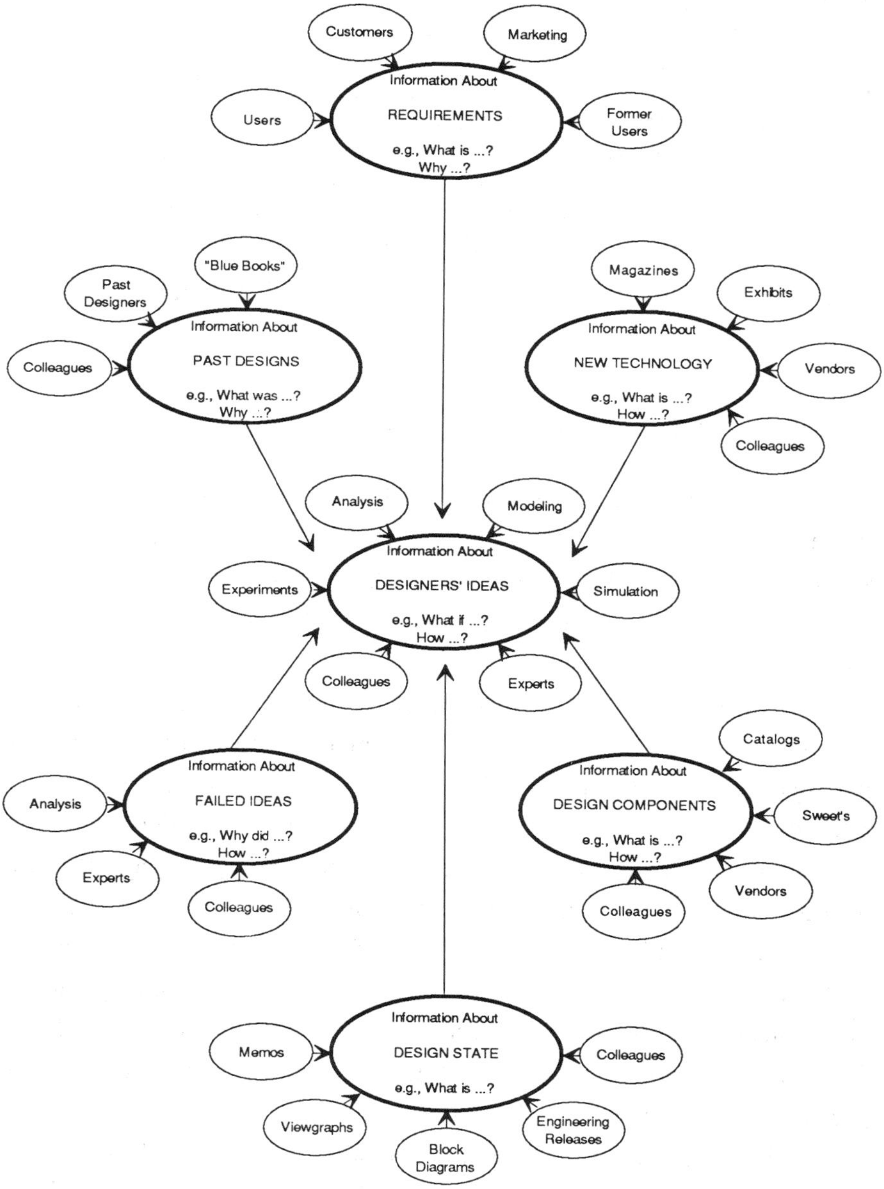

Fig. 7. Information Environment of Designers

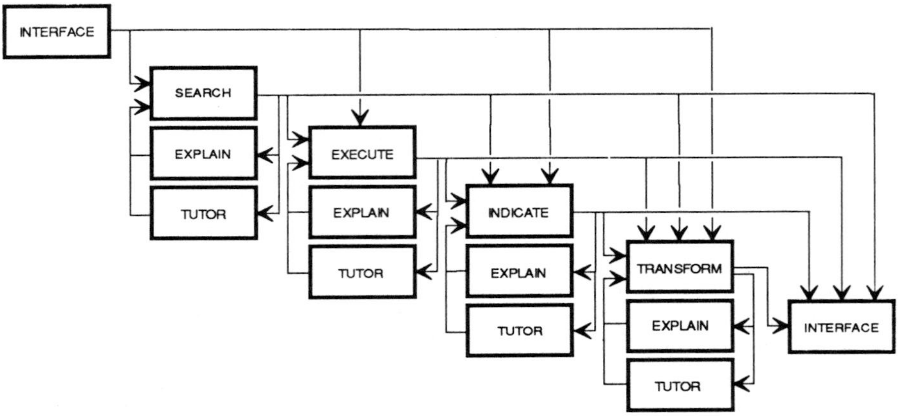

Fig. 8. Overall architecture for design support

A third noteworthy aspect of Fig. 8 is the interface, which is shown twice in this diagram. The interface provides functionality that goes substantially beyond input devices and display pages. We envision an intelligent interface of the type advocated by Rouse and his colleagues [46].

The interface should have three central functions: information management, error monitoring, and adaptive aiding. The purpose of <u>information management</u> is to deal with the "overhead" associated with retrieving, creating, editing, and storing information in the process of utilizing the functionality shown in Fig. 8. The role of <u>error monitoring</u> is to observe the data stream generated by the designer, looking for inconsistencies and anomalies indicative of inadequate, inappropriate, or incorrect use of the functionality in Fig. 8. The purpose of <u>adaptive aiding</u> is automatic prompting, and occasionally initiation, of use of the functionality in Fig. 8 to enhance information access and utilization in the process of pursuing designers' goals.

Past experience in developing intelligent interfaces [46, 47] has shown that the key to providing the above types of support in an intelligent interface lies in being able to know what the human is doing and how these activities relate to goals and plans. This requirement can be satisfied for design support by recalling that our overall analysis is premised on design behavior being described as trajectories, or sequences of tasks, in the design space depicted in Figs. 1 and 2.

If the intelligent interface knows where the designer is in the design space, in terms of task and levels of abstraction and aggregation, as well as the likely trajectory of the designer, then it should be quite feasible to provide the type of intelligent interface described above. In past experiences in aviation and process control applications, we found that much of humans' activities could be described, at least qualitatively, in terms of informal scripts and plans that emerge repeatedly and can provide the basis for inferring goals and plans. We expect that "standard" trajectories in the design space can be identified and serve the same purpose [43].

Our analysis thus far has considered design in general rather than any particular design domain. However, in order to provide a much more concrete image of the type of support that we envision for a "Designers' Associate," it is necessary to consider the implications of the foregoing discussion for a particular design domain, namely, human-machine systems.

V. DESIGN OF HUMAN-MACHINE SYSTEMS

A. Motivation for Choosing Human-Machine System Design

Three reasons underlie our selecting this particular domain to illustrate a design information system that supports information access and utilization.

First, the need for innovative design supports in this domain is undeniable. Technological advances continue to enable ever-more complex systems. Examples that immediately come to mind include spacecraft, automated process control and manufacturing facilities, and new transportation systems. However, there is ample evidence to show that the personnel involved with such systems often encounter serious problems which end up reducing system effectiveness and safety. Cases of operators becoming overwhelmed by number of subsystems and modes, maintainers by convoluted failure sequences and repair requirements, and managers by volume and variety of data, seem to be reported daily.

Presumably, these problems are sown inadvertently during design, which suggests that the job of human-machine design is itself a difficult enterprise. Thus, new supports that help designers to avoid human-related problems appear needed and should have wide appeal.

The second motivation is the challenge that developing a support system for this domain represents. Human-machine system design is different from other engineering disciplines for which support systems might be developed. For one thing, it is multi-disciplinary. Information needed to make human-related decisions is distributed across several specialties. In aerospace, for example, over 40 specialties contribute to crew system design decisions (see Fig. 9). Thus, the domain poses a significant design support challenge on the basis of information volume and distribution alone.

Beyond volume and distribution, however, the larger challenge is in translating among and reconciling the disparate points of view. Due to its multi-disciplinary makeup, human-machine system design admits a wide variety of representational schemes, methods and analytical tools. These have been borrowed from mechanical and electrical engineering, operations research, computer science as well as the behavioral sciences. Unlike aerodynamicists, for instance, who can converse in the singular representation of the Navier-Stokes equations, human-machine system designers can appear more like the participants in the Tower of Babel as they try to share representations [5].

Information transfer has a subtle side as well. Studies of engineers' and technologists' information seeking habits have shown that searching for and using information from outside of one's expertise is quite different from inside [16, 31]. Search is less comprehensive, less effective, and interpretation of what is obtained is sometimes quite muddled. This phenomenon is no where better illustrated than in the use of behavioral information by nonspecialists in human-machine system design [30].

Acoustics	Industrial Design
Aeronautical Engineering	Logistics Support Analysis
Anthropometry	Maintenance Engineering
Artificial Intelligence	Manufacturing & Fabrication
Atmospheric Physics	Mathematics
Cartology	Mechanical Engineering
Chemical Engineering	Meteorology
Climatology	Navigation
Colorimetry	Optics
Computer Science	Optimization
Control Systems Engineering	Personnel Management
Cost Effectivess Modeling	Photometry
Cybernetics	Physiology & Madicine
Display System Engineering	Piloting
Dynamics	Pneumatics
Education & Training	Procurement
Electrical Engineering	Psychology
Electronic Engineering	Quality Assurance & Control
Ergonomics	Safety
Facilities Engineering	Sensors
Flight Simulation	Simulation Modeling
Fluidics & Hydraulics	Speech & Linguistics
Gunnery & Weaponry	Statistics & Probability
Human Factors	Systems Engineering
Heating & Air Conditioning	Thermodynamics

Fig. 9. Aircrew system design involves many disciplines

In part, cross-disciplinary information seeking may be hindered by people's judging the risks associated with issues they don't know much about differently from those they know a lot about [3]. Gemunden [16] has reviewed the evidence which demonstrates that risk judgments are related to information seeking (although the relationship is complex). For purposes of design support, the lesson is that the system must both enable users to access unfamiliar information sources and provide help in the proper use of what they obtain.

The third motivation is that the human-machine system design domain provides a strong test for the analysis method used here to develop the design information system. If the system that emerges is roundly criticized by its intended users then, assuming that an implementation delivers the intended functionality, the development method is brought into question. On the other hand, if the system is an appealing one, this would encourage the use of the method for deriving other support systems.

B. Nature of Human-Machine System Design

It is useful to examine briefly the nature of the problem that human-machine system designers face to better understand their tasks and information needs. (More detailed accounts of this design area are available [1, 29, 31]). This domain can be distinguished from other design disciplines in terms of the system attributes that fall into its jurisdiction. These attributes can be grouped into *micro* and *macro* human-machine properties which together govern the role and well-being of humans in the system as well as system cost [21].

Micro human-machine attributes include comfort, safety, habitability, ease with which tasks are learned, effectiveness with which they are performed, error potential, tolerance of the system to human misuse, maintainability, and others. The designer controls these attributes by manipulating several variables. These include the physical compatibility of equipment and environmental conditions with human input-output characteristics, the cognitive demands of the structure and content of user-system dialogues, personnel characteristics, training regimen, various types of online and offline aid, and protection equipment.

Macro human-machine properties are the manpower, personnel, training and organizational requirements of the system. These are determined partly by design choices at the micro level but by economic and political factors as well.

Fig. 10, adapted from Rouse [37], illustrates the nature of the design problem. This depiction suggests that designers control micro and macro system properties through selections they make with respect to six key design factors (left-most column). These factors govern human performance determinants such as physical and intellectual properties,

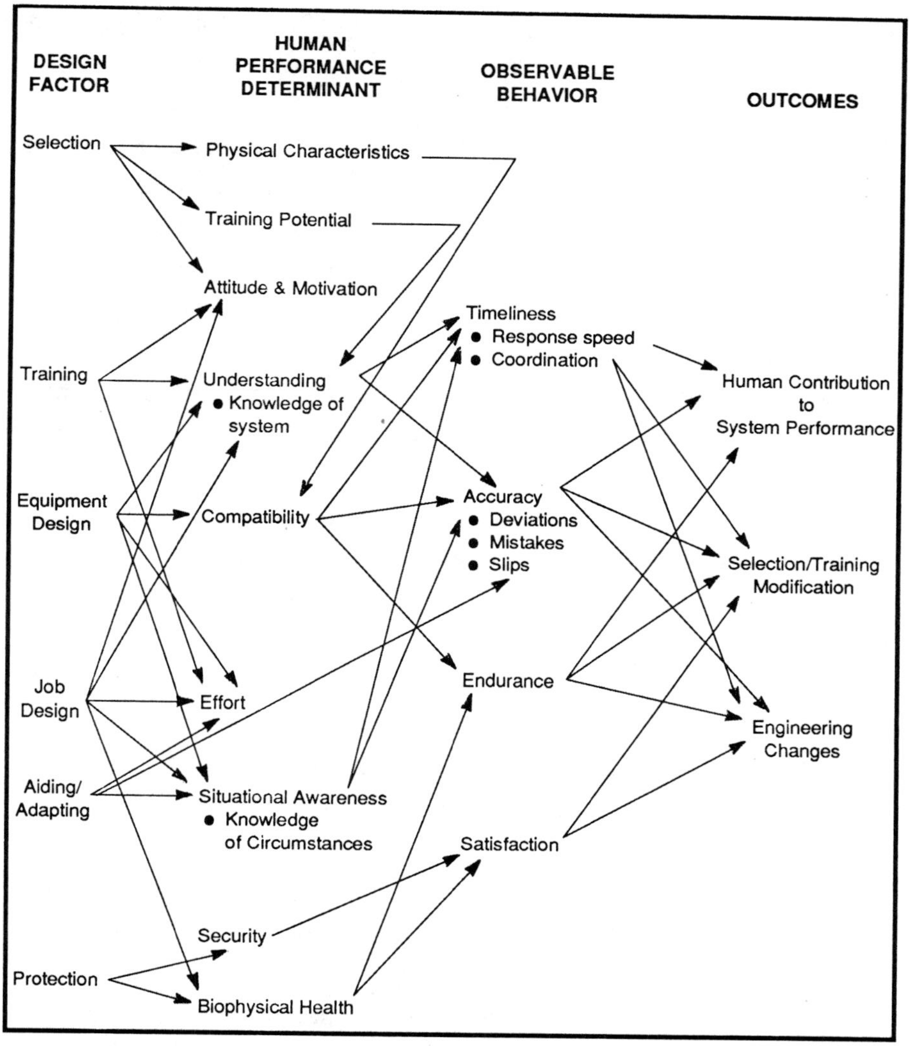

Fig. 10. A model of the crew system design problem

numbers of people required to do the job, knowledge of the system, human-equipment physical compatibility, effort or workload during system operations, biophysical health and perception of safety. In turn, these intervening variables govern the observable human contributions to the system and, ultimately, the adequacy of system performance. As Fig. 10 suggests, the design problem is challenging for two reasons. Many of the variables and relationships are difficult to define and quantify. They also interact in sometimes intractable ways.

Fig. 10 also helps to illustrate the multi-disciplinarity of this domain. In aerospace, for example, personnel selection practices are dominated by individuals with behavioral sciences backgrounds. Training curricula are designed by subject matter experts and specialists in education and psychology. In contrast, training devices (e.g., simulators) are the province of control engineering, electronics, computer science and optics. For equipment design, job design and aiding, designers include specialists in avionics, sensors, computer hardware and software, human engineering, life support, safety, operational analysis, and piloting. Protection is the bailiwick of physiologists, biomedical engineers and other engineering specialties that contribute to life support equipment.

In addition to the complexities that are fundamental to all human-machine system design, we currently are witnessing a major shift in human-machine interaction paradigms that is changing the nature of designers' information needs. In the older paradigm, people are conceptualized as input-output devices that close loops between displays and controls in several discrete subsystems. The designer's information requirements are dominated by the need to predict how the human will perform as a system module. Sensory, perceptual and motor-control performance information helps with equipment design. Attention switching rates and vigilance characteristics support task scheduling. Skill acquisition rates support training systems design. Speed and accuracy of recalling things from memory help with procedures design.

Advances in technology are changing all of this. With respect to the human's role, the emergence first of reliable automatic controls allowed the designer to shift the human from component to "supervisor." In this capacity, the human is more oriented toward setting initial conditions, monitoring overall system performance, making adjustments, and intervening as needed.

Now as intelligent systems technology is maturing, humans and their machine facsimiles can swap the supervisor and subordinate roles.

At times, the human can monitor and correct the automation; at other times, it makes sense for a computer to monitor and correct the human. Rouse and colleagues [46, 47] have termed this arrangement "collaborative control."

Supervisory and collaborative control have radically expanded the designer's needs for behavioral information. Knowing the engineering properties of the human controller is necessary but no longer sufficient. How humans seek information, integrate disparate facts, make decisions, judge risks, etc., are becoming central to advanced systems design. Indeed, the primary information requirements in design are shifting from what the human component *can* do to what the supervisor or collaborator *will* do.

C. Tasks in Human-Machine System Design

What tasks does human-machine system design involve, and to what extent do the behaviors listed in Fig. 2 cover these tasks? Their correspondence is important because the design information system was derived to service the behaviors in Fig. 2.

Despite a voluminous literature on the topic of design, there is surprisingly little information about what designers do. As far as we know, no one has reported a detailed "task analysis" of the human-machine design job. Many prescriptions for how design *ought* to be done have appeared, but there are very few studies on how design actually *is* done. Some information about what designers do can be gleaned from descriptions of roles and responsibilities that most engineering organizations prepare for technical staff members [9]. These descriptions help to define what designers are supposed to accomplish but not how they actually go about it.

To address this issue, we extracted a set of key human-machine design problems from literature sources [1, 2, 29, 31], field interviews that we conducted in the aerospace domain [11, 12, 43, 44], and our own design experiences. Fig. 11 shows a "top ten" problems list along with samples of the associated inputs and outputs. While resolving just these

HUMAN-MACHINE SYSTEM DESIGN PROBLEMS	DESIGN OBJECTS INPUTS	DESIGN OBJECTS OUTPUTS
Identify Functional Requirements	o Statements of need o Current system deficiencies o Engineering change proposals o Performance & cost constraints	o Function definitions o Functional flow diagrams o Function timeline o Plan/Goal graphs
Allocate Functions to Human & Machine Elements	o Function timeline o Human limits & abilities o Technology assessment	o Allocation timeline o Task definitions, performance criteria & estimate o Manpower, personnel & training estimates
Establish Information & Control Requirements	o Function definitions o Allocation timeline o Scenarios	o Information dictionary o Information flow diagram
Design Equipment, Displays, & Controls	o Human engineering properties o Constraints (size, power, ...) o Information flow diagrams	o Media specifications o Contents & format specifications o Training requirements
Design Procedures & Dialogue	o Functional & information flow o Human communication phenomena o Performance criteria	o Task definitions (OSDs, IBDs) o Training requirements o Manuals o Workload & error predictions
Layout Workspace	o Human anthropometry/biomechanics o Constraints (electrical, mechanical) o Human spatial perception & memory phenomena	o Geometry drawings & specs o Equipment arrangement o Link analysis & performance estimates
Design Aids & Adaptive Interfaces	o Allocation timeline o Workload & error predictions o AI technology o Performance criteria	o Control switching specifications o Job aids o Adaptive interface software
Design Training Systems	o Personnel aptitude o Knowledge & skill requirements o Human learning & memory phenomena	o Curriculum design o Courseware o Training device specifications o Training time/cost predictions
Design Protection & Life Support Systems	o Expected operational requirements o Human physiology o Biomedical technology	o Work environment specifications o Personal gear & equipment o Escape, survival, & recovery systems
Evaluate Human & System Performance	o Artifact descriptions o Artifact o Scenario of expected use o Performance criteria	o Data collection plans o Modeled or measured performance o Deviations from expectations o Explanations of deviations

Fig. 11. Problems, inputs and outputs in human-machine system design

problems may not exhaust the designer's responsibilities, most people would agree that they are salient issues and critical to system success.

For each problem, we identified what the designer would be expected to do to achieve a resolution, and then compared these behaviors with the general tasks listed in Fig. 2. This exercise produced ten trajectories, one per problem, across the tasks in the general design task space. Our goal was to identify any glaring omissions or unnecessary tasks in the general list from the perspective of human-machine system design.

While the analysis identified no outright omissions, a few activities in human-machine system design seemed to be underemphasized in the list of general tasks. In particular, activities that involve projecting the consequences of design changes to system properties and making explicit specification of design parameters are central to resolving the design problems in Fig. 11, but seem underemphasized in Fig. 2. No doubt, the apparent lack of emphasis is due to our concentrating on information access and utilization.

Regarding unnecessary tasks, none of the twelve general tasks emerged as superfluous. Although not all tasks were required in the resolution of each problem, each task was sampled across the ten design problems.

D. Limitations in Designers' Abilities to Perform Human-Machine Design Tasks

Given this correspondence, we then sought evidence in the human-machine design domain for the types of limitation that our support system is aimed at ameliorating. Recall that seven limitations were identified (Fig. 4), and that these could be clustered into two groups. Four limitations pertain to design objects (i.e., not knowing about useful information, where to obtain it, how to obtain it, and not being able to obtain it). Three limitations pertain to performing design tasks (i.e., not know what to do, how to do it, and not being able to do it).

Using the primary and modifying objects in Fig. 6 as a starting point, we generated a detailed list of objects that typically are used, are produced, or should be in the course of resolving the ten design problems. Basically, "objects" refer to task inputs and outputs and can be the product of performing an information search, using a design tool, conducting an experiment, etc. Several are listed in Fig. 11. We then reviewed several critiques of design practice in human-machine systems to establish the nature and prevalence of both object-related and task-related limitations in conjunction with these objects.

Beevis [2] assessed the quality of human engineering activities in ten military system design projects. He found a variety of organizational barriers to human-centered design. These included the low status of human engineering and consequent lack of time, staffing, and resources devoted to operability issues; inertia of former solutions; and the sequential nature of design in complex organizations that tends to shortchange the potential influence of disciplines that contribute near the tail end of the process (i.e., human engineering). Perrow [33] has reported similar findings.

In terms of the seven limitations in Fig. 4, these barriers manifest themselves as limitations in designers' abilities to obtain information both inside and outside of their organizations and in their not be being able (permitted) to perform certain tasks.

Beyond organizational barriers, Beevis [2] also found deficiencies in the approaches taken by individual designers to nearly all of the problems listed in Fig. 10. This suggests that not knowing what or how to do some human-related design task stand in the way of the individual designer. Examples include using unrealistic mission scenarios for functional requirements analyses, allocating functions on a strictly ad hoc basis, designing procedures and tasks that clearly exceed human capabilities, and others.

In a similar vein, Promisel and his colleagues [34] "reverse engineered" several design decisions related to human use in four case studies of military system design. They uncovered various deficiencies in how the human user was considered during design which led to subsequent operational problems. For example, the design of one procedural sequence in an anti-tank weapon was so complicated that most members of the intended soldier population were unable to operate the system well enough to meet the performance requirements. Evidently, the designers had difficulty taking personnel capabilities into account and judging the training requirements of their creation.

Meister and his colleagues [29] and Lintz et al. [27] conducted a series of studies to establish why human engineering information in particular is introduced late or not at all into systems design. They found that engineers did not oppose using this class of information in design tradeoffs if it was readily available and if the statement of work required them to do so. However, these individuals showed unexpectedly high rates of misinterpreting the information and making design decisions completely at odds with what the information would recommend. The investigators concluded that the manner in which this class of information has traditionally been prepared for nonspecialists in human performance is the culprit, not that the engineers are dull. From our perspective as support system developers, these findings imply that accessing technical information is not strictly a matter of gaining physical custody of a source. Rather, inaccessibility is as much a function of cognitive factors as physical ones.

In the realm of software engineering, Gould and Lewis [19], Hammond et al. [20] and Dagwell and Weber [14] all have studied designers' "common sense psychology" of users and their tasks. These investigators have shown that designers' beliefs about users can interfere with access and use of pertinent information. For instance, when asked to explain specific interface design decisions, designers often provided theories of what users will know or be able to learn. These theories usually contained specific assumptions about human cognition for which empirical evidence may or may not exist, let alone agree. A particularly striking feature about these personal theories is that they are stated as broad generalizations, not subject to variation over tasks or individuals. "All users will...."

The extent to which such theories contribute to design decisions rather than justify them after the fact is not clear. However, their pervasiveness suggests that many designers are not knowledgeable about evidence from the cognitive sciences that bear on their choices.

Moreover, since personal opinion is so readily available compared with more formal sources of information, the competition they pose is substantial if it cannot be demonstrated that formal sources yield more effective or less expensive inputs.

Rouse and Cody [43] and Cody [11] interviewed over 60 aerospace system designers as to their daily activities, design practices and information needs especially regarding human-related issues. Interviewees recounted several vignettes of personal design experiences to exemplify how they identified and resolved design problems in general. Three pervasive strategies for resolving human-related issues emerged from analysis of these vignettes. First, designers often resorted to personal models and theories of end-user psychology to justify design choices, a finding that corroborates results from the software engineering domain. Second, they "role played" the part of the user, considering themselves to be representative surrogates for all members of the user population. Chatelier [10] has reported a similar phenomenon based on twenty years of observing the military system acquisition process. Third, interviewees described cases in which they basically avoided the human-related problem altogether by discounting the risk associated with them in comparison with engineering and economic matters.

Each of these strategies suggests one or more of the seven task limitations. Using personal theories and role playing as opposed to accessing accurate information suggests that designers may not have known about the information or could not obtain it for any of several reasons. Similarly, discounting risks when the risks are real may indicate a lack of knowledge about human-related problems and how to address them.

Rouse and Cody [44] developed a fictional scenario, based on the theory presented above in Section II, in which an engineer re-designs a system that tends to induce human error. In the scenario, the designer must access a variety of information sources, models, and theoretical explanations for human behavior in a complex work context, and clearly, has difficulty knowing what to ask for and occasionally how to use what he gets. In interviews, practicing engineers who had read the scenario said that it was a credible portrait of design, especially the sleuthing component of the engineering change process. Although they felt that a support system depicted in the story was unrealistic (too futuristic), the interviewees had no objections with the tasks that the fictional designer performed; these seemed natural and familiar.

Regarding limitations, where the fictional designer had difficulties obtaining information and performing tasks, the practitioners both empathized with his predicament and recounted episodes from their own experiences in which they had very similar problems. This concurrence was particularly important to the present line of reasoning. It suggests that the types of limitation identified in Fig. 4 pertain in particular to human-machine system design. Furthermore, designers' assent to the story provide some confidence in the theory (Fig. 2) which is the basis for the design information system.

Finally, Cody and Rouse [12] polled over 100 system designers who attended a recent workshop on human performance models regarding their use of these tools in actual practice. Responses showed that, prior to the workshop, attendees were largely unaware of the existence of most of the thirty tools and approaches that were presented, let alone know how to use them. Following the workshop, attendees were attracted to several general modeling packages (e.g., MicroSAINT, SLAM) and conceptual frameworks for interpreting human behavior in systems contexts. They showed only lukewarm interest, however, in obtaining and using most of the other tools that had been described. General modeling packages permit the user to create context-specific tools that are perceived to be worth the effort to design and build. In contrast, the remaining offerings, which appeared to lack generality across contexts, were perceived to cost more to learn and use than the practical benefits they could deliver.

What is important about the poll findings is that, as Boff [3, 4] has discussed, the value of information in clearly in the eye of the beholder and is governed by two independent factors: usefulness and usability. Usefulness refers to pragmatic value (applicability, credibility, etc).

Usability refers to the appropriateness of form and presentation to the problem at hand. Thus, regardless of their actual value, information sources can elicit reactions in their intended consumers which encourage or discourage use for two very different reasons; reactions are governed by content as well as by "packaging." The lesson is that object- and task-related limitations result not only from deficiencies with the information sources that are external to the designer, but also from designers' perceptions and beliefs about these external sources.

To summarize, there is clear evidence from a variety of sources that the design of human-machine systems exhibits the types of limitation listed in Fig. 4. Although questionnaires, interview data and examination of designed artifacts themselves often cannot distinguish which limitation(s) was operative in a particular case, it is safe to conclude that all of the limitations emerged in one context or another. In addition, reasons for the limitations were seen to be a complex mix of information source characteristics, designer characteristics, and environmental forces such as the organizational milieu. These findings suggest that 1) a support system which helps the designer overcome all of the limitations should have wide appeal, and b) there are many challenges to its implementation beyond the obvious technological ones.

Given that the types of support we envision will be relevant in human-machine design, the next question was the extent to which designers can find the necessary functionality in existing tools and supports.

VI. SUPPORTING HUMAN-MACHINE SYSTEM DESIGN

A. Objectives Of A Support System For This Domain

There is no question that designers want to produce high-quality operational systems for those who purchase and use them. In general, a system is higher quality the easier it is to learn, easier it is to use, the fewer people it requires to operate, the easier it is to maintain, the more tolerant it is of misuse, and so on in endless variety. Given this to be the human-machine designer's goal, it is useful to reflect on the distribution of responsibility in design and on the objectives that a support system can and cannot be expected to achieve.

People who request systems decide what purpose the system shall help them or their constituents to accomplish. In turn, designers have the authority, implicitly or explicitly, to decide how the system shall behave and what it will contain to achieve these purposes. This includes system characteristics that influence end-users. Therefore, the requester bears responsibility for the purposes while the designer bears responsibility for how those purposes are accomplished and how well. These roles are patently human ones. We currently know of no way to assign such value-laden responsibilties to machines.

A design support system can do several things, but it cannot assume responsibility. It can provide relevant information in usable forms, explain the information, give advice on sound uses of the information, and even reduce the administrative overhead of doing design. The architecture described above in Section IV provides for these functions and more. It cannot, however, do the designer's job for him by, for instance, choosing one design alternative over another. Nor can the support system guarantee or be held accountable for the quality of the designer's output. This implies that even if the support system delivers its intended functions flawlessly, the designer ultimately can reject the information, its implications, and produce a poor product.

With these observations made, the following objective for a support system that aids human-machine system design emerged: *to ensure that system design choices reflect consideration of relevant information about the humans who are expected to use the system.*

Four elements of this statement merit a closer look. First, "consideration of relevant information about humans" means that designers need only take this class of information into advisement, not necessarily accept and use it. The support system could be judged a success if it encouraged designers to formulate and resolve human-related design questions more frequently than they currently do.

Second, the objective statement does not tie the support system to a specific subset of the people who contribute to design. It applies to all contributors whose decisions affect end-users, which include the engineer, human factors engineer, project manager, training device designer, etc. Rather than concentrate on who might use the design information system, the primary focus is on what design decisions tend to affect end-users regardless of who makes them.

Third, for present purposes, relevant information refers to reliable phenomena and concepts about humans (e.g., anthropometric properties, physical tolerances, performance, learning, reasoning, cognitive characteristics, etc.) These phenomenon may or may not have scientifically acceptable explanations. Theoretical accounts may be useful, but are not necessary for supporting design efforts. The reliability with which a given phenomenon occurs under specified conditions is the crucial dimension for designers.

Relevant information can reside in a variety of places and forms (see Fig. 6). The implication for the support system is that, at a minimum, it must "know" where information resides and, even better, be able to connect the designer directly to the source. Hence, in addition to the obvious archived sources, the support system would know about and network users to other people (experts), tools, methods, and facilities that are capable of producing facts and concepts that are not available in the archives.

As Boff [4] has discussed, much of the human-related information in archived sources that is relevant to system design was not produced with designers' needs in mind. For such information to contribute to a design choice, it often must be interpreted or transformed in some manner to make its import apparent and valued. Thus, to achieve its objective, the support system needs either to perform these interpretations and transformations or support the designer in making them himself.

Finally, the objective is to "ensure" that the designer considers relevant human-related information. This is a tall order and implies that the support system would be capable of the following functions:

- Help designers construct questions that lead to relevant information even if it resides outside their areas of expertise. Hence, the system should support specialists and nonspecialists alike.

- Help the designer interpret and transform information into forms appropriate for design decision making.

- "Understand" what a design decision is in general, understand which decisions affect end-users, and recognize when such decisions are about to be made in actual practice.

- Determine whether the designer is or is not considering relevant human-related information that the support system is capable of providing.

- Persuade the designer to consider the information if in fact he shows signs of not doing so.

Referring to the system architecture discussed in Section IV, the first two of these requirements are met through the primary functions (Search, Execute, Indicate, etc.) which are aimed at understanding designers' questions, connecting them to pertinent information, explaining information, etc.

The latter three requirements exemplify the envisioned functionality of the interface. Information management, error monitoring, and adaptive aiding reduce the overhead involved with doing design as opposed to resolving specific design issues. These functions will rely on various computer-based representations of the design situation including models of the artifact being designed, of the design process (e.g., high-probability trajectories in the design task space), and of the designer's intentions and activities.

This objective poses challenging technological, conceptual and organizational problems. To determine the extent to which solutions to these problems are in hand, we reviewed and characterized the available support that existing human-machine design tools provide.

B. State of the Art Review of Supports

A literature search uncovered 135 design and evaluation tools that appear to constitute the state of the art in this domain. The list is reproduced in Appendix A. We emphasized computer-based products and adopted a fairly liberal criterion with regard to tool maturity. All tools that were working prototypes or further along in development were included. Fig. 12 shows the distribution of the 135 tools by application.

No.	Application
6	General model building packages (e.g., SAINT)
8	Specialized human-machine system modeling packages
43	Biomechanical CAD
14	Displays construction
5	Display/control prototyping tools
22	Timeline construction and analysis
17	Spreadsheets for manpower, personnel & training costs
16	Electronic retrieval of human engineering standards
2	Human workload evaluation tools
135	

Fig. 12. Distribution of human-machine system design tools

To map the functionality of these tools to needs, we first decomposed each tool into its basic support verb(s), designer verb(s), and object(s) using the grammar summarized in Figs. 5 and 6. Then for each of the ten human-machine problem areas (Fig. 11), we identified applicable supports from among the 613 theoretically-derived supports. Finally, we determined how many of the 135 tools were relevant to each problem and the proportion of 613 supports they supplied.

C. Problem-Independent Design Support

Looking first at supports independently of specific problem areas, the 135 tools and methods supply a total of 2,247 instances of support. The number of supports provided by individual tools range from 1 to 35 with a median of 12. When duplicate supports across tools are removed, however, the tools supply only 70 of the 613 needed supports, or just over 11%. Figs. 13 and 14 show the distribution of these 70 in terms of our designer and support verb vocabularies, respectively.

DESIGNER VERB	EXPECTED OCCURRENCE	PERFORMED WITH TOOLS	PERCENT COVERAGE
Create	92	9	9.8
Evaluate	170	11	6.5
Select	114	7	6.2
Identify	54	1	1.9
Locate	42	1	2.4
Obtain	84	11	13.1
Access	22	12	54.5
Construct	9	5	55.5
Measure	3	1	33.3
Monitor	16	8	50.0
Run	3	2	66.7
[ ]	4	2	50.0
Total	613	70	11.4

Fig. 13. Distribution of existing supports

SUPPORT	NUMBER NEEDED	NUMBER DELIVERED	PERCENT COVERAGE
Search	109	3	2.7
Execute	78	34	43.6
Indicated	21	8	38.1
Transform	2	1	50.0
Explain	325	24	7.4
Tutor	78	0	0.0
Total	613	70	11.4

Fig. 14. Distribution of existing supports

Using the expected occurrences of designer verbs as a rough yardstick of his activity, Fig. 13 reveals the difference between what our theory suggests the designer spends his time doing and what the available support tools would have him do. In theory, the designer's activities cluster into three groups. Primary activities focus on creating the artifact, evaluating it and confirming its specification (i.e., selecting). Secondary activities involve information gathering (identifying, locating, and obtaining it) which provide the needed inputs to the primary ones. The remaining activities are, in a sense, design overhead. Accessing tool functionality in order to construct representations of the artifact, running studies, monitoring tool operations, and measuring system behavior are necessary, but their accomplishment is not the designer's chief purpose.

Examination of the "Percent Coverage" column in Fig. 13 shows that, when using existing tools, the designer is largely engaged in overhead tasks! His primary activities and their more immediate information support tasks are less well represented.

Results shown in Fig. 14 reinforce this image. Existing tools provide nearly 44% of the procedure execution needs. However, the majority of these procedures are associated with overhead tasks such as constructing formal representations, accessing and running tools, and measuring system properties.

The original analysis of the designer's tasks suggested relatively few needs for the support functions of indicating and transforming as defined in Fig. 5. In part, this explains their apparently deeper coverage through existing tools. Furthermore, the few instances of these two supports that emerged basically amount to on-line help facilities and data highlighting following computation, respectively. Thus, the percentages in Fig. 14 should not be taken to mean that these supports are well in hand.

Large deficits appear in the remaining three support areas. Some explanation facilities exist (24), but the coverage is meager relative to the need. Similar disparities are associated with information searching. Finally, we uncovered no tutoring supports among the 135 tools that we reviewed.

D. Support for the Human-Machine Problem Areas

Turning to the support coverages in each of the ten human-machine design areas, Fig. 15 summarizes how well these needs appear to be met. As shown under the Applicable Services column, design problem areas exhibit needs for between 190 and 355 of the 613 general supports. The number of relevant tools out of 135 range from none (aiding and adaptive interface design) to 94 (evaluation). Finally, based on the number of supports out of 70 that the tools actually deliver, at best, about 25% of the needs in these ten areas can be met with existing tools.

Based on the number of services, human-system evaluation and workspace layout enjoy somewhat greater levels of support than the remaining areas. All tools that support evaluation, regardless of the primary problem area for which they are intended, were included under

HUMAN-MACHINE SYSTEM DESIGN PROBLEMS	# APPLICABLE SUPPORT SERVICES	# RELEVANT TOOLS	# SERVICES SUPPLIED ACROSS TOOLS	% COVERAGE
Identify Functional Requirements	267	7	18	6.7
Allocate Functions to Human & Machine Elements	355	10	38	10.7
Establish Information & Control Requirements	190	18	39	20.5
Design Equipment, Displays, & Controls	355	25	51	14.3
Design Procedures & Dialogue	318	54	54	17.0
Layout Workspace	355	48	64	18.0
Design Aids & Adaptive Interfaces	318	0	0	0
Design Training Systems	355	17	35	9.8
Design Protection & Life Support Systems	355	31	37	10.4
Evaluate Human & System Performance	269	94	69	25.6

Fig. 15. Design supports provided via existing tools

human-machine evaluation. Hence, the 94 tools and 69 services duplicate several tools from the other problem areas.

For workspace layout, there are over 40 biomechanical CAD packages available (e.g., SAMMIE, COMBIMAN) and together they provide 64 forms of support. These tools are used primarily for detailed design to visualize physical interactions between human and machine and evaluate designs in terms of reach, visual restrictions, and safety.

Anthropometric properties of the simulated human are typically under user control, and a few of the tools contain various human engineering standards on-line. The conceptual bases for these tools reside in mechanical engineering and, hence, information that they produce transfers well to other disciplines that share the language and methods of this discipline.

The next most supported areas are display/control design and procedures design. Tools for the former area include rapid prototyping facilities and several interface drawing packages. Prototyping facilities often contain functions for defining display/control requirements and checking actual designs for violations. For the most part, drawing packages are strictly computer-based equivalents of the drafting table. PC paint packages typify this form of support.

Procedures and dialogue design rely on the computerized descendents of paper-and-pencil tools for timeline analysis, information flow analysis, operational sequence diagramming, link analysis, and decision analysis. We located 20 such packages. Basically, they help the designer to construct task networks, compute time and identify task scheduling conflicts. In addition, there are several general simulation modeling packages (e.g., SAINT, SLAM II) and packages specialized to human-machine modeling (e.g., PROCRU, HOS) for this area. These tools rest on concepts and methods adopted from network and queuing theory, stochastic process modeling and control theory.

Function allocation is supported by the same general purpose and specialized modeling packages used for procedures design. Information requirements analysis also uses these packages but gains additional support from several "automated guidelines" that can be used to select appropriate mappings between the functions allocated to humans in systems and the classes of information they will need to accomplish the functions.

We uncovered 17 tools that helped the designer to forecast the manpower, personnel and training requirements for a given design. These tools are rudimentary computational models that convert descriptions of jobs, skill requirements, and training regimes into training and operational cost estimates.

There is a long history of biomedical experimentation and modeling on which to represent and analyze the human health consequences of design options. We did not attempt to review tools developed exclusively for this problem area and, therefore, our estimates of support coverage for protection and life support system design are likely to be conservative. Tools that were mapped to this area included the biomechanical CAD packages that support computer representation and analysis of protective gear and physiological consequences of equipment and environment design.

Only 7 tools were found that support functional requirements identification and analysis. These tools supply 18 needed functions which consist for the most part of automatic retrieval of human engineering standards and requirements that the designer had developed himself and stored away for future reference.

Finally, we found no explicit supports among the 135 tools for the design of aiding and adaptive interfaces. These features have emerged only recently with the advent of knowledge-based systems and intelligent computing technology [46, 47]. Currently, their *design* appears to rely strictly on the creative capabilities of the human designer. Among existing design tools, supports that have been envisioned for constructing expert systems come closest to the specialized needs in this area.

E. Summary and Assessment

The picture that emerges from this review shows that existing tools fall far short of meeting the objectives that we laid out for a design information system for this domain. Gaps are apparent for all six of the primary types of support listed in Fig. 5 as well as in the specific content areas of human-machine system design.

With regard to the functions indicated by support verbs, available tools concentrate on executing procedures associated with the designer's tertiary or overhead tasks. Indeed, support is quite strong where syntactic features of the designer's task map directly from older ways of performing the task to newer, electronically mediated ways. Examples that support this observation include the following: 1) drafting done with paper and pencil has been transformed by CAD into drafting with light pen and screen, but the syntax of the task has not changed; 2) computations formerly done by hand are now performed much more quickly and accurately by computer; and 3) retrieving papers from hardcopy filing systems is being replaced with electronic media. (Hypermedia may be an exception to the strategy of converting manual methods to computerized versions of the same thing.) However, execution support for procedures that are more central to the designer's goals are meager. Such supports rely on deeper semantic understanding of the nature of design problem solving.

There is some support for search and retrieval, but designers still obtain most of their technical information from personal sources and social exchange with colleagues. Although difficult to estimate with any precision, the costs of these retrieval deficits to design innovation and improvement are probably considerable assuming that valuable information exists that is not being tapped.

Facilities that explain or provide tutoring about the operation of the support system or the problem domain are practically nonexistent in design support. These facilities depend on reasonably robust models of what the asker knows and does not know about the topic under consideration. They also depend on a pedagogical model of how to teach something to someone. For the few tools that do provide some explanation and tutoring, tool developers have attempted to anticipate their users' needs through question and answer checklists and "See Also" documentation. As helpful as they can be, these vehicles rarely provide comprehensive explanation and tutoring because they cannot anticipate the myriad individual differences among designers' knowledge, skills and particular design problems. That is, they possess no "model" of the individual being instructed.

Intelligent interface functions that could help manage information, monitor design activities for error occurrences, and adapt support as the designer's needs shift have not yet been applied to design information systems.

For human-machine design content areas, existing tools: 1) replicate their manual predecessors without significant advances in the conceptual treatment of human-machine interaction; 2) focus on the micro human-machine issues to the near exclusion of macro issues; and 3) are not integrated in any coherent manner.

In addition, tools for display/control design and procedures/dialogue design in particular tend to focus the designer's attention on the end-user's sensory-motor engineering properties and leave consideration of his learning, reasoning and decision making properties up to designers' intuitions and personal theories. Similarly, the absence of adaptive interface design tools is symptomatic of problems in capturing the end-user's cognitive contributions to human-machine systems. Thus, to the extent they are used, the tools themselves appear to promote the older paradigm of human-machine interaction. (See McMillan et al. [28] for the state of the art in human performance modeling technology and its usefulness in system design.)

VII. PROSPECTS

When one ponders these results, comprehensive support for information access and utilization in human-machine system design looks to be a long way off. In fact, our analyses convince us that the obstacles to *comprehensive* support realistically cannot be overcome any time soon. The technological, conceptual and organizational difficulties are simply too great.

However, the analyses also encourage optimism over the prospects for significant improvement even in the near-term. Trends in several key technologies and ongoing efforts to support the design, engineering and manufacturing processes are providing some of the enabling technology as well as the market demand for the supports that we have in mind.

The final section of this chapter takes a realistic look at the prospects for developing practical design information systems in the image of Fig. 8. The discussion is organized around three challenges that face the support system developer: 1) constructing the data and knowledge bases; 2) implementing the functionality; and 3) transitioning the support system into practice. Our discussion continues to focus on human-machine system design, but we would be surprised if the points did not apply equally well to other engineering disciplines.

A. Data Bases

In the conventions of software design, the proposed support system can be decomposed into two parts -- data and programs (functions). The relevant data are represented by Fig. 7 and support functions by Fig. 8.

Alone, the task of compiling and integrating the pertinent data could be a daunting one to say the least. Volume, growth rate, syntactic differences (e.g., query languages) and semantic differences (e.g., terminology) across databases all discourage simple solutions to constructing useful bases.

Some sense of the magnitude of the bases represented by Fig. 7 can be gleaned from Licklider's [26] estimates made in 1965. At that time, just the text associated with science and technology that was stored in all the world's libraries approximated 10^{13} bits. This could be divided into 1,000 subfields (40 of which we suggested in Fig. 9 contribute to human-machine system design), each with 10^{10} bits. In addition, he projected that the corpus would double every 15-20 years.

When other types of information are added, this estimate expands rather dramatically. Speech, sounds, pictures, maps, engineering drawings, dynamic images, and process simulations that can be accessed electronically would easily double the estimate. Further, if we include information that practitioners carry about in their heads that is not duplicated in formal sources (e.g., "when we used the XL4912 widget back in 1988,..."), the base may double again. A simple armchair calculation backs this statement. For human-machine system design, 10^1 subfields contribute and each subfield may have 10^5 experienced practitioners. If each practitioner were to learn non-redundant facts and relationships at the (very generous) rate of 5 items/hour, then every 10 years the corpus of experiential knowledge would expand by another 10^{11} items. Electronic bulletin boards, e-mail and voice lines are the channels for accessing these data.

Whether Licklider's estimates and our additions are accurate is not of major consequence to the present point. Volume and growth rate are simply enormous. However, two observations give the support system developer hope: storage and communications media are maturing rapidly; and it is not necessary to access all the world's technical data to provide valuable support.

First, the practical difficulties of compiling the amount of information depicted in Fig. 7 notwithstanding, solutions to physical storage and communication at least seem plausible. Data format standards have begun to emerge from industrial and government R&D programs devoted specifically to the digital exchange of engineering information, e.g., the Computer-aided Acquisition and Logistics Support (CALS) effort. Advanced optical media already permit construction of massive electronic archives. Currently available CD ROMs (compact disk, read only memory), with a capacity of over 600 megabytes, can store the text of about six hundred 400-page books [23, 24]. The burgeoning local- and wide-area networks are laying the necessary voice and data communication paths. The integrated services digital network (ISDN) will be capable of high-bandwidth transmission of digitally encoded information of any type (voice, data, facsimile, video), and is expected to be in operation in some form in the next two decades [22].

Second, design information systems that support relatively narrow, but still critical problem areas, could be constructed from small fractions of the technical data in Fig. 7. Boff and Lincoln [6], for instance, have compiled in hardcopy form a wide variety of human perception and performance phenomena that pertain to display and control system design. Currently, we are writing this database to a CD ROM and augmenting its value with interactive simulations of several phenomena. Also, existing tools for display design that include human engineering guidance (e.g., KADD, RAPID, RIPL) are forebears to more powerful products of the sort we have in mind.

The stickier side of large database construction centers on how to organize and encode information to support timely and intelligent retrieval. Existing indexing and retrieval schemes rely on users' abilities to formulate questions and recognize useful returns. These schemes are inherently limited to the expressive power of language. Efforts to model the conceptual structure and pragmatic import of a data set from a particular point of view, like design, are in their infancy.

Beyond structural considerations, content difficulties that plague would-be human information sources need to be remedied. Card [8] and Boff [4, 5] have summarized the designers' chief complaints when trying to use the general sources that are currently available: suspicious content validity; lack of functional relationships between operational conditions and human behavior as opposed to point estimates from comparative evaluation studies; verbal as opposed to formal models, especially in the realms of human learning and reasoning; and lack of a coherent framework within which to integrate existing models of narrow psychological phenomena.

As alluded to earlier, existing human-machine design tools appear to be more appropriate to the older interaction paradigm than to newer supervisory control and collaborative paradigms. Hence, to support function allocation, procedures design and adaptive interface design, advances in modeling end-user cognition will be important.

It is difficult to estimate how much effort will be required to overcome the content problems of human-related databases for design in general. However, the task clearly seems to be more manageable if approached by developing a series of mini support systems, one mini-system per problem area. Chief differences from present day tools of such mini-systems would be the range of information types (Fig. 7) that would be compiled to provide thorough support and the requirement to supply all of the primary and interface support functions. Present day expert systems come closest to this image, although they do not provide the range of functionality that we propose for design support, to which we now turn.

B. Implementing System Functions

Assuming that databases could be constructed for narrow problems, the next question centers on the extent to which technology exists or can be developed to implement the support system functions (Fig. 8). Each of the six primary modules and the three interface functions poses distinct conceptual and technological demands.

With regard to "Search," our ability to store massive amounts of information is way ahead of our ability to find and interact with it in natural and productive ways. To offload the overhead from the designer of retrieving information, a useful search function, even for a mini support system, will depend on encoded knowledge about where information resides as well as the data formats, query languages and communication protocols for getting to it. More importantly, it will require a capacity to interpret what the designer is looking for, perhaps through conversational interaction, to map from his specific problem to the many sources that might pertain. Filtering relevant information from noise is a totally different matter from navigation that, along with a model of the designer's needs, requires a robust comparison facility. A fair analogy to this last capability is found in the pattern recognition properties of human vision.

Alternative implementations of the Search function are imaginable and hinge on whether to put the "smarts" into the databases or into the search mechanism. Along with their primary content, smart databases would have to contain information about applicability, limitations, constraints, etc. with regard to specific design issues. Such a requirement would place extraordinary (and probably unacceptable) demands on information producers. Alternatively, the Search function might be given general knowledge about the information bases and, at execution time, be provided with problem-specific expertise by the human designer to augment these general capabilities. Both solutions face complex and difficult problems.

"Execute" can include both mundane operations and sophisticated procedures that are context-specific. For instance, the particular manner in which an evaluation is performed might vary with the state of the design and purpose of the evaluation. We suspect that a reasonably sophisticated model of *how* design typically is performed underlies most procedure execution.

"Indicate" and "Transform" supports can be thought of as facilities that recommend to and remind the designer, respectively, about procedures, open issues, observations, etc. Like procedure execution, these functions will depend on models of the designer's immediate goals, nominal design procedures, plus extensive knowledge about the "psychology of designing" [41]. The latter topic refers to human reasoning, biases, risk assessment, and decision processes under all of the conditions that typify design problems, e.g., conflicting requirements, ambiguous and partial information, etc.

The "Explain" and "Tutor" functions each admit to several potential levels of sophistication depending on the goal of support. At the extreme, if the goal were to enable all design contributors to behave as if they were expert in all relevant disciplines, then both modules would have to contain very sophisticated modeling and knowledge-based computing -- more sophisticated than anyone currently knows how to implement. On the other hand, if the more modest goal of improving novice design behavior were adopted, then the types of advanced educational technology that enable the "electronic book" offer feasible approaches. Automated definition, elaboration, maps, interactive simulation, and question/answering facilities exemplify these technologies.

Finally, the three interface functions were discussed in regard to support system objectives (see Section VI, B). Suffice it to say that information management, error monitoring and adaptive aiding are technologies that have been applied successfully to the operations and maintenance domains [46, 47]. What humans and systems can do in these domains exhibit much greater levels of proceduralization and predictability than occur in the design domain. This distinction prevents a wholesale adoption of existing intelligent interface implementations into design. However, the conceptual groundwork as well as the tough lessons that these implementation attempts have taught should transfer [47].

In summary, Fig. 16 attempts to illustrate along two dimensions the current frontier and what lies in store for the design support community. Technological challenges center around measurable system properties such as size, speed, completeness, and reliability. Conceptual difficulties pertain more to the presence or absence of enabling theories or world views. No doubt, the exact placement of the example technologies along these dimensions could be debated.

C. Technology Transition

Implementation issues are significant; there is no question about that. However, there are also constraints and preferences, from the level of the individual designer through the organization, that must be anticipated to ensure the support system is purchased and used [39].

One particularly knotty problem is the question of proprietary information that enables profit-making organizations to remain competitive. It is very unlikely that these organizations could be persuaded to grant to open databases which they believe sustain them. However, everything has its price. Under this philosophy, a support system might function to enable a free market for information purchases between buyers and sellers. The requirement is that the system be given sufficient match-making information by those who wish to participate. Incentives for such behavior are quite imaginable.

There are also "pushes" and "pulls" inside and out of human-machine system design that affect the rate of progress toward a functioning support system. On the "push" side, the tool development community has been criticized for producing data and instruments that

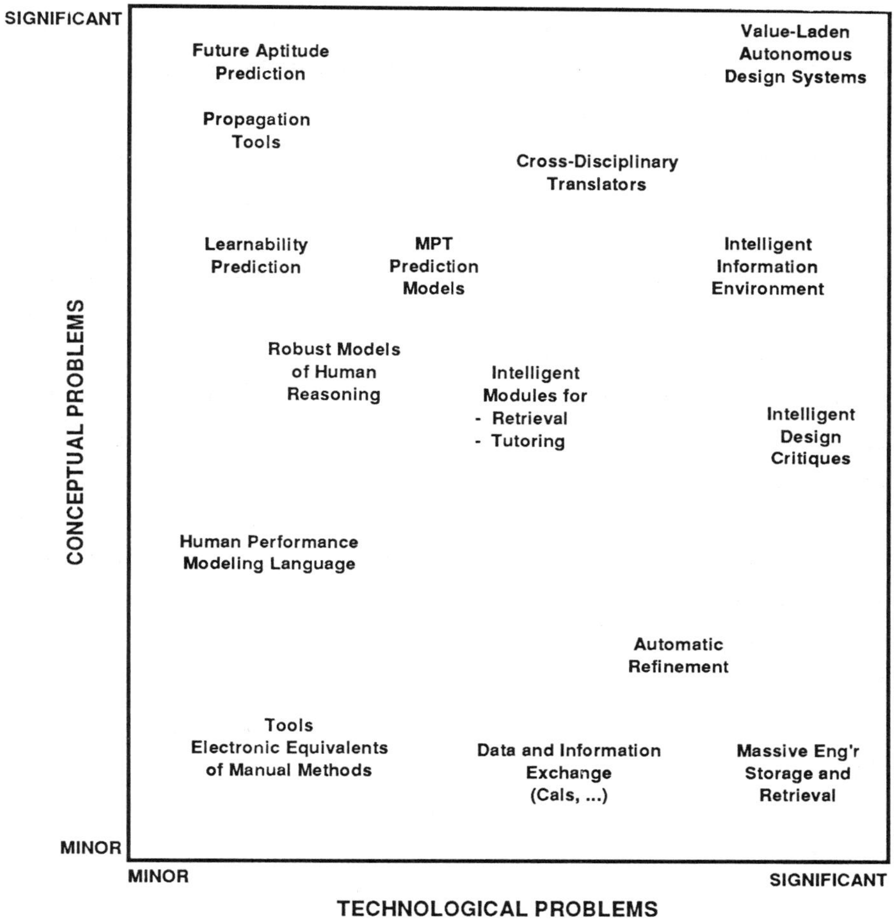

Fig. 16. The frontier of comprehensive design support

have little practical value to system developers [3, 8, 38]. This may be because tool developers simply do not know or care to know what system designers need. Indeed, some model developers hold that the goal of modeling is foremost to produce better theoretical explanations of human behavior. These explanations may have utility for other endeavors, including system design. However, for these individuals, this utility is a by-product and not the primary motivation for their efforts.

As Card [8] recommends, one way to get human-related instruments and data sources that are usable for design is specifically to fund research activities that produce tools for engineering use. This approach may also help to warm the cool attitude that practitioners have traditionally demonstrated toward these products.

On the "pull" side, human-machine system design has suffered from its multi-disciplinarity and dispersion when it comes to responsibility and authority for design decisions. Managerial intervention through efforts such as the Army's MANPRINT program may improve this status problem [7]. Unfortunately, prior efforts with similar intent have had little lasting impact [10, 21]. To counter criticisms of its weak inputs to design, the human-machine design domain must develop ways to measure the magnitude of the benefits that human-related improvements make to system effectiveness and cost. As our review of macro human engineering tools showed, there is yet a long way to go in this regard.

Finally, how the envisioned support system might alter the individual designer's job and whether the changes are acceptable ones are at the heart of whether the effort ever turns out to be a successful one. Rouse and Morris [49] have discussed the issues of technology transition from the end-user's perspective and recommend several ways to avoid disaster.

D. Next Steps

Our discussion of a Designers' Associate has focused along three dimensions: 1) the system user is the *individual designer*; 2) user's tasks concentrate on *seeking and using information*; and 3) the application problems pertain to *human-machine system design*. How should the effort proceed?

Based on our analysis and experiences with similar support system technology, the judicious approach seems to be to start by tackling a single problem area within human-machine design from among the ten identified in Fig. 11. Using field studies to establish the range and types of design problems, collecting and integrating the data required for rich, if not comprehensive support, would then become the first order of business. In concert with this effort, all six primary functions and three interface functions should be included both to test the feasibility of the entire concept as well as solicit user feedback about what is useful and what is not.

It is possible that a successful effort to implement design support in the image that we have proposed could bring to design support a measure of standardization that has so far been missing. More specifically, should the primary and interface functions prove to enhance designers' productivity as we believe they will, then to the extent that the supports are fully adopted by other developers, designer limitations and needs would be driving design support. This thrust is comparable to those that underlie interface standards for the Macintosh and data exchange standards in the CALS efforts that have proven to be so successful.

Looking beyond a specialized Designers' Associate for individuals' needs in human-machine system design, the next steps will be to enhance supports for collaborative design, activities beyond information seeking, and domains in addition to human-machine design. While the challenges are substantial, the utility of the present theory and analysis method along with the success we expect to achieve in human-machine systems encourage the journey.

ACKNOWLEDGEMENT

This chapter is based upon research supported by the U.S. Air Force Armstrong Aerospace Medical Research Laboratory under Contract No. F33615-86-C-0542. Dr. Boff of AAMRL is the technical director of this project, which is entitled "Automated Data Management for Designers: A Designer's Associate."

REFERENCES

1. R.W. Bailey, "Human Performance Engineering: A Guide for System Designers", Prentice Hall, New Jersey, 1982.

2. D. Beevis, "Experience in the integration of human engineering effort with avionics systems development", *in* "AGARD Conference Proceedings, 417, The Design Development and Testing of Complex Avionics Systems", Neuilly Sur Seine, France, 1987.

3. K.R. Boff, "Meeting the challenge: Factors in the design and acquisition of human-engineered systems", *in* H.E. Booher (ed.), "People, Machines and Organizations: The MANPRINT Approach to System Integration", Van Nostrand Reinhold, New York, 1990.

4. K.R. Boff, "The value of research is in the eye of the beholder", *Human Factors Society Bulletin,* **31**, 1-4 (1988).

5. K.R. Boff, "The Tower of Babel revisited: On cross-disciplinary chokepoints in systems design", *in* W.B. Rouse and K.R. Boff (eds.), "System Design: Behavioral Perspectives on Designers, Tools, and Organizations", North Holland, New York, 1987.

6. K.R. Boff and J.E. Lincoln, "Engineering Data Compendium: Human Perception and Performance", Wright Patterson Air Force Base, OH, Harry G. Armstrong Aerospace Mideical Research Laboratory, 1988.

7. H.E. Booher (ed.), "People, Machines and Organizations: The MANPRINT Approach to System Integration", Van Nostrand Reinhold, New York, 1990.

8. S.K. Card, "Theory-Driven Design Research", *in* G.R. McMillan, D. Beevis, E. Salas, M.H. Strub, R. Sutton, and L. van Breda (eds.), "Application of Human Performance Models to System Design", Plenum Press, New York, 1989.

9. G.J. Chambers, "What is a systems engineer?", *IEEE Transactions on Systems, Man and Cybernetics*, **SMC-15**(4), 517-521 (1985).

10. P.R. Chatelier, "Psychology or reality", *in* W.B. Rouse & K.R. Boff (eds.),"System Design: Behavioral Perspectives on Designers, Tools, and Organizations", North Holland, New York, 1987.

11. W.J. Cody, "Designers as users: Design supports based on crew system design practices", *in* "Proceedings of the American Helicopter Society 45th Annual Forum", Boston, 1989.

12. W.J. Cody and W.B. Rouse, "A test of criteria used to select human performance models", *in* G.R. McMillan, D. Beevis, E. Salas, M.H. Strub, R. Sutton, and L. van Breda (eds.), "Application of Human Performance Models to System Design", Plenum Press, New York, 1989.

13. B. Curtis, H. Krasner, and N. Iscoe, "A field study of the software design process for large systems", *Communications of the ACM,* **31**, 1268-1287 (1988).

14. R. Dagwell and R. Weber, "Systems designers' user models: A comparative study and methodological critique", *Communications of the ACM*, **26**, 987-997 (1983).

15. M.M. Gardiner and B. Christie (eds.), "Applying Cognitive Psychology to User-Interface Design", John Wiley & Sons, New York, 1987.

16. H.G. Gemunden, "Perceived risk and information search: A systematic meta-analysis of the empirical evidence", *International Journal of Research in Marketing*, **2**, 79-100 (1985).

17. P.G. Gerstberger and T.J. Allen , "Criteria used by research and development engineers in the selection of an information source", *Journal of Applied Psychology*, **52**, 272-279 (1968).

18. V. Goel and P. Pirolli, "Motivating the notion of generic design within information processing theory: The design problem space", *AI Magazine*, Spring, 18-36 (1989).

19. J.D. Gould and C. Lewis, "Designing for usability: Key principles and what designers think", *Communications of the ACM*, **28**, 300-311 (1985).

20. N.V. Hammond, A.H. Jorgenson, A. MacLean, P.J. Barnard, and J.B. Long, "Design practice and interface usability: Evidence from interviews with designers", *in* "Proceedings of the 1983 CHI Conference on Human Factors in Computing Systems", ACM, New York, 1983.

21. E.M. Johnson, "The role of man in the system design process: The unresolved dilemma", in W.B. Rouse and K.R. Boff (eds.), "System Design: Behavioral Perspectives on Designers, Tools, and Organizations", North Holland, New York, 1987.

22. R.E. Kahn, "Networks for advanced computing", Scientific American, **257**, 136-143 (1987).

23. M.H. Kryder, "Data-storage technolgies for advanced computing", Scientific American, **257**, 116-125 (1987).

24. S. Lambert and S. Ropiequet (eds.), "CD ROM: The New Papyrus", Microsoft Press, Redmund, WA., 1986.

25. L. Leifer, "On nature of design and an environment for design", *in* W.B. Rouse & K.R. Boff (eds.), "System Design: Behavioral Perspectives on Designers, Tools, and Organizations", North Holland, New York, 1987.

26. J. Licklider, "Libraries of the Future", MIT Press, Cambridge, MA, 1965.

27. L.M. Lintz, W.B. Askren, and W.J. Lott, "System Design Trade Studies: The Engineering Process and Use of Human Resources Data", Report No. AFHRL-TR-71-24. Wright Patterson AFB, Ohio, (1971).

28. G.R. McMillan, D. Beevis, E. Salas, M.H. Strub, R. Sutton, and L. van Breda (eds.), "Application of Human Performance Models to System Design", Plenum Press, New York, 1989.

29. D. Meister, "Behavioral Analysis and Measurement Methods", John Wiley and Sons, New York, 1985.

30. D. Meister and D.E. Farr, "The utilization of human factors information by designers", *Human Factors*, **9**(1), 71-87 (1987).

31. J. Moraal and K.F. Kraiss (eds.), "Manned Systems Design: Methods, Equipment and Applications", Plenum Press, New York, 1981.

32. C.A. O'Reilly, "Variations in decision makers' use of information sources: The impact of quality and accessibility of information", *Academy of Management Journal*, **25**, 756-771 (1982).

33. C. Perrow, "The organizational context of human factors engineering", *Administrative Science Quarterly*, **28**, 521-541 (1983).

34. D.M. Promisel, C.R. Hartel, J.D. Kaplan, A. Marcus, and J.A. Whittenburg, "Reverse Engineering: Human Factors, Manpower, Personnel, and Training in the Weapon System Acquisition Process", Technical Report No. 659, Army Research Institute, Alexandria, Virginia, (1985).

35. J.R. Rasmussen, "Information processing and human-machine interaction: An approach to cognitive engineering", North Holland, New York, 1986.

36. J.R. Rasmussen, "A cognitive engineering approach to the modeling of decision making and its organization in process control, emergency management, CAD/CAM, office systems, and library systems", *in* W.B. Rouse (ed.), "Advances in Man-Machine Systems Research", JAI Press, Greenwich, CT, 1988.

37. W.B. Rouse, "Optimal allocation of system development resources to reduce and/or tolerate human error", *IEEE Transactions on Systems, Man, and Cybernetics*, **SMC-15**(5), 620-630, 1985.

38. W.B. Rouse, "On the value of information in system design: A framework for understanding and aiding designers", *Information Processing and Management*, **22**, 217-228 (1986).

39. W.B. Rouse, "Designers, decision making, and decision support", *in* W.B. Rouse & K.R. Boff (eds.), "System Design: Behavioral Perspectives on Designers, Tools, and Organizations", North Holland, New York, 1987.

40. W.B. Rouse, "Intelligent decision support for advanced manufacturing systems", *Manufacturing Review,* **1**, 236-243 (1988).

41. W.B. Rouse and K.R. Boff, "System design: Behavioral perspectives on designers, tools, and organizations", North Holland, New York, 1987.

42. W.B. Rouse and W.J. Cody, "Functional allocation in manned system design", *in* "Proceedings of the 1986 IEEE International Conference on Systems, Man, and Cybernetics", 1600-1606, 1986.

43. W.B. Rouse and W.J. Cody, "On the design of man-machine systems: Principles, practices, and prospects", *Automatica,* **24**, 227-238 (1988).

44. W.B. Rouse and W.J. Cody, "A theory-based approach to supporting design decision making and problem solving", *Information and Decision Technologies*, to appear.

45. W.B. Rouse, W.J Cody, and K.R. Boff, "The human factors of system design: Understanding and enhancing the role of human factors engineering", to appear.

46. W.B. Rouse, N.D. Geddes, and R.E. Curry, "An architecture for intelligent interfaces: Outline of an approach to supporting operators of complex systems", *Human Computer Interaction,* **3**, 87-122 (1987).

47. W.B. Rouse, N.D. Geddes, and J.M. Hammer, "Computer-aided fighter pilots", *IEEE Spectrum,* **27**, 38-41 (1990).

48. W.B. Rouse, R.A. Kisner, P.R. Frey, and S.H. Rouse, "A method for analytical evaluation of computer-based decision aids", Oak Ridge National Laboratory, Technical Report NUREG/CR-3655 (1984).

49. W.B. Rouse and N.M. Morris, "Understanding and avoiding potential problems in implementing automation", *in* "Proceedings of the 1985

IEEE International Conference on Systems, Man, and Cybernetics", 787-791, 1985.

50. W.B. Rouse and S.H. Rouse, "A framework for research on adaptive decision aids", Aerospace Medical Research Laboratory, Report TR-83-082 (1983).

51. A.P. Sage, "Organizational and behavioral considerations in the design of information systems and processes for planning and decision support", *IEEE Transactions on Systems, Man, and Cybernetics,* **SMC-11**, 640-678 (1981).

52. D.A. Schon, "The reflective practitioner: How professionals think in action", Basic Books, New York, 1983.

53. D.G. Ullman, T.G. Dietterich and L.A. Stauffer, "A model of the mechanical design process based on empirical data", Oregon State University, 1988.

54. W. Zachary, "A cognitively based functional taxonomy of decision support techniques", *Human-Computer Interaction,* **2**, 25-63 (1986).

Appendix A

List of human-machine system design tools

General model building packages

EXTEND
GENSAW (Generic Systems Analysis Workstation)
Micro SAINT (Micro-Systems Analysis of Integrated Networks of Tasks)
SAINT (Systems Analysis of Integrated Networks of Tasks) I & II
SLAM II (Simulation Language for Alternative Modeling)
STELLA (Structural Thinking, Experimental Learning Laboratory with Animation)

Specialized human-machine system modeling packages

DMS (CAFES)
HOS (Human Operator Simulator)
Interops
PROCRU
PROFILE
Siegal-Wolf
SIMNET (Simulation Network)
SIMWAM (Simulation for Workload Assessment and Modeling)

Biomechanical CAD

ATB Model
BIOMAN
BOEMAN (CGE)
BUFORD
CAD (CAFES) (Computer Aided Design)
CADAM (ADAM and EVE)
CADET (Computer Aided Design and Evaluation Techniques)
CALSPAN 3D CVS
CAPABLE (Controls And Panel Arrangement By Logical Evaluation)
CAPE (Computer Accommodated Percentage Evaluation)
CAR (CAFES/CHESS) (Crew Assessment of Reach)
CGE (CAFES) (Crewstation Geometry Evaluation)
CINCI-KID
COM-GEOM
COMBIMAN (Computerized Biomechanical Man-Model)
CONSOLE (CAFES)
CORELAP (Computerized Relationship Layout Planning)
CRAFT (Computerized Relative Allocation of Facilities)
CREW CHIEF
CVAS (Crewstation Vision Analysis System)
CYBERMAN
ERGOMAN
GEOMOD (Geometric Modeling Tool)
GRASP (Graphical Robot Applications Simulation Package)
HECAD (Human Engineering Computer-Aided Design)
HSRI Models
IRA (CHESS)
LAYGEN (LAYout GENerator)

NUDES
PLAID (Panel Layout Automated Interactive Design)
SAMMIE (System for Aiding Man-Machine Interaction Evaluation)
SFU Model
SIMUL (Simulation of CAD)
SIMULA/PROMETHEUS
SPRINGMAN
STICKMAN
TEMPUS
TTI Models
TX-105 (Operator/Crew Workload Assessment Technique)
UCIN
WOLAG (Workstation Layout Generator)
WOLAP (Workspace Optimization and Layout Planning)
WORG (Workspace ORGanizer)

Display construction

ADM (A Dialog Manager)
CHIPS
COUSIN (COoperative USer INterface)
CUBITS (Criticality/Utilization/Bits of Information)
Dan Bricklin's Demo Program
FLAIR (Functional Language Articulated Interactive Resource)
IDL (Interface Design Language)
IDSA (Input Device Selection Assistant)
Interviews
ITS
MENULAY
MMP (MicroMain Prototyper)
Serpent
SIMKIT

Display/control prototyping tools

KADD (Knowledge Aided Display Design)
RAPID
RIPL
Univ. of Michigan Control Panel Design Evaluation Tool
USIPS (User-System Interface Prototyping System)

Timeline construction and analysis

BEMOD (Behavior Modification)
CAPRA (Computer Aided Probabilistic Risk Assessment)
CRAWL
DART (Data Analysis and Retrieval Technique)
EDGE (Ergonomic Design using Graphical Evaluation)
ERGONOGRAPHY
ESAT (Expert System for Applied Task Taxonomy)
FAM (CAFES) (Function Allocation Model)
Function Allocation Decision Aid
HOMME (Human Operator Model Management Environment, with HOS IV)
ICAM (Interactive Control Assessment Methodology)
MOPSIE (Multiple Operator Parallel Systems Evaluation)

ORACLE (Operations Research and Critical Link Evaluation)
OWLES (Operator Workload Evaluation System)
Penn State Ergonomic Expert System
TASCO (Timeline Analysis of Significant Coordinated Operations)
TEPPS (Technique for Establishing Personnel Performance Standards)
TLA (TimeLine Analysis Program)
TLE (CHESS) Time Line Evaluation
WAA (Workload Assessment Aid - MAN-SEVAL - MANPRINT)
WAM (CAFES) (Workload Assessment Model)
WOSTAS (Workstation Assessor)

Spreadsheets for manpower, personnel, and training costs

AMCOS (Army Manpower Cost System)
ASSET (Acquisition of Supportable Systems Evaluation Technology)
CRDS (Crew Requirements Definition Subsystem and Methodology)
EAM (Electronic Aids to Maintenance) Impact on Weapon System Availability
ETAS (Essex Training Analysis System)
HARDMAN II
JASS (Job Assessment Software System)
Knowledge-based HFE Document Preparation System
M-CON (Manpower Constraints Aid - MANPRINT)
MAN-SEVAL (Manpower Systems Evaluation Aid - MANPRINT)
MDT (Mission Decomposition Tool)
Methodologies for Planning Unit and Displaced Equipment Training
MIST (Man-Integrated Systems Technology) see HARDMAN II
MMAA (Maintenance Manpower Analysis Aid - MAN-SEVAL - MANPRINT)
Operations and Maintenance Requirements Simulation Methodology and Model
P-CON (Personnel Quality Constraints Aid - MANPRINT)
PER-SEVAL (Personnel-based System EVALuation aid - MANPRINT)
SPARC (System Performance and RAM Criteria Estimation Aid - MANPRINT)
Supply Support Methodology and Model

Electronic retrieval of human engineering standards

AUTOSPEC
AWARE Design Tool
CACHE (Computer-Aided Checklist for Human Engineering)
COPE (Contract Preparation Environment - part of Knowledge-based HFE Document Preparation System
DAP (Display Analysis Program)
DRUID (Dynamic Rules for User Interface Design)
Graphical Marionette
HEED (Human Engineering Equipment Design - part of Knowledge-based HFE Document Preparation System
HF-ROBOTEX (Human Factors-Robotics Expert System)
HIMS (Helicopter Inflight Monitoring System) II
IDEA (Integrated Engineering/Decision Aid)
NaviText SAM
OWLKNEST (Operator Workload Knowledge-based Expert System Tool)
POSE (Program Organization and Scheduling Environment - part of Knowledge-based Document Preparation System
POSIT
SCOPE (Smart Contract Preparation Expediter)

Human workload evaluation tools

SWAT (CHESS) (Subjective Workload Assessment Technique)
ZITA (Zero Input Tracking Analyzer)

Formulation of a Minimum Variance Deconvolution Technique for Compensation of Pneumatic Distortion in Pressure Sensing Devices

Stephen A. Whitmore, Aerospace Engineer

NASA Ames Research Center

Dryden Flight Research Facility

1 Introduction

Recent advances in aircraft performance and maneuver capability have dramatically complicated the problem of flight control augmentation. With increasing regularity, aircraft system designs require that aerodynamic parameters derived from pneumatic measurements be used as control system feedbacks. These requirements necessitate that pneumatic data be measured with accuracy and fidelity. To date this has been a difficult task. The primary difficulty in obtaining high frequency pressure measurements is pressure distortion due to frictional attenuation and pneumatic resonance within the sensing system. Typically, most of the distortion occurs within the pneumatic tubing used to transmit pressure impulses from the surface of the aircraft to the measurement transducer.

To avoid pneumatic distortion, experiment designers have sought to mount the pressure sensor at the surface of the aircraft; a process referred to as *in situ* mounting. In some cases this mounting technique provides a viable solution. In other cases, however, as when many pressures must be measured in a small surface area, designers simply cannot crowd enough pressure transducers into the available space. As a result, typical installations require that sizeable lengths of pneumatic tubing be run from the surface of the aircraft to the pressure transducer.

In addition to space limitations, other operational difficulties occur with in-flight use of *in situ* sensors. Since the *in situ* transducer is in direct contact with the external aircraft enviorment, insuring survivability is extremely difficult. Additionally, since the temperature enviorment cannot readily be controlled, transducer calibration scale factors will tend to drift causing transducer readings to wander off-scale.

In the long term, new non-intrusive pressure sensing technologies such as $piezo - film$ or optical techniques such as $laser/Doppler$ must be developed, if the pressure sensing field is to continue to grow. Unfortunately, these technologies are still in their infancy, and will not be reliably proven for at least the remainder of this century. Additionally, these advanced technologies are quite expensive. In light of these operational and cost limitations, this chapter develops a technique for measuring unsteady pressure data using conventional pressure sensing technology. Preliminary results of this development have already been published in references [15], [16], and [17].

In this chapter a pneumatic distortion model will be formulated and reduced to a low order state-variable model which retains most of the dynamic characteristics of the full model. The reduced order model will be coupled with standard results from minimum variance estimation theory to develop an algorithm to compensate for the effects of pneumatic distortion. Both post-flight and real-time algorithms will be developed. Algorithm characterisitcs will be illustrated using both real and simulated data.

2 List of Symbols and Mathematical Operators

Symbol	Definition
A_c	Cross-sectional Area of Pneumatic Tubing
A_{k+1}	Rauch Smoother Gain Matrix
B	State Equation Input Geometry Matrix
C	Observation Matrix
c	Sonic Velocity
C_{f_w}	Skin Friction Coefficient at Tubing Wall
C_0	Polytropy Coefficient
D	Pressure Tubing Cross-Sectional Diameter
dx	Spatial Coordinate Differential
F_w	Damping Force Acting at Tubing Wall
G	Stationary Measurement Error Variance
G_{k+1}	Measurement Noise Covariance Matrix
$G(1)$	First Diagonal Component of G_{k+1}
$G(2)$	Second Diagonal Component of G_{k+1}
g_{k+1}	Measurement Noise Function
I	Identity Matrix
i	Spatial Index
J	Quadratic Cost Index
k	Data Frame Index
L	Length of Pneumatic Tubing
M	Number of Spatial Gridpoints

Symbol	Definition
N	Number of Data Frames Available
K_{k+1}	Kalman Filter Gain Matrix
P_a	Ambient Pressure
$P_{k+1/k}$	Prediction Error Covariance Matrix
$P_{k+1/k+1}$	Filter Error Covariance Matrix
$P_{k+1/N}$	Smoother Error Covariance Matrix
P_L	Downstream Pressure Response
P_0	Input Pressure Function
$P(x,t)$	Pressure Function
Q_{k+1}	State Noise Covariance Matrix
$Q_{U_{k+1}}$	Prescribed Input Function Error Variance
R	Acoustic Resistance of Pneumatic Configuration
R_e	Reynold's Number
R_{k+1}	Residual Error Vector
t	Time Coordinate
U_{k+1}	Input Vector
$U(x,t)$	Velocity Function
V	Enclosed Transducer Volume
V_e	Effective Sensor Volume, $V + \frac{A_c L}{2}$
x	Spatial Coordinate
X	State Vector
Z	Observation Vector
$\tilde{Z}$	Filter Response Residual

Symbol	Definition
Γ_{k+1}	Prescribed Input Vector
γ	Ratio of Specific Heats
Δ_{k+1}	Recursive Residual Error Matrix
δ_{k+1}	Input Error Vector
$\delta\rho$	Density Perturbation
Δt	Discretization Time Interval
μ	Dynamic Viscosity
ξ	Polytropy Heat Transfer Parameter
ξ	Damping Ratio of Reduced Order Model
$\rho(x,t)$	Density Function
ρ_0	Linearization Density
τ_w	Pressure Tubing Shear Stress at Wall
Φ	State Equation Transition Matrix
ω_n	Natural Radian Frequency of Reduced Order Model
$E[...]$	Expectation Operator
$\hat{[...]}$	Estimated Quantity
$\frac{\partial(...)}{\partial[...]}$	Partial Derivative of (...) with Respect to [...]
$\frac{d(...)}{d[...]}$	Total Derivative of (...) with Respect to [...]

3 Background

As mentioned in the introduction, for most flight test applications pressure sensing devices must be located remotely from the aircraft surface; usually somewhere within the internal cavity of the aircraft. To transmit pressure changes at the surface to the pressure transducer, a length of connective tubing is used. Pressure variations at the surface propogate as waves from the upstream end through the connective tubing to the transducer. The wave propogation is damped by frictional attenuation along the walls of the tubing. This damping causes spectral attenuation of the pressure response and produces both a magnitude attenuation and a phase lag. As the wave reaches the downstream end of the pneumatic tubing, it is reflected back up the tube. Depending upon the frequency of the incoming waves and the length of the tubing, the reflected waves either cancel or reinforce incoming pressure waves. If the waves cancel, then further spectral attenuation occurs; however, if the waves reinforce, the power of the incoming wave will be amplified and acoustic resonance occurs. Only for zero volume configurations, in which the transducer diaphragm is mounted exactly flush to the external surface (*in situ* mounting), will spectral distortion not occur.

4 Derivation of Mathematical Model

A considerable body of information concerning the effects of pneumatic distortion is available. Early analyses modeled the tubing dynamics for specialized inputs (Refs. [2], [8] , [9], [10], [13], and [14]), assuming either highly damped or undamped measurement configurations. While accurately predicting pressure responses for their prescribed types of inputs, these methods were unable to predict the behavior of pneumatic systems subjected to arbitrary pressure inputs. A more general model is required.

This section presents the idealized configuration to be analysed first. Next, the governing equations will be presented. These equations will be linearized assuming small amplitude inputs and coupled to give a linear partial differential equation for pressure in terms of position and time. Initial and boundary conditions will be developed and the complete model will be stated. Numerical techniques for solving the full mathematical model will be developed. Once the full model has been developed, it will be reduced to a simple second order filter which retains most of the dynamic characteristics of the complete model. The reduced order model will be integrated to give a discrete-time state variable model. Using minimum variance deconvolution techniques, this state variable model will be used to develop the pressure distortion compensation algorithm in section 5.

4.1 Idealized Configuration Geomtery

The sensor configuration is modeled as a straight cylindrical tube with a constant volume attached to its downstream end. The tube represents the transmission line from the surface to the sensor, and the attached volume, V, represents the internal volume of the pressure transducer. The tube is considered to be of constant radius and has length, L. The tubing cross-sectional area is A_c. A longitudinal coordinate, x, is measured from the upstream end of the tube, and a time coordinate, t, is measured forward from some initial time.

For this analysis a small pressure pulse is introduced at the upstream end of the pressure tubing and propogates downstream in the form of a longitudinal wave. Since the sensor is closed at one end, flow velocities are small and air density within the sensor may be assumed to experience only small changes from the initial value., i.e.

$$\rho(x,t) = \rho_0 + \delta\rho(x,t).$$

If little heat transfer is assumed to occur, temperature variations within the sensor may be assumed to be negligible. A one-dimensional analysis will be performed, thus, the flow velocity is approximated by its radial average. The idealized configuration to be analysed is depicted in figure 1.

The governing equation will now be presented. The Navier-Stokes equations of momentum and continuity will be presented first. The equations of state will then be presented. All thermodynamic parameters are assumed to be constant in this analysis.

4.2 Continuity Equation

The longitudinal component of the Navier-Stokes equation of continuity is given by reference [5] as

$$\frac{\partial\rho(x,t)}{\partial t} + \frac{\partial[\rho(x,t)U(x,t)]}{\partial x} = 0, \tag{1}$$

where, $\rho(x,t)$ is the flow density, and $U(x,t)$ is the flow velocity.

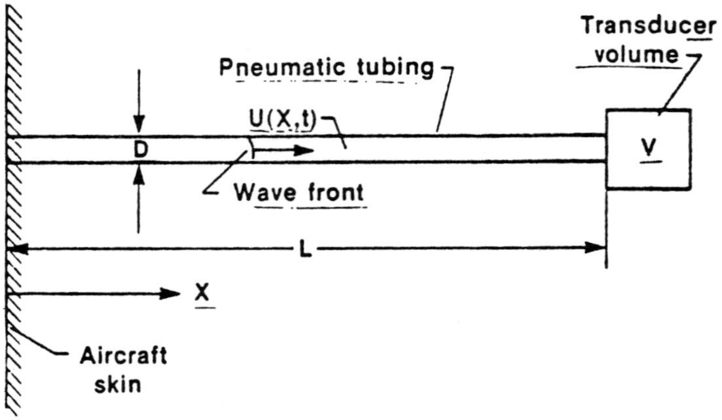

Figure 1: Schematic of Pressure Sensing Device

4.3 Momentum Equation

The general form of the longitudinal component of the Navier-stokes momentum equation is given by reference [5] as

$$\rho\frac{\partial U(x,t)}{\partial t} + \frac{\partial P(x,t)}{\partial x} + \mu\frac{\partial^2 U(x,t)}{\partial x^2} = 0. \tag{2}$$

where, $P(x,t)$ is the local pressure within the tubing. The last term on the left hand side, $\mu\frac{\partial^2 U(x,t)}{\partial x^2}$, represents the viscous force per unit volume acting against the flow. For flow in a cylindrical channel, the viscous force per unit volume acting upon an elemental cylindrical volume with diameter, D, and length, dx, is

$$F_w = \frac{\pi D dx}{A_c dx}\tau_w = \frac{\pi D \tau_w}{A_c},$$

where, τ_w, is the shear stress at the wall, and F_w is the resistive force per unit volume. This can be expressed in terms of a shear-stress coefficient as

$$F_w = \frac{\pi D C_{f_w}(\frac{1}{2}\rho U^2)}{A_c} = \frac{2R_e C_{f_w}\mu U}{D^2},$$

where, C_{f_w} is the shear stress coefficient at the wall and R_e, is the Reynold's number. Defining the acoustic resistance as

$$R = \frac{2R_e C_{f_w}\mu}{D^2}, \tag{3}$$

equation 2 reduces to

$$\rho\frac{\partial U(x,t)}{\partial t} + \frac{\partial P(x,t)}{\partial x} + RU(x,t) = 0. \tag{4}$$

For contemporary sensing devices in which internal volumes are low, the flow is laminar (Refs. [1], [5], [13]), and the coefficient for shear stress at the wall is

$$C_{f_w} = \frac{16}{R_e}. \tag{5}$$

Substituting equation 5 into equation 3, the expression for the acoustic resistance reduces to

$$R = \frac{32\mu}{D^2} . \tag{6}$$

4.4 Equation of State

For types of fluid flow in which temperature and density changes are small, the flow may be assumed to behave as a *polytropic* process (Ref. [1]). In this type of thermodynamic process changes in pressure and density are related according to the equation

$$\frac{P}{\rho^{\xi}} = C_0, \tag{7}$$

where, ξ, is the coefficient of polytropy and C_0 is a constant. Conservation of energy requires that

$$1 \leq \xi \leq \gamma,$$

where, γ is the ratio of specific heats. A technique for evaluating ξ for flow in channels is presented in reference [16]. Simulations indicate that for pressure sensors in which flow velocities are small and little heat transfer occurs, $\xi \approx \gamma$. This assumption will be used throughout this paper. Rates of change of pressure and density may be related by differentiating equation 7 with respect to the appropriate independent variable.

4.5 Reduction of Governing Equations to Wave Model

The governing equations will now be linearized and coupled to produce a single damped wave equation. The equation of continuity will be linearized first. Next the equation of state will be used to write the continuity equation in terms of pressure and velocity. The result will then be substituted into the equation of momentum to give a single damped wave equation.

As mentioned earlier, small amplitude inputs are assumed and equation 1 is written as

$$\frac{\partial[\rho_0 + \delta\rho(x,t)]}{\partial t} + \frac{\partial[\rho_0 + \delta\rho(x,t)]}{\partial x}U(x,t) + [\ \rho_0 + \delta\rho(x,t)\ ]\frac{\partial[U(x,t)]}{\partial x} = 0.$$

Neglecting all second order perturbation terms, (remembering that velocity is a perturbation in this case) the above equation is written in terms of pressure by substituting the equation of state to give

$$\frac{\partial P(x,t)}{\partial t} + c^2\rho_0\frac{\partial U(x,t)}{\partial x} = 0, \tag{8}$$

where, c is the local sonic velocity and

$$c^2 = \frac{\gamma P}{\rho}.$$

The equations of momentum and continuity are coupled by differentiating equation 4 with respect to x, equation 8 with respect to t, and equating the results.

$$\frac{\partial^2 P(x,t)}{\partial t^2} - c^2\rho_0[\frac{R}{\rho_0}\frac{\partial U(x,t)}{\partial x} + \frac{1}{\rho_0}\frac{\partial^2 P(x,t)}{\partial t^2}] = 0. \tag{9}$$

But from equation 8

$$\frac{\partial U(x,t)}{\partial x} = -\frac{1}{c^2\rho_0}\frac{\partial P(x,t)}{\partial t}. \tag{10}$$

Substituting 10 into 9 the result is

$$\frac{\partial^2 P(x,t)}{\partial t^2} + \frac{R}{\rho_0}\frac{\partial P(x,t)}{\partial t} = c^2 \frac{\partial^2 P(x,t)}{\partial x^2}. \tag{11}$$

Equation 11 is a form of the classical wave equation. Pressure variations within the tube propogate as longitudinal compression waves with undamped wave speed, c–sonic velocity.

4.6 Initial and Boundary Conditions

In order to complete the mathematical model, the initial and boundary conditions must still be defined. The initial and upstream boundary conditions will be prescribed. The downstream boundary condition will be defined by satisfying the equations of momentum and continuity at $x = L$.

4.6.1 Initial Condition

The initial conditions are prescribed by assuming that at the initial time the system is at rest, thus,

$$P(x,0) = P_a, \quad \frac{\partial P(x,t)}{\partial t}\Big|_{t=0} = 0, \tag{12}$$

4.6.2 Upstream Boundary Condition

The upstream boundary condition at $x = 0$ is prescribed as a known input function,

$$P(0,t) = P_0(t). \tag{13}$$

$P_0(t)$ will change depending upon what sort of input is analysed.

4.6.3 Downstream Boundary Condition

The governing equations must also hold at the downstream boundary, thus continuity must be satisfied,

$$V\frac{d\rho_L(t)}{dt} = \rho_L(t)A_cU_L(t), \tag{14}$$

and similarly, momentum must be satisfied,

$$\rho\frac{dU_L(t)}{dt} + RU_L(t) + [\frac{\partial P(x,t)}{\partial x}]|_{x=L} = 0. \tag{15}$$

As assumed during the derivation of the wave model, input amplitudes and velocities are small. Dropping all second order perturbations and applying the equation of state to write density in terms of pressure, the linearized continuity constraint is written as

$$\frac{dP_L(t)}{dt} = \frac{c^2A_c}{V}\rho_0U_L(t). \tag{16}$$

If equation 16 is differentiated with respect to time, and equations 14 and 15 are substituted to eliminate $U_L(t)$, the downstream boundary condition may be written in terms of pressure as

$$\frac{d^2P_L(t)}{dt^2} + \frac{R}{\rho_0}\frac{dP_L(t)}{dt} + \frac{c^2A_c}{V}[\frac{\partial P(x,t)}{\partial x}]|_{x=L} = 0. \tag{17}$$

4.7 Numerical Solutions of Complete Model

Equations 11, 12, 13, and 17 collected together represent a completely posed linear boundary value problem with a unique solution for every finite input function (Ref. [16]). The model may be solved numerically using an implicit differencing technique where the pressure function is approximated by a series of discrete gridpoints, with partial derivative operators being approximated by finite differences with first derivatives approximated by

$$\frac{\partial P(x,t)}{\partial t} = \frac{p_{i_{k+1}} - p_{i_k}}{\Delta t}, \tag{18}$$

$$\frac{\partial P(x,t)}{\partial x} = \frac{p_{i+1_{k+1}} - p_{i_{k+1}}}{\Delta x}, \tag{19}$$

and second derivatives approximated by

$$\frac{\partial^2 P(x,t)}{\partial t^2} = \frac{p_{i_{k+1}} - 2p_{i_k} + p_{i_{k-1}}}{\Delta t^2}, \tag{20}$$

$$\frac{\partial^2 P(x,t)}{\partial x^2} = \frac{p_{i+1_{k+1}} - 2p_{i_{k+1}} + p_{i-1_{k+1}}}{\Delta x^2}, \tag{21}$$

where, the index, i, represents the i'th spatial gridpoint, and the index, k, represents the k'th temporal gridoint; while, Δx and Δt, represent the spatial and temporal distances between the discrete gridpoints. The difference operators as defined in equations 18, 19, 20, and 21 are substituted into equation 11 and rearranged to give

$$-\lambda p_{i+1_{k+1}} + (\delta + 2\lambda)p_{i_{k+1}} - \lambda p_{i-1_{k+1}} = -(1+\delta)p_{i_k} - p_{i_{k-1}}, \tag{22}$$

where,

$$\lambda = (\frac{c\Delta t}{\Delta x})^2,$$

and

$$\delta = 1 + [\frac{R\Delta t}{\rho_0}].$$

Equation 22 is a recursive formula which must be satisfied for

$$i = 1,\ 2,\ 3,\ \dots M-1,\ \ M = \frac{L}{\Delta x},$$

at every time frame.

The downstream boundary condition is discretized in a similar manner, with finite differences being taken over the last finite element (from $M-1$ to M). The result is

$$-g\lambda p_{M-1_{k+1}} + (\delta + g\lambda)p_{M_{k+1}} - (1+\delta)p_{M_k} - p_{M_{k-1}}, \tag{23}$$

where, λ and δ are defined as above, and [1]

$$g = \frac{A_c \Delta x}{V_t}\left(\frac{c\Delta t}{\Delta x}\right)^2.$$

Equation 23 may now be augmented to the set of equations described by equation 22 and the entire spatial grid at temporal gridpoint, k+1, may be written in matrix form as the non-symmetric tridiagonal system

$$\begin{bmatrix} \delta+2\lambda & -\lambda & 0 & \dots & 0 & 0 & 0 \\ -\lambda & \delta+2\lambda & -\lambda & \dots & 0 & 0 & 0 \\ 0 & -\lambda & \delta+2\lambda & \dots & 0 & 0 & 0 \\ \dots & \dots & \dots & \dots & \dots & \dots & \dots \\ 0 & 0 & 0 & \dots & -\lambda & 0 & 0 \\ 0 & 0 & 0 & \dots & \delta+2\lambda & -\lambda & 0 \\ 0 & 0 & 0 & \dots & -\lambda & \delta+2\lambda & -\lambda \\ 0 & 0 & 0 & \dots & 0 & -g\lambda & \delta+g\lambda \end{bmatrix} \begin{bmatrix} p_{1_{k+1}} \\ p_{2_{k+1}} \\ p_{3_{k+1}} \\ \dots \\ p_{M-3_{k+1}} \\ p_{M-2_{k+1}} \\ p_{M-1_{k+1}} \\ p_{M_{k+1}} \end{bmatrix}$$

[1] In the discretization of the downstream boundary condition, $V_t = V + \frac{A_c \Delta x}{2}$. The addition of the $\frac{A_c \Delta x}{2}$ term accounts for the fact that some finite volume has been neglected by assuming that the pressure gradient at the downstream end is approximated by the average pressure gradient over the last finite element.

$$= \begin{bmatrix} (1+\delta)p_{1_k} \\ (1+\delta)p_{2_k} \\ (1+\delta)p_{3_k} \\ \cdots \\ (1+\delta)p_{M-3_k} \\ (1+\delta)p_{M-2_k} \\ (1+\delta)p_{M-1_k} \\ (1+\delta)p_{M_k} \end{bmatrix} - \begin{bmatrix} p_{1_{k-1}} \\ p_{2_{k-1}} \\ p_{3_{k-1}} \\ \cdots \\ p_{M-3_{k-1}} \\ p_{M-2_{k-1}} \\ p_{M-1_{k-1}} \\ p_{M_{k-1}} \end{bmatrix} + \begin{bmatrix} \lambda p_{0_{k+1}} \\ 0 \\ 0 \\ \cdots \\ 0 \\ 0 \\ 0 \\ 0. \end{bmatrix} \tag{24}$$

Given estimates of the pressure vector at times $k-1$ and k, and the prescribed pressure input, $P_{0_{k+1}}$, equation 24 is now simply an algebraic system in which the vector of pressures $[p_{1_{k+1}}, \ \ldots \ p_{M_{k+1}}]^T$, are the only unknowns. The system of linear equations may be solved using elimination techniques. Unfortunately, because equation 24 is non-symmetrical, fast symmetric factorization techniques such as Choleski factorization cannot be used to solve the system. In order to insure stability and accuracy, some form of pivoting is required. For this analysis simple *L-U* factorization with row pivoting was used (Ref. [7] pp. 100-108.) The tridiagonal structure of equation 24 was exploited.

The model was extensively verified by comparisons to special-case analytical solutions, laboratory data, and flight data. Results of this verification are presented in references [15] and [16]. Detailed descriptions of the experimental tests and procedures are presented in references [15], [16], and [17]. Results presented in these references indicate that predictions of the wave model are accurate. For well controlled experimental data, virtually no differences between the model predictions and the experimental results were found to exist.

4.8 Reduction of Wave Model to a Second Order Filter

The full wave model will now be reduced to a lower order model which retains most of the dynamic characteristics. The reduced order model is more amenable to computation and will be used to develop the compensation algorithm. Solutions of the full wave model will be used to evaluate the validity range of the reduced order model.

If the pressure gradient at the downstream boundary is approximated by the spatial average of the gradient within the pressure tubing, then the wave model reduces to a second order filter of the form (Ref. [16])

$$\frac{d^2 P_L(t)}{dt^t} + 2\xi\omega_n \frac{dP_L(t)}{dt} + \omega_n^2 P_L(t) = \omega_n^2 P_0(t), \tag{25}$$

where, ξ is the equivalent damping ratio, and ω_n is the natural frequency. For this model

$$\omega_n^2 = \frac{c^2 A_c}{LV_e},$$

$$2\xi\omega_n = \frac{R}{\rho_0},$$

where, V_e the effective volume (that volume entrapped downstream of the pressure tubing midpoint) is, $V_e = V + \frac{LA_c}{2}$.

As mentioned earlier, the single-input/single-output filter of equation 25, retains most of the frequency and time response characteristics of the original wave model up through the first wave harmonic. This fact is illustrated in figures 2 and 3 where frequency response comparisons of the full and reduced order models are presented.

Figure 2 displays frequency response results for an overdamped sensor with a tubing length of 8 ft, a tubing diameter of 0.06 $in.$, and a transducer of negligible volume. The operating altitude is $40,000$ $feet$. The equivalent damping ratio of the reduced order model is 1.93, and the natural frequency is 26.75 $hertz$. The solid line represents the frequency response of the full wave model, while the dashed line represents the frequency response of the reduced order model. Analysis of the phase angle data reveals that at the natural frequency, 26.75 hertz, the reduced order model introduces a time lag of 9.3 milliseconds, the full order model introduces a time lag of 12.9 milliseconds. The time lag difference, 3.6 milliseconds, is approximately $\frac{1}{10}^{th}$ of a cycle–a negligible difference. At the natural frequency the magnitudes comparisons differ by approximately 2.0 decibels. This difference is also considered negligible.

At frequencies much beyond the first harmonic the reduced order model may not be acceptable for lightly damped sensing systems. This fact is illustrated in figure 3. Here the sensor is lightly damped having 2 ft. of 0.06 in. diameter tubing and negligible transducer volume. The altitude of operation is 20,000 ft. For this configuration, the equivalent damping ratio of the reduced order system is 0.25 and the natural frequency is 108.93. The comparisons are good up through 100 hertz, however, beyond this frequency magnitude comparisons rapidly diverge. The reason for the divergence is clear–the second order approximation does not allow for higher-order wave harmonics, whereas the wave model does. In lightly

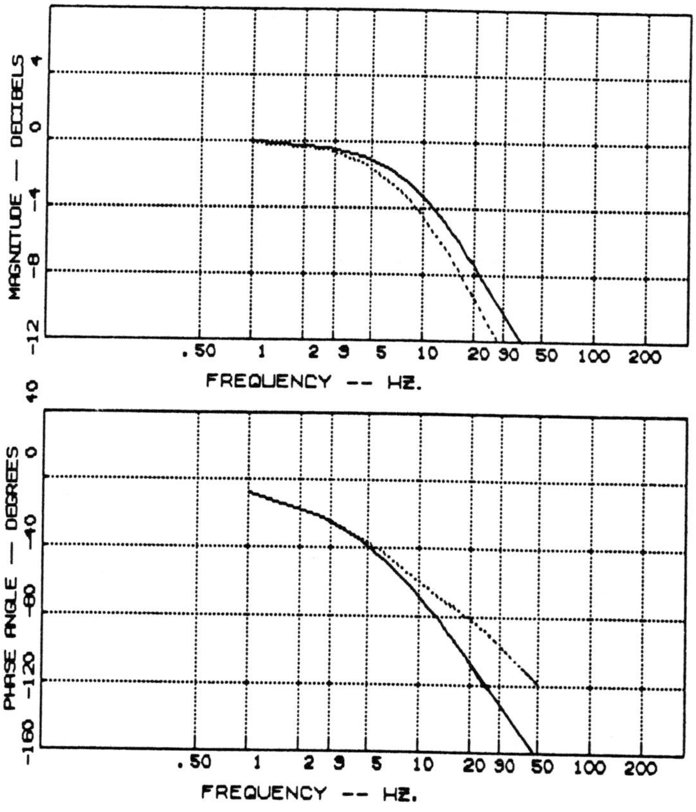

Figure 2: Frequency Response Comparisons: Full and Reduced Order Models for Overdamped Sensor Configuration.

damped configurations these harmonics are not eliminated by frictional attenuation. 3.

For analysis on lightly damped sensing systems, a higher order model which allows for additional wave harmonics should be used. One approach attempts to match the frequency response of a fourth order model to the frequency response of the full wave equation up through the second harmonic using system identification techniques. Results from this area of research will not be presented in this chapter.

4.9 Expression of Reduced Order Model as a Discrete State Variable Model

If equation 25 is integrated using the Implicit Euler Method (Ref. [6]) by performing the discrete approximations

$$\frac{d^2 P_L(t)}{dt^2} = \frac{P_{L_{k+1}} - 2P_{L_k} + P_{L_{k-1}}}{\Delta t^2}, \quad and \quad \frac{dP_L(t)}{dt} = \frac{P_{L_{k+1}} - P_{L_{k-1}}}{\Delta t},$$

where, Δt is a small time interval, the resulting difference equation is

$$P_{L_{k+1}} = \frac{2(1 + \xi\omega_n\Delta t)P_{L_k} - P_{L_{k-1}} + \omega_n\Delta t^2 P_{0_{k+1}}}{1 + 2\xi\omega_n\Delta t + (\omega_n\Delta t)^2}. \tag{26}$$

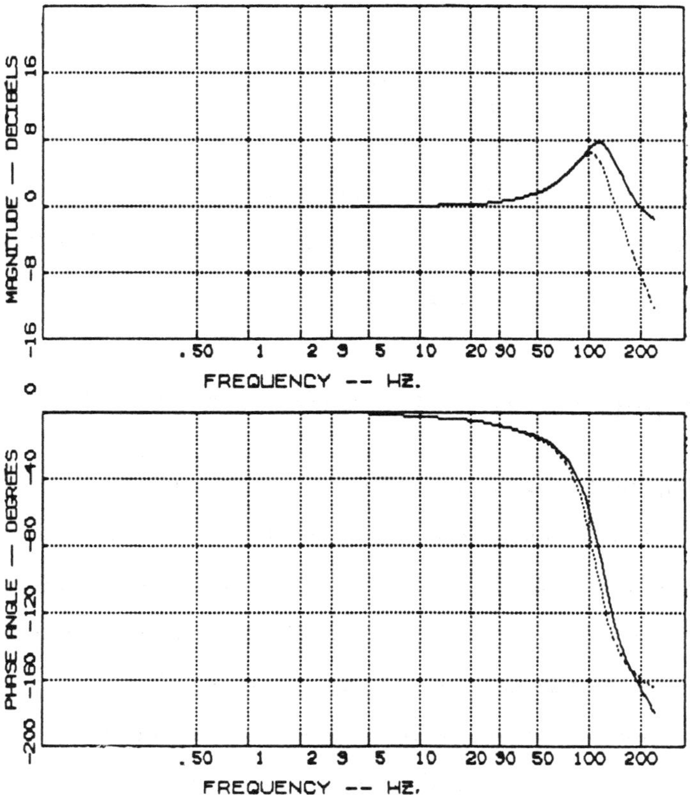

Figure 3: Frequency Response Comparisons: Full and Reduced Order Models for Underdamped Sensor Configuration.

Equation 26 may be written in state variable form as the matrix equation

$$\begin{pmatrix} P_{L_k} \\ \\ P_{L_{k+1}} \end{pmatrix} = \begin{pmatrix} 0 & 1 \\ \\ \frac{-1}{1+2\xi\omega_n\Delta t+(\omega_n\Delta t)^2} & \frac{2(1+\xi\omega_n\Delta t)}{1+2\xi\omega_n\Delta t+(\omega_n\Delta t)^2} \end{pmatrix} \begin{pmatrix} P_{L_{k-1}} \\ \\ P_{L_k} \end{pmatrix} \tag{27}$$

$$+ \begin{pmatrix} 0 \\ \\ \frac{(\omega_n\Delta t)^2}{1+2\xi\omega_n\Delta t+(\omega_n\Delta t)^2} \end{pmatrix} P_{0_{k+1}}.$$

Equation 27 is the reduced order dynamics model which will be used to develop the compensation algorithm.

5 Derivation of Compensation Algorithm

The compensation algorithm will be developed in this section. The specific objective is to develop an algorithm to infer the external pressure input to the measurement system based upon observations of the pressure response at the transducer. In general the technique of infering a system input given knowledge of the system structure and a measurement of the system response is known as *deconvolution*. First, a general deconvolution algorithm will be developed using the techniques of minimum variance estimation theory. This development will parallel results presented in references [3], [4], and [12]. Once the general algorithm has been formulated, it will be applied to the pressure compensation problem. Both post-flight and real-time versions of the algorithm will be developed.

5.1 Background on Minimum Variance Deconvolution

If equation 27 is solved for $P_{0_{k+1}}$ in terms of $P_{L_{k-1}}$, P_{L_k}, and $P_{L_{k+1}}$, the resulting equation is numerically ill-conditioned, i.e. small changes in P_{L_k} produce large changes in $P_{0_{k+1}}$. As a result noise in the pressure measurements at the transducer end of the sensor is greatly over-amplified and overwhelms any estimate of the input pressure. The over-amplified error must be identified and controlled as a part of the estimation loop. A minimum variance deconvolution algorithm will be derived in order to accomplish this objective. The basic approach is to prescribe an input to the state variable model and then identify the residual between the prescribed input and the actual input using minimum variance estimation techniques. The deconvolution algorithms will be derived using standard results from minimum variance estimation theory. These standard results will be stated without proof; complete derivations of these results may be found in reference [11].

5.1.1 Assumed Model Form

The assumed form of the model is

$$X_{k+1} = \Phi X_k + B[\Gamma_{k+1} + \delta_{k+1}], \tag{28}$$

and

$$Z_{k+1} = CX_{k+1} + g_{k+1}, \tag{29}$$

where,

$$U_{k+1} = \Gamma_{k+1} + \delta_{k+1}. \tag{30}$$

In Eqn. 30 U_{k+1} is the actual input to the system, Γ_{k+1} is the prescribed (non-random) input to the system model, and δ_{k+1} is a random process which models the residual between the actual and prescribed inputs. The covariance of δ_{k+1} is assumed to be known *a priori* as

$$COV[\delta_{k+1}] = Q_{k+1}.$$

Measurements, Z_{k+1}, of the state vector are assumed to be observed in the presence of noise. The measurement noise is assumed to have a covariance that is given *a priori* as

$$COV[g_{k+1}] = G_{k+1}.$$

For notational convenience the state transition matrix, Φ, and the input and measurement geometry matrices, B and C, are written as constants. Results to be derived are equally valid if these matrices are allowed to vary as a function of the time index. Notational conventions in this derivation follow the conventions of reference [11].

5.1.2 Fundamental Lemma of Minimum Variance Estimation Theory

The goal of the estimation process is to identify an estimate of δ_{k+1} so that the cost functional

$$J = E\{[U_{k+1} - (\Gamma_{k+1} + \delta_{k+1})]^T [U_{k+1} - (\Gamma_{k+1} + \delta_{k+1})]\}$$

is minimized with respect to δ_{k+1}. The minimization is to be conditioned upon the measured observations and constrained by the state equations.

Based on the fundamental lemma of minimum variance estimation theory (Ref. [11]), the minimizing estimate is given by the conditional expectation

$$\hat{\delta}_{k+1/N} = E\{\delta_{k+1}/(Z_1, ..., Z_N)\},$$

where, $Z_1, ..., Z_N$ is the set of observations. This result will be applied later in section 5.1.3.

5.1.3 Derivation of Residual Error Recursive Equation

The minimization is performed by defining a residual vector, $R_{k/N}$, where

$$P_{k/k-1}R_{k/N} = \hat{X}_{k/N} - \hat{X}_{k/k-1}, \tag{31}$$

and $P_{k/k-1}$ is the (Kalman) filter error covariance matrix,

$$P_{k/k-1} = COV[X_k - \hat{X}_{k/k-1}].$$

Proceed by substituting the Kalman filter,

$$\hat{X}_{k/k} = \hat{X}_{k/k-1} + K_k\tilde{Z}_{k/k-1},$$

where the Kalman Gain Matrix is

$$K_k = P_{k/k-1}C^T[CP_{k/k-1}C^T + G_k]^{-1},$$

and the predicted filter output error (innovations process) is

$$\tilde{Z}_{k/k-1} = Z_k - C\hat{X}_{k/k-1},$$

into the Rauch Smoother,

$$\hat{X}_{k/N} = \hat{X}_{k/k} + A_k[\hat{X}_{k+1/N} - \hat{X}_{k+1/k}],$$

where the Rauch Gain Matrix is,

$$A_k = P_{k/k}\Phi^T[P_{k+1/k}]^{-1}.$$

The result is

$$\hat{X}_{k/N} - \hat{X}_{k/k-1} = A_k[\hat{X}_{k+1/N} - \hat{X}_{k+1/k}] + K_k\tilde{Z}_{k/k-1}.$$

Proceeding further by substituting the residual (equation 31) vector into the above result

$$P_{k/k-1}R_{k/N} = A_k[P_{k+1/k}R_{k+1/N}] + K_k\tilde{Z}_{k/k-1},$$

But since

$$A_k = P_{k/k}\Phi^T[P_{k+1/k}]^{-1}.$$

then

$$P_{k/k-1}R_{k/N} = P_{k/k}\Phi^T R_{k+1/N} + K_k\tilde{Z}_{k/k-1}.$$

Inverting $P_{k/k-1}$ then

$$R_{k/N} = [P_{k/k-1}]^{-1}P_{k/k}\Phi^T R_{k+1/N} + [P_{k/k-1}]^{-1}K_k\tilde{Z}_{k/k-1},$$

and since (from Kalman filtering results)

$$K_k = P_{k/k-1}C^T[CP_{k/k-1}C^T + G_k]^{-1},$$

and

$$P_{k/k} = (I - K_kC)P_{k/k-1} = P_{k/k-1}(I - K_kC)^T,$$

then

$$R_{k/N} = (I - K_kC)^T\Phi^T R_{k+1/N} + C^T[CP_{k/k-1}C^T + G_k]^{-1}\tilde{Z}_{k/k-1}.$$

Defining

$$\Delta_k^T = (I - K_k C)^T \Phi^T = [\ \Phi(I - K_k C)\ ]^T,$$

then,

$$R_{k/N} = \Delta_k^T R_{k+1/N} + C^T [C P_{k/k-1} C^T + G_k]^{-1} \tilde{Z}_{k/k-1}. \tag{32}$$

Now apply the fundamental lemma of estimation

$$B\hat{\delta}_{k+1/N} = BE\{\delta_{k+1}/(Z_1, ..., Z_N)\} = E\{(X_{k+1} - \Phi X_k - B\Gamma_{k+1})/(Z_1, ..., Z_N)\}$$

$$= \hat{X}_{k+1/N} - \Phi \hat{X}_{k/N} - B\Gamma_{k+1}.$$

But from the defininiton of the residual vector

$$\hat{X}_{k+1/N} = \hat{X}_{k+1/k} + P_{k+1/k} R_{k+1/N},$$

and

$$\hat{X}_{k/N} = \hat{X}_{k/k-1} + P_{k/k-1} R_{k/N},$$

then

$$B\hat{\delta}_{k+1/N} = \hat{X}_{k+1/k} + P_{k+1/k} R_{k+1/N} - B\Gamma_{k+1} - \Phi[\hat{X}_{k/k-1} + P_{k/k-1} R_{k/N}].$$

Regrouping gives

$$B\hat{\delta}_{k+1/N} = \hat{X}_{k+1/k} - \Phi \hat{X}_{k/k-1} - B\Gamma_{k+1} + P_{k+1/k} R_{k+1/N} - \Phi P_{k/k-1} R_{k/N}.$$

But

$$\hat{X}_{k+1/k} = \Phi \hat{X}_{k/k} + B\Gamma_{k+1},$$

therefore

$$\hat{X}_{k+1/k} - \Phi \hat{X}_{k/k-1} - B\Gamma_{k+1} = \Phi(\hat{X}_{k/k} - \hat{X}_{k/k-1}) = \Phi K_k \tilde{Z}_{k/k-1}$$

As a result

$$B\hat{\delta}_{k+1/N} = \Phi K_k \tilde{Z}_{k/k-1} + P_{k+1/k} R_{k+1/N} - \Phi P_{k/k-1} R_{k/N}.$$

But as derived earlier (Eqn. 32)

$$R_{k/N} = \Delta_k^T R_{k+1/N} + C^T [C P_{k/k-1} C^T + G_k]^{-1} \tilde{Z}_{k/k-1},$$

therefore,

$$B\hat{\delta}_{k+1/N} = \Phi K_k \tilde{Z}_{k/k-1} + P_{k+1/k} R_{k+1/N} -$$

$$\Phi P_{k/k-1} \Delta_k^T R_{k+1/N} - \Phi P_{k/k-1} C^T [C P_{k/k-1} C^T + G_k]^{-1} \tilde{Z}_{k/k-1}.$$

Applying the definition of the Kalman Gain Matrix, clearly the first and last terms drop and the resulting expression is

$$B\hat{\delta}_{k+1/N} = P_{k+1/k} R_{k+1/N} - \Phi P_{k/k-1} \Delta_k^T R_{k+1/N}. \tag{33}$$

Furthermore applying Kalman Filter results for the error covariance propogation (Ref. [11])

$$P_{k+1/k} = \Phi P_{k/k} \Phi^T + COV[B\delta_{k+1}] = \Phi P_{k/k} \Phi^T + B Q_{k+1} B^T =$$

$$\Phi (I - K_k C) P_{k/k-1} \Phi^T + B Q_{k+1} B^T.$$

But

$$\Delta_k = \Phi (I - K_k C),$$

hence,

$$P_{k+1/k} = \Phi P_{k/k-1} \Delta_k^T + B Q_{k+1} B^T.$$

and

$$P_{k+1/k} R_{k+1/N} = B Q_{k+1} B^T R_{k+1/N} + \Phi P_{k/k-1} \Delta_k^T R_{k+1/N}.$$

Substituting this expression into equation 33, the result is

$$B\hat{\delta}_{k+1/N} = BQ_{k+1}B^T R_{k+1/N} + \Phi P_{k/k-1}\Delta_k^T R_{k+1/N} - \Phi P_{k/k-1}\Delta_k^T R_{k+1/N} =$$

$$BQ_{k+1}B^T R_{k+1/N}.$$

If B is full column rank then the resulting estimate for δ, based upon N observations, is given by

$$\hat{\delta}_{k+1/N} = Q_{k+1}B^T R_{k+1/N}. \tag{34}$$

By summing the result from equation 34 with the prescribed input at each data frame, minimum variance estimates of the model input may be achieved. Two complete estimation algorithms which use the result of equation 34 will now be presented. The first algorithm, intended for post-flight data reconstruction, is implemented as a two-pass smoother. The second algorithm, intended for real-time operation, is implemented as a time-recursive filter.

5.1.4 Post-Flight Smoothing Algorithm

The post-flight algorithm relies on the use of all available data measurements–both past and future. Assuming that observations of the system response are available for data frames $k = 0,\ 1,\ ...,\ N$, the follwing sequence of computations are required for the smoothing algorithm:

Kalman Filter Loop,

performed for $k = 0,\ 1,\ ...,\ N$,

$$\hat{X}_{k+1/k} = \Phi\hat{X}_{k/k} + B\Gamma_{k+1},$$

$$P_{k+1/k} = \Phi P_{k/k}\Phi^T + BQ_{k+1}B^T,$$

$$\tilde{Z}_{k+1/k} = Z_{k+1} - C\hat{X}_{k+1/k},$$

$$K_{k+1} = P_{k+1/k}C^T[CP_{k+1/k}C^T + G_{k+1}]^{-1},$$

$$\hat{X}_{k+1/k+1} = \hat{X}_{k+1/k} + K_{k+1}\tilde{Z}_{k+1/k},$$

$$P_{k+1/k+1} = [I - K_{k+1}C]P_{k+1/k}.$$

Resulting estimates for the state vector, the error covariances, and the innovations process must be stored for each recursion, $k = 0,\ 1,\ ...,\ N$.

Input Residual Estimation Loop,
performed backwards for $k = N,\ N-1,\ ...,\ 0$, using stored data

$$R_{k/N} = \Delta_k^T R_{k+1/N} + C^T[CP_{k/k-1}C^T + G_k]^{-1}\tilde{Z}_{k/k-1},$$

$$\hat{\delta}_{k/N} = Q_k B^T R_{k/N},$$

and finally,

$$\hat{U}_{k/N} = \Gamma_{k/N} + \hat{\delta}_{k/N}.$$

The final value for the residual vector, $R_{N+1/N}$, is given by the original definition (equation 31) with k replaced by N,

$$R_{N+1/N} = [P_{N+1/N}]^{-1}[\hat{X}_{N+1/N} - \hat{X}_{N+1/N}] = 0.$$

5.1.5 Real Time Filtering Algolrithm

The real-time algorithm relies only on past and present data to perform the estimation. Assuming that observations of the system response are available for data frames $k = 0,\ 1,\ \ldots,\ k+1$, then the residual estimation is performed with N replaced by $k+1$

$$\hat{\delta}_{k+1/k+1} = Q_{k+1} B^T R_{k+1/k+1},$$

But

$$R_{k+1/k+1} = [P_{k+1/k}]^{-1}[\hat{X}_{k+1/k+1} - \hat{X}_{k+1/k}],$$

therefore,

$$\hat{\delta}_{k+1/k+1} = Q_{k+1} B^T [P_{k+1/k}]^{-1}[\hat{X}_{k+1/k+1} - \hat{X}_{k+1/k}],$$

and substituting in the Kalman Filter,

$$\hat{\delta}_{k+1/k+1} = Q_{k+1} B^T [P_{K+1/K}]^{-1} K_{k+1} \tilde{Z}_{k+1/k}.$$

Applying the definition of the Kalman Gain Matrix,

$$\hat{\delta}_{k+1/k+1} = Q_{k+1} B^T C^T [C P_{k+1} C^T]^{-1} \tilde{Z}_{k+1/k}.$$

The input may be estimated as a part of the Kalman filtering loop according to the sequence of computations

Kalman Filter Step,

perform for $k = 0,\ 1,\ \ldots,\ N$

$$\hat{X}_{k+1/k} = \Phi \hat{X}_{k/k} + B\Gamma_{k+1},$$

$$P_{k+1/k} = \Phi P_{k/k} \Phi^T + B Q_{k+1} B^T,$$

$$\tilde{Z}_{k+1/k} = Z_{k+1} - C\hat{X}_{k+1/k},$$

$$K_{k+1} = P_{k+1/k} C^T [C P_{k+1/k} C^T + G_{k+1}]^{-1},$$

$$\hat{X}_{k+1/k+1} = \hat{X}_{k+1/k} + K_{(k+1} \tilde{Z}_{k+1/k},$$

$$P_{k+1/k+1} = [I - K_{k+1}C]P_{k+1/k}.$$

Residual Estimation Step,

$$\hat{\delta}_{k+1/k+1} = Q_{k+1}B^T C^T [C P_{k+1} C^T]^{-1} \tilde{Z}_{k+1/k},$$

and

$$\hat{U}_{k+1/k+1} = \Gamma_{k+1/k+1} + \hat{\delta}_{k+1/k+1}.$$

5.2 Application of Minimum Variance Deconvolution Algorithms to Pressure Compensation Problem

The above algorithms will now be applied to the pneumatic distortion problem. Equation 27 may be written as the first order matrix difference equation

$$X_{k+1} = \Phi X_k + BU_{k+1}, \tag{35}$$

and if the downstream pressure is observed in the presence of noise, then

$$Z_{k+1} = X_{k+1} + g_{k+1}. \tag{36}$$

If the substitution

$$U_{k+1} = \Gamma_{k+1} + \delta_{k+1},$$

is performed, then equations 35 and 36 are of the same form as the assumed deconvolution model, equations 28 and 29, and the deconvolution algorithms may be directly applied.

The input function, Γ_{k+1}, is prescribed by solving equation 26 for the input pressure in terms of the downstream pressures, and evaluating the result using state estimates from previous time recursions of the Filter Loop. Over-amplified noise in Γ_{k+1} is estimated and controlled by the deconvolution algorithm through judicious selection of the *a priori* state and measurement noise covariance matrices.

Derived from equation 27, the state noise covariance matrix is given by the expression

$$BQ_{k+1}B^T = \begin{pmatrix} 0 & 0 \\ 0 & Q_{U_{k+1}}\left\{\frac{(\omega_n\Delta t)^2}{1+2\xi\omega_n\Delta t+(\omega_n\Delta t)^2}\right\}^2 \end{pmatrix} \tag{37}$$

where, $Q_{U_{k+1}}$ is the variance in the residual, δ_{k+1}.

Similarly, assuming that the measurement noise is white, the measurement noise covariance matrix is given by the expression

$$G_{k+1} = \begin{pmatrix} G(1)_{k+1} & 0 \\ 0 & G(2)_{k+1} \end{pmatrix} \tag{38}$$

where, $G(1)_{k+1}$ is the error variance of the measured downstream pressure at data frame k, and $G(2)_{k+1}$ is the error variance in the measured downstream pressure at data frame $k+1$. For stationary measurement processes

$$G(1)_{k+1} = G(2)_{k+1} \equiv G.$$

6 Results of Deconvolution Analysis

Selected results of the minimum variance deconvolution analysis will now be presented. Illustrations of filtering tuning will be presented first. These illustrations will be performed using simulated data. Next the deconvolution algorithms will be applied to flight data. Analyses will be performed using both the smoothing and filtering algorithms. Comparisons with reference pressure data will be performed.

6.1 Filter Tuning

A priori selection of the proper values for the error covariance matrices is critical. This practice, usually referred to as "filter tuning," is typical of all minimum variance estimators. It requires considerable experience and more often than not, the appropriate values for the covariance matrices are unique to each individual configuration. Proper "tuning" of the deconvolution algorithms will be illustrated using simulated data generated from numerical solutions of the full wave model. Simulated data is used to illustrate the tuning process because the actual measurement and input error covariances can be pre-defined. For this illustration simulated data were generated using the full wave model for a sensor configuration with a tubing length of 8 ft, a tubing diameter of 0.06 $in.$, a transducer with negligible volume, and operating at an altitude of $40,000$ $feet$. For this configuration and altitude, the equivalent damping ratio of the reduced order system is 1.93; the natural frequency is 26.75. Simulated zero-mean white noise with a variance of 0.1 $(\frac{lbf}{ft^2})^2$ was superimposed on the output time history; simulated zero-mean white noise with a variance of 1.0 $(\frac{lbf}{ft^2})^2$ was superimposed on the input time history.

The minimum variance deconvolution analysis was performed using the full smoothing algorithm. Figures 4 – 5 demonstrate the effects that selected values of G have on the estimates of U. In these figures the value of Q_U was fixed at the appropriate value of 1.0, while the value for G was varied. Recall that the actual measurement error variance is 0.1 $(\frac{lbf}{ft^2})^2$.

Figure 4 depicts the deconvolution results for $G = 10$, a value which is much too large. Two time history graphs are presented. Depicted in the lower graph are time history overplots of the 'measured' downstream pressure response (solid line), the smoother estimate of the downstream response (short dashes), and the smoother estimate of the input pressure (long dashes). Depicted in the upper graph are time history overplots of the estimated input (solid line) and the 'measured' input (dashed line). Comparisons are poor. Clearly this set of weights does not allow the estimation algorithm to extract enough information from the measured pressure data; resulting estimates are overly smoothed.

The above situation cannot be improved by arbitrarily lowering the value for G. This point is illustrated in figure 5. Depicted are similar time history comparisons; however, in this instance the value for G is set to be much too small–0.001. The end result is that the smoother tries to identically match the measured response–noise and all. As discussed earlier, this results in over-amplification of the measurement noise and the estimate is overwhelmed. The selected value for G did not allow the estimator to identify and control noise in the measured response. The filter is essentially trying to perform an open-loop deconvolution of the state-equation.

Figure 6 demonstrates the effect of Q_U on the estimation of U. In this figure the value of G was fixed at the appropriate value of 0.10,

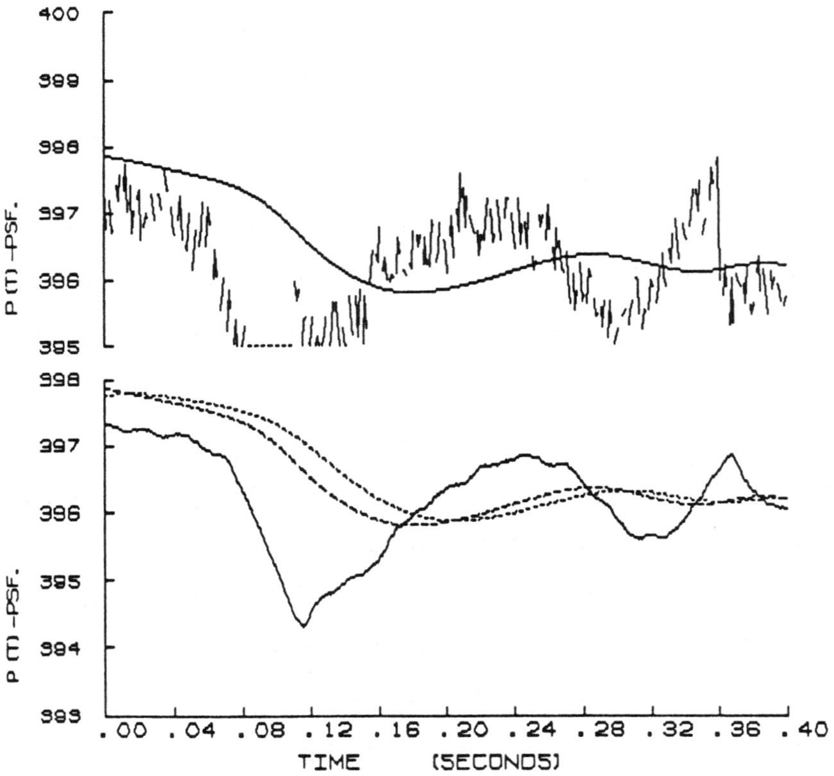

Figure 4: Deconvolution Filter Tuning Example: Effect of Improperly Selected Measurement Error Covariance, $Q_U = 1.0$, $G = 10.0$

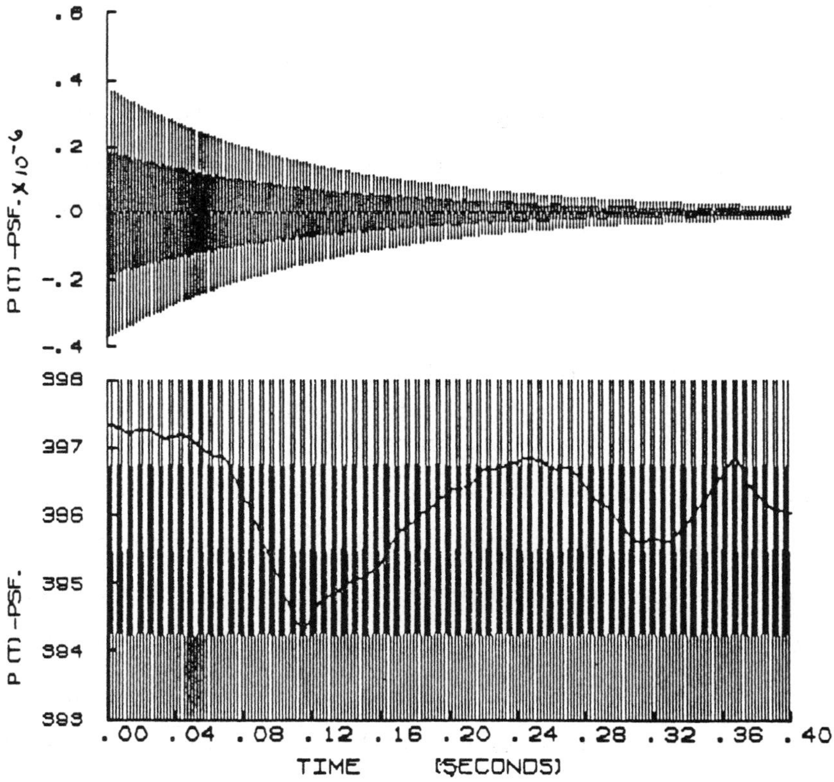

Figure 5: Deconvolution Filter Tuning Example: Effect of Improperly Selected Measurement Error Covariance, $Q_U = 1.0$, $G = 0.001$

while the value for Q_U was set at $Q_U = 10$. The resulting estimate contains extraneous harmonics not present in the original input signal. The assumed value for Q_U allows the input error to oscillate in an overly random manner, indicating that the the dynamic constraints of the model have not been enforced strongly enough.

The appropriate level of smoothing and compensation is as presented in figure 7. Here the proper values for the measurement and input error covariances are used. In this case the estimated input matches the actual input very well and much of the superimposed noise has been removed.

6.2 Application to Flight Data

The deconvolution algorithms will now be applied to real data. Using flight data derived from the experiment described in references [15], [16], and [17], the deconvolution analysis was performed. Pressure data obtained from a conventionally mounted pressure sensing device was compensated using both the smoothing and filtering algorithms. The resulting input estimates will be compared with data obtained from a specially and carefully mounted *in situ* pressure transducer. Two time history data cases will be presented; one for an overdamped configuration, and one for an underdamped configuration.

Deconvolution results for an overdamped flight maneuver obtained at 42000 ft. altitude with a sensor configuration having 8 ft. of 0.06 in. diameter tubing and negligible transducer volume are presented in figure 8 for

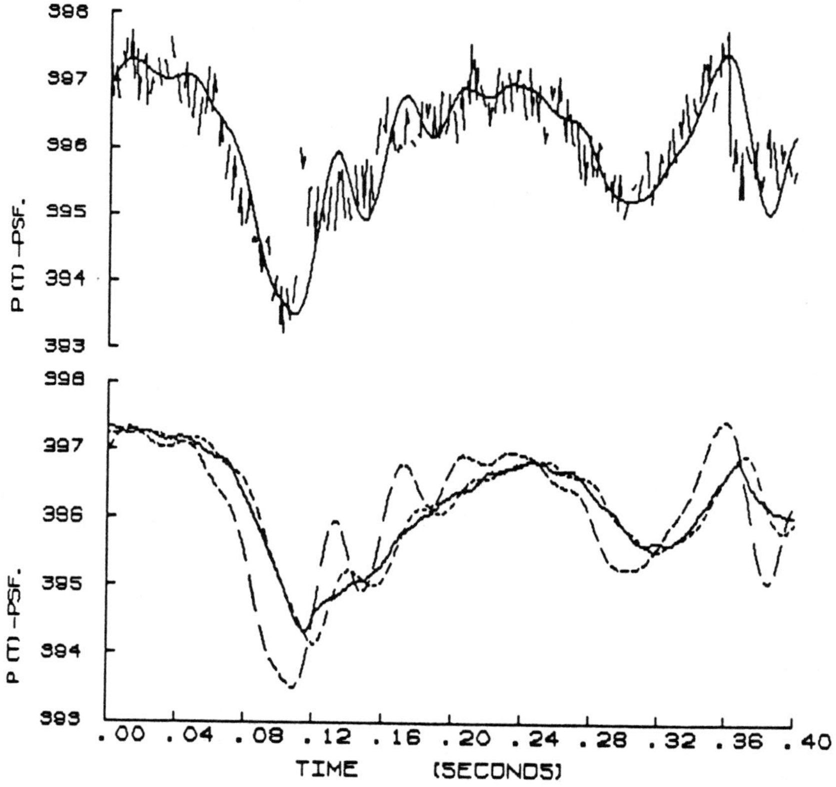

Figure 6: Deconvolution Filter Tuning Example: Effect of Improperly Selected Input Error Covariance, $Q_U = 10.0$, $G = 0.1$

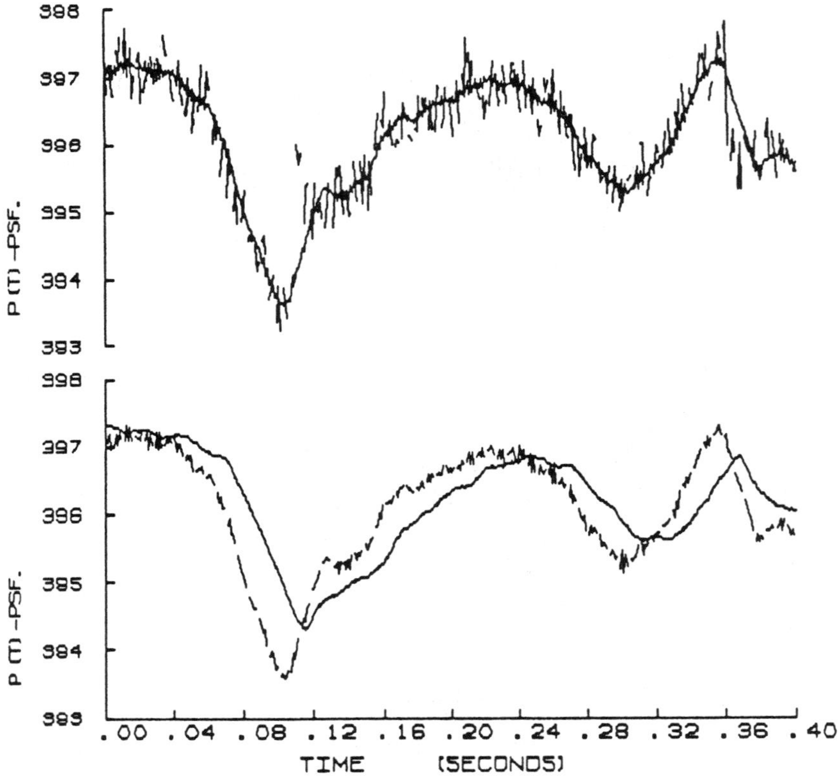

Figure 7: Deconvolution Filter Tuning Example: Effect of Properly Selected Input and Measurement Error Covariances, $Q_U = 1.0$, $G = 0.1$

the smoothing algorithm and figure 9 for the filtering algorithm. For this configuration and altitude, the equivalent damping ratio of the reduced order system is 2.125; the natural frequency is 26.75. The comparisons are excellent for both the smoothing and filtering algorithms. The filtering algorithm does not smooth the estimate quite as much as does the smoothing algorithm; this result is as expected.

Deconvolution results for an underdamped flight maneuver obtained at 24000 ft. altitude with a sensor configuration having 2 ft. of 0.06 in. diameter tubing and negligible transducer volume are presented in figure 10 for the smoothing algorithm and figure 11 for the filtering algorithm. For this configuration and altitude, the equivalent damping ratio of the reduced order system is 0.279; the natural frequency is 107.18. Both the smoothing and filtering algorithms work well. The smoothing algorithm removes nearly all of the resonant harmonic, ≈ 100 *hertz*, from the measured pressure response. The filtering algorithm removes a sizable portion of the resonance. Both algorithms account for phase lag very well.

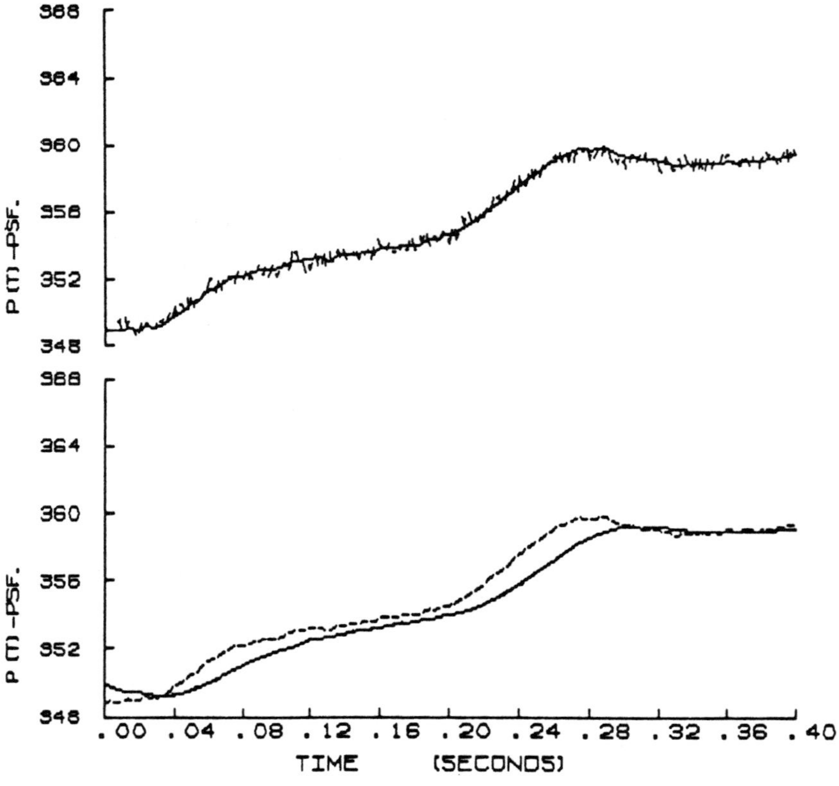

Figure 8: Flight Maneuver Time History Comparison I: Output From the Minimum Variance Deconvolution Smoothing Algorithm for Overdamped System

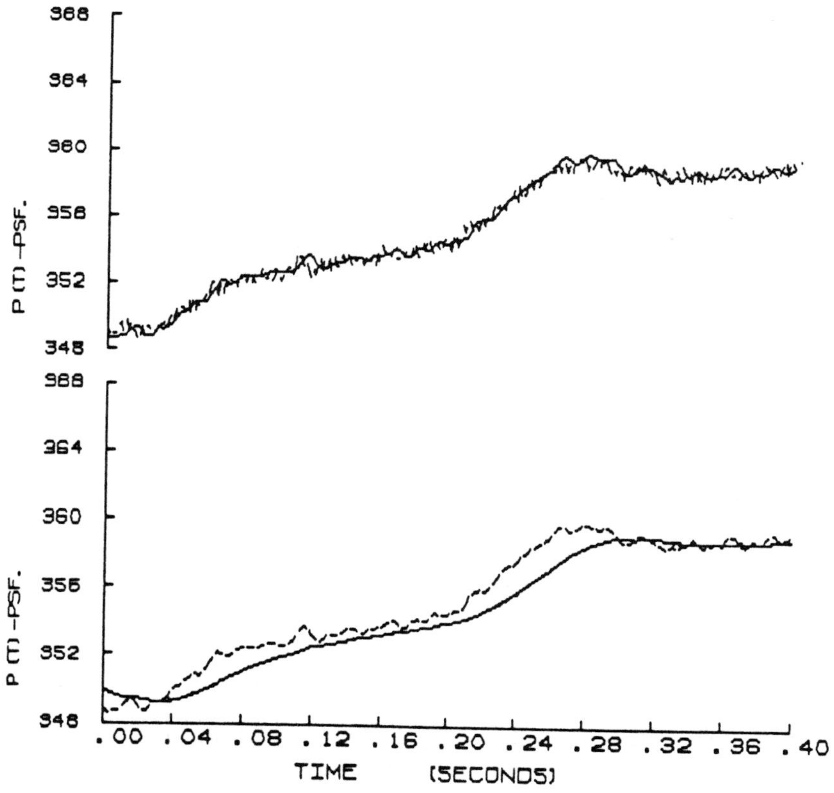

Figure 9: Flight Maneuver Time History Comparison I: Output From the Minimum Variance Deconvolution Filtering Algorithm for Overdamped System

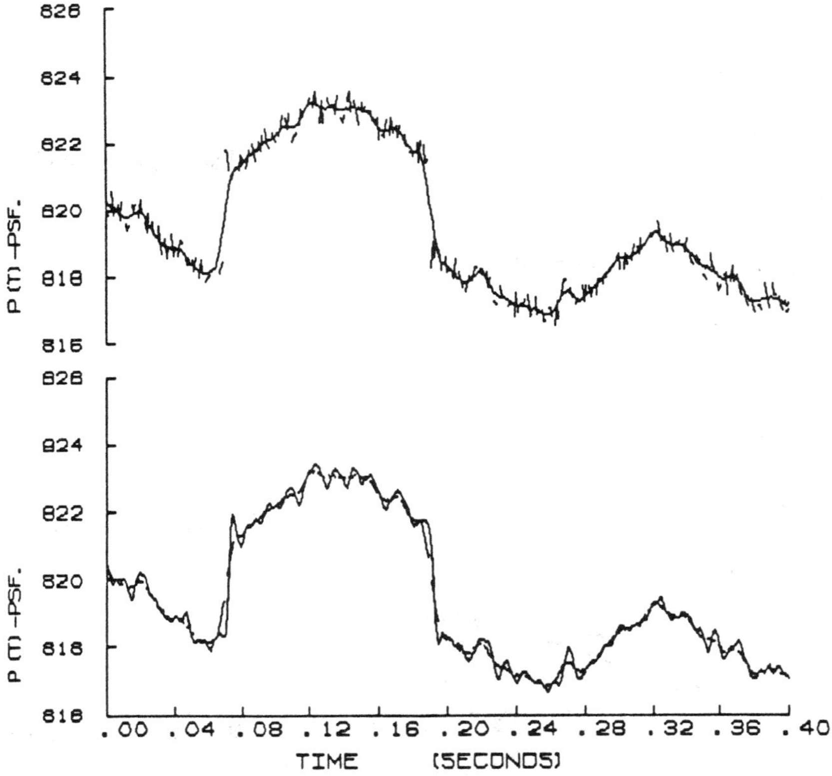

Figure 10: Flight Maneuver Time History Comparison II: Output From the Minimum Variance Deconvolution Smoothing Algorithm for Underdamped Sensor

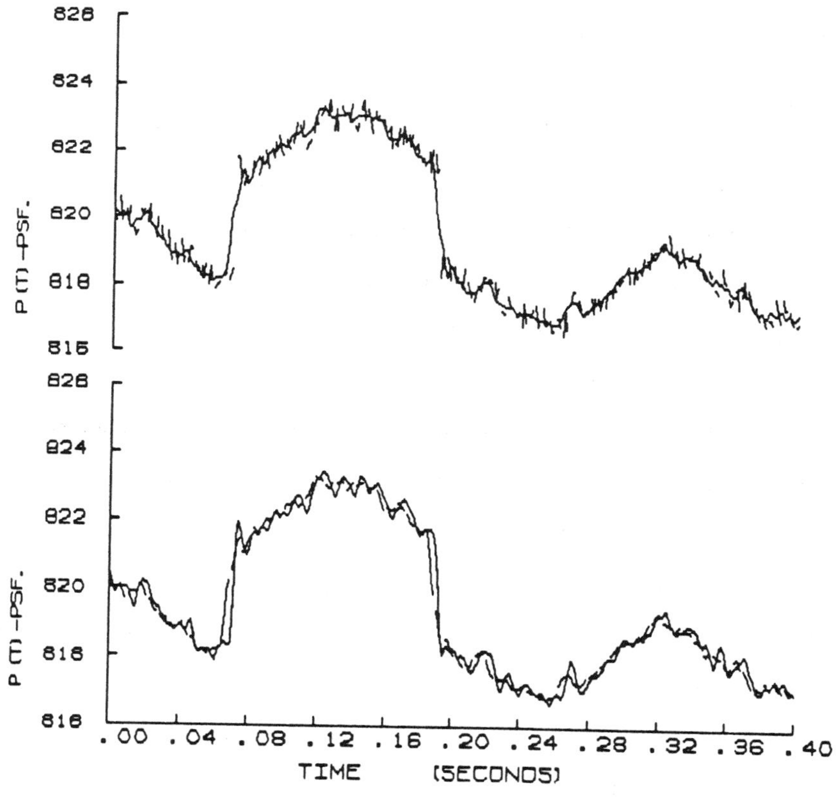

Figure 11: Flight Maneuver Time History Comparison II: Output From the Minimum Variance Deconvolution Filtering Algorithm for Underdamped Sensor

7 Concluding Remarks

The emphasis of this chapter was on the development of a general numerical technique with which unsteady pressure measurements may be obtained using conventional pressure sensing technology. A mathematical model, based on the Navier–Stokes equations of momentum and continuity was developed. The model, essentially a damped wave model, was reduced to second order and discretized to give a simple state variable model. Areas of applicability of the reduced order model were discussed.

Using a collection of standard results from minimum variance estimation theory, the reduced order model was incorporated into a deconvolution algorithm which estimates the system input based upon observations of the system output. Both post-flight and real-time algorithms were developed. Over-amplification of measurement error is controlled by judicious selection of the elements of the algorithm state and measurement error covariance matrices.

Performance characteristics and implementation approaches were discussed. Filter tuning characterisitcs were illustrated using simulated data. Applications of the techniques to real data were presented. Both the smoothing and filtering algorithms exhibited excellent performance for overdamped and underdamped sensor configurations.

The results presented in this chapter display considerable promise for the modeling and compensation techniques as developed. These techniques provide a means by which the effects of pneumatic distortion in pressure sensing devices can be predicted and accounted for. Application of these techniques offers a reliable and cost effective means of measuring unsteady pressure data. These techniques will help to bridge-the-gap until new pressure sensing technologies can be sufficiently matured.

References

[1] The International Dictionary of Applied Mathematics, Van Nostrand Company Inc., Princeton, New Hersey, 1960.

[2] Berg, H., Tijdeman, H., Theoretical and Experimental Results for the Dynamic Response of Pressure Measuring Systems, NLR report F.238 (1965)

[3] Chi, Chong-Yung, A Further Analysis of Minimum-Variance Deconvolution Filter Performance, IEEE Transactions on Acoustics, Speech, and Signal Processing, Vol ASSP-35, June 1987, p. 888-889.

[4] Chi, Chong-Yung, Mendel, J.M. A Fast Approach to Identification Using Deconvolution, Institute of Electrical and Electronics Engineering Inc., Proccedings of the Conference on Decision and Control, Volume 3, 1983, P. 1347-1352.

[5] Currie, I. G., Fundamental Mechanics of Fluids, McGraw-Hill Book Company, New York, 1974.

[6] Franklin, Gene F., and Powell, J. David, Digital Control of Dynamic Systems, Addison Wesley Publishing Co., Reading Mass., 1980.

[7] Golub, Gene, H., and Van Loan, Charles, F., Matrix Computations, John Hopkins University Press, Baltimore, 1983.

[8] Huston, W.B., The Accuracy of Airspeed Measurements and Flight Calibration Procedures, NACA Report 919, 1948.

[9] Iberall, A.,S., Attenuation of Oscillatory Pressures in Instrument Lines, National Bureau of Standards Research Paper, RP 2115, July, 1950.

[10] Lamb, J. P., The Influence of Geometry Parameters Upon Lag Error in Airborne Pressure Measurement Systems, WADC TR 57-351, Wright-Patterson AFB, Ohio, July 1957.

[11] Meditch, J. S., Stochastic Optimal Linear Estimation and Control, McGraw Hill Book Company, New York, 1969. Optimal Seismic Deconvolution, An Estimation Based Approach, The Academic Press, New York, 1983.

[12] Mendel Jerry M., Optimal Seismic Deconvolution, An Estimation Based Approach, The Academic Press, New York, 1983.

[13] Schuder, C.,B., and Binder, R., C., The Response of Pneumatic Transmission Lines to Step Inputs, Journal of Basic Engineering, ASME Transactions, December, 1959.

[14] Stephens, R. W. B., and Bate, A. E., Acoustics and Vibrational Physics, St. Martin Publishing Co., New York, 1966.

[15] Whitmore, Stephen A., Formulation of a General Technique for Prediciting Pneumatic Attenuation Errors in Airborne Pressure Sensing Devices, NASA Technical Memorandum 100430, May 1988.

[16] Whitmore, Stephen A., Formulation and Verification of a Techique for Compensation of Pneumatic Attenuation Errors in Airborne Pressure Sensing Devices, PhD Dissertation, University of California, Los Angeles, University Microfilms International, 1989.

[17] Experimental Characterization of the Effects of Pneumatic Tubing on Unsteady Pressure Measurements, Proposed NASA Technical Memorandum, Currently Awaiting Publication, 1989.

SYNTHESIS AND VALIDATION OF FEEDBACK GUIDANCE LAWS FOR AIR-TO-AIR INTERCEPTIONS

JOSEF SHINAR
Faculty of Aerospace Engineering,
Technion-Israel Institute of Technology,
Haifa, Israel.

and

HENDRIKUS G. VISSER
Faculty of Aerospace Engineering,
Delft University of Technology,
Delft, The Netherlands.

This Chapter is dedicated to the memory of Prof. H.J. Kelley,
whose ideas have inspired both authors.

I. INTRODUCTION

Interception of adversary aircraft with the objective to disrupt a hostile mission has been a basic air-to-air task since the very first years of air warfare. Not surprisingly, this task has spurred a continuous research effort as a part of evaluating air-to-air combat performance of fighter aircraft. The assumption of a known adversary trajectory allows to formulate the interception as a minimum-time optimal control problem. Simplified analysis [1,2] indicates that the optimal trajectory of the interceptor has three distinct phases:

(i) an initial turn-climb-acceleration phase,
(ii) a steady-state cruise at maximum speed,
(iii) an end-game terminating with capture as determined by the weapon of the interceptor.

Such a decomposition is valid only if the initial conditions of the encounter allow the interceptor to reach its maximum speed before the end-game is initiated, a condition generally satisfied in long-range interceptions. For shorter ranges, phase (ii) disappears and the initial and terminal parts of the trajectory are merged together. The optimal control solution associated with such a trajectory is generally obtained in an open-loop form by solving a high dimensional nonlinear two-point-boundary-value problem using some iterative algorithm.

For a real-time airborne application, as well as for a systematic performance assessment, a reasonably accurate feedback approximation of the optimal control solution is more useful. This challenge drew a considerable interest for investigations in the last decade. The first works [3, 4] were oriented towards long-range interceptions of low flying targets with the requirement of point capture. Later extensions included medium-range scenarios, characterized by the absence of the dash segment [5], validation by ground based pilot-in-the-loop simulations [6] and some flight testing [7]. These last activities [6,7] provided an encouraging proof of feasibility for such algorithms. Unfortunately, however, the accuracy of the guidance algorithm derived in [5] by applying singular perturbation methods, has never been tested.

The objective of this Chapter is to summarize a multi-year joint effort of the authors to develop and validate a feedback algorithm for medium-range air-to-air interceptions with realistic capture conditions compatible with the deployment of advanced guided weapons in a future air combat. This effort was based, similarly to [3-7], on a singular perturbation approach, but used a different multiple time-scale model. Moreover, at each milestone during the development of the algorithm, a special effort was undertaken to compare the outcome of the feedback approximations to the optimal control solution. In the open literature only a single similar study of limited scope [8] is known. In the first phase, horizontal [9,10] and vertical [11,12] interceptions were analyzed separately, leading to an enhanced insight into the problems associated with each planar maneuver. The feedback guidance law for a three-dimensional interception was then synthesized based on a vectorial combination of the two-dimensional controls and compared to the results of open-loop optimal control solutions [13,14].

This Chapter includes a review of previous works [9-14], as well as some new results and is organized as follows. In Section II the detailed mathematical formulation of the problem is presented and the formal optimal control solution is derived. In Section III modeling considerations and the application of singular perturbation theory for approximating the optimal control in feedback form are outlined. In Section IV the interception in the horizontal plane is analyzed and a uniformly valid feedback guidance law is derived. In section V the problems of the interception in a vertical plane are discussed and the synthesis of the resulting feedback guidance law is presented. It is followed in Section VI by the description of the problems involved in the synthesis of a three-dimensional feedback control strategy and a new numerical example, in which the simulated feedback approximation is compared with the open-loop optimal solution, using an aircraft model representative of a state-of-the-art high-performance fighter interceptor.

II. PROBLEM FORMULATION

A. MATHEMATICAL MODEL

The air-to-air interceptions analyzed in this chapter are characterized by the following features:

(i) The adversary aircraft (target) is assumed to fly at a fixed altitude and direction with constant speed.

(ii) The initial distance of separation is large compared with the turning radius of the interceptor, but not large enough to allow the interceptor to reach its maximum speed. This last statement defines the domain of "medium-range" interceptions.

(iii) The interception terminates when the distance of separation becomes equal to the effective firing range of the interceptor weapon (assumed to be an air-to-air missile).

In a Cartesian coordinate system, centered at the target (T) and with the x-axis is aligned with its velocity vector, the equations of relative motion (see Fig. 1 for the definition of the variables) are given by:

$$\dot{x} = V\cos\gamma\cos\chi - V_T \overset{\Delta}{=} F_x \quad , \quad x(t_o) = x_o \tag{1}$$

$$\dot{y} = V\cos\gamma\sin\chi \overset{\Delta}{=} F_y \quad , \quad y(t_o) = y_o \tag{2}$$

$$\Delta\dot{h} = \dot{h} = V\sin\gamma \overset{\Delta}{=} F_h \quad , \quad \Delta h(t_o) = h(t_o) - h_T = \Delta h_o \tag{3}$$

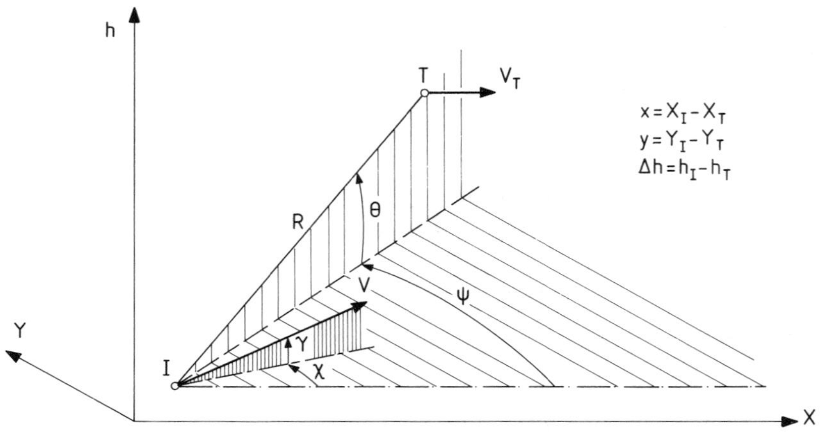

Fig. 1 Three-dimensional Interception Geometry.

The dynamic model of the interceptor assumes a flat non-rotating earth, point-mass approximation and thrust aligned with the velocity vector. The corresponding equations of motion are:

$$\dot{V} = g[(T-D)/W - \sin\gamma] \stackrel{\Delta}{=} F_V \quad , \quad V(t_o) = V_o \tag{4}$$

$$\dot{\gamma} = (g/V)[(L/W)\cos\mu - \cos\gamma] \stackrel{\Delta}{=} F_\gamma \quad , \quad \gamma(t_o) = \gamma_o \tag{5}$$

$$\dot{\chi} = (g/V\cos\gamma)[(L/W)\sin\mu] \stackrel{\Delta}{=} F_\chi \quad , \quad \chi(t_o) = \chi_o \tag{6}$$

The aerodynamic forces (lift and drag) and the maximum available thrust are functions of speed and altitude:

$$L = 0.5\ \rho(h)V^2SC_L \tag{7}$$

$$D = 0.5\ \rho(h)V^2SC_D \tag{8}$$

$$T = \eta T_{max}\ (h,V) \tag{9}$$

Based on the definition of the aerodynamic load factor:

$$n \triangleq (L/W) \tag{10}$$

and assuming a parabolic drag polar, the total drag force can be expressed as:

$$D = D_o + n^2 D_i = D(h,V,n)\ , \tag{11}$$

where the zero-lift drag and the induced drag in level flight (n=1) are defined respectively by:

$$D_o = 0.5\ \rho(h)V^2SC_{D_o} \tag{12}$$

$$D_i = 2KW^2/(\rho V^2 S)\ , \tag{13}$$

The non-dimensional coefficients C_{D_o} and K are functions of Mach number. In this mathematical model the controls are:

(i) The throttle parameter η constrained by:

$$0 \leq \eta \leq 1 \tag{14}$$

(ii) The bank angle μ which determines the direction of the lift force.

(iii) The aerodynamic load factor n defined by Eq. (10). It is subject to two different constraints: a structural limit which is effective at high speeds:

$$|n| \leq n_{max} \tag{15}$$

and a limit imposed by the maximum lift coefficient:

$$|n| \leq n_L(h,M) \triangleq \frac{0.5\, \rho\,(h)\, V^2}{(W/S)}\, C_{L_{max}}(M) \tag{16}$$

Quite often it is convenient to use as controls the horizontal and vertical components of the load factor, defined by:

$$n_h \triangleq n \sin\mu \tag{17}$$

$$n_v \triangleq n \cos\mu \tag{18}$$

The interceptor's trajectory has to be confined, as any other maneuver, to the flight envelope of the aircraft, determined by the following state constraints:

$$0 < h_{min} \leq h \leq h_{max} \tag{19}$$

$$V_{min}(h) \leq V \leq V_{max}(h) \tag{20}$$

B. OPTIMAL CONTROL FORMULATION

For the sake of a concise formulation let us define the state vector of the three-dimensional interception problem as

$$\underline{X}^T \triangleq (x, y, \Delta h, V, \gamma, \chi) \tag{21}$$

The optimal control problem to be solved is to determine the control vector $\underline{U}^{*T} \triangleq (\eta^*, \mu^*, n^*)$ that brings the interceptor from a given set of initial conditions $\underline{X}(t_o)$ to a terminal manifold, - which depends on the effective weapon firing envelope, - in the shortest possible time, subject to the state and control constraints. The performance index is therefore the final time:

$$J = \int_{t_o}^{t_f} dt = t_f \ , \tag{22}$$

defined by the terminal manifold:

$$t_f = \arg \{\Phi[t_f, \underline{X}(t=t_f)] = 0\} \tag{23}$$

The variational Hamiltonian of this problem is:

$$H = -1 + \underline{\Lambda} \cdot \dot{\underline{X}} + \text{constraints} = H(\underline{X}, \underline{\Lambda}, \underline{U}) \quad , \tag{24}$$

where the costate vector $\underline{\Lambda}$ has to satisfy the adjoint equations:

$$\dot{\underline{\Lambda}} = - \frac{\partial H}{\partial \underline{X}} \tag{25}$$

and the corresponding transversality conditions:

$$\underline{\Lambda}(t_f) = \nu \left. \frac{\partial \Phi}{\partial \underline{X}} \right|_{t=t_f} \quad , \tag{26}$$

where ν is a Lagrange multiplier.

According to the Maximum Principle the optimal controls have to maximize the Hamiltonian:

$$U^* = \arg \max_{\underline{U}} H(\underline{X}, \underline{\Lambda}, \underline{U}) \tag{27}$$

Moreover, since time does not appear explicitly in the equations and the final time is not specified:

$$H^* \triangleq H(\underline{X}, \underline{\Lambda}, \underline{U}^*) = H \Big|_{t_f} = 0 \tag{28}$$

This formulation requires to solve a nonlinear two-point-boundary-value problem of 12 dimensions resulting in $\underline{X}^*(t)$, $\underline{\Lambda}^*(t)$ and consequently $\underline{U}^*(t)$. A feedback approximation of the optimal control can be obtained if one can approximate the adjoint variables by explicit functions of the state variables. One approach to carry out such an operation is the application of singular perturbation theory, as outlined in the next section.

III. APPLICATION OF SINGULAR PERTURBATION THEORY

A singular perturbed dynamic system is characterized by a small parameter ε, multiplying the time derivatives of some components of the state vector. These components behave as "fast" variables compared to the other part of the state vector. This indicates that the mathematical structure of singularly perturbed systems is always associated with the physical phenomenon of time-scale separation. Singularly perturbed mathematical models are frequently encountered in celestial mechanics, fluid dynamics, physical chemistry etc. and have been subjects of thorough investigations [15-19]. The basic approach to solve singular perturbation problems, formulated as initial value problems, has been the method of Matched Asymptotic Expansions [20, 21]. This method was also extended to deal with two-point boundary-value problems and adapted to solve optimal control problems as well [22-26].

If one sets in a singular perturbation problem $\varepsilon = 0$, the "fast" dynamics are neglected and the order of the dynamic system is reduced. The solution of the "reduced-order" system may serve as an approximation, though it cannot satisfy the initial and terminal conditions imposed on the "fast" variables. This deficiency is corrected by initial and terminal "boundary-layer" solutions computed on a stretched time-scale. Expanding all variables in both the original and the boundary-layer problems in asymptotic power series of ε and matching the corresponding terms lead to a uniformly valid "composite solution" of the problem. Using only a finite number of terms of the expansions yields an approximation. The optimal control solution approximated by this method is obtained in an open-loop form. It was, however, found [23, 27-28] that if the terminal constraints do not involve the "fast" variables, the terminal "boundary-layer" may disappear and in this case the optimal control approximation can be expressed in a feedback form. This result, which is very attractive for real-time on-line applications, was recently confirmed [29] by demonstrating that under certain conditions the optimal feedback control solution of a singularly perturbed system can be constructed, - by a recursive solution of the Hamilton-Jacobi-Bellman equation, - as a single and uniformly valid expansion of the parameter ε.

The application of singular perturbation theory to optimize aircraft trajectories is based on identifying the actual time-scale separations between the state variables. In a linear time-invariant system such time-scale separation can be expressed by the ratio of the respective eigenvalues, which in some case is a small parameter. An appropriate scaling transformation results in multiplying the derivative of the "fast" variable by this small parameter. In the strongly nonlinear problems encountered in flight mechanics the identification of a small parameter associated with the observed time-scale separation is, unfortunately, rather difficult. For this reason it was proposed [30] to insert the singular perturbation parameter ε artificially by multiplying the observed "fast" time derivatives. This approach, called in some papers [9-14] a "forced singular perturbation" technique (FSP), has been used in most aircraft performance optimization studies [3-14, 27]. In [28] it was formally demonstrated that the zeroth-order feedback control approximation of an FSP problem is identical to the solution of a similar classical SP problem obtained by a scaling transformation. Moreover, this result was later extended to higher order corrections [31,10].

The application of the FSP technique for obtaining a zeroth-order feedback control approximation can be summarized by the following steps:

1. Order the state variables according to their relative rate of change (the faster following the slower). This first step is of major importance and the key to a satisfactory result.
2. Transform the set of original state equations to a singularly perturbed multiple time-scale dynamic model by multiplying the time derivatives of the "fast" variables of the same time-scale by increasing powers of ε. It is preferable to have only a single (active) variable on the same time-scale.
3. Set $\varepsilon = 0$ and solve the resulting "reduced-order" optimal control problem. In this "reduced-order" solution the "fast" variables play the role of pseudo-controls, constrained to an "integral manifold". Thus, the controls, as well as the respective costate variables are functions of the "slow" state variables only.
4. Solve the first "boundary layer" problem by a time-scale stretching transformation. In this stretched time-scale the "slow" variables are

frozen to their initial values. The major element in the solution is to express the active costate variable as a function of the active state variable and the frozen initial values of the "slow" states.

5. Repeat this step for all "boundary layers". It will finally result in approximating all costate variables by a feedback type expression as functions of the "frozen" initial states and the active state.
6. Find the expression for the control variables in the last (fastest) "boundary layer" at the initial time. This expression will be a function of the initial conditions only.
7. Since any current state can be considered as a new set of initial conditions, a uniformly valid feedback control law can be synthesized by replacing the initial values of the state variables with the current values.

The method for higher order corrections is similar, however the matching process is far more elaborate. For details the reader is referred to [31].

Since the application of the above outlined method to the problem of a three-dimensional air-to-air interception involves very complex modeling considerations, in the next sections the more simple planar geometries are analyzed.

IV. INTERCEPTION IN A HORIZONTAL PLANE

A. FORMULATION OF THE HORIZONTAL INTERCEPTION PROBLEM

1. DYNAMIC EQUATIONS

For the restricted case of flight in a horizontal plane ($F_h = 0 \Rightarrow \gamma = 0$), it is sometimes more convenient to describe the relative motion between the two aircraft in polar coordinates using the distance of separation R and the line-of-sight angle ψ:

$$\dot{R} = V_T \cos\psi - V\cos(\psi - \chi) \triangleq f_R(\psi, V, \chi) \tag{29}$$

$$\dot{\psi} = [-V_T \sin\psi + V\sin(\psi - \chi)]/R \triangleq f_\psi(R, \psi, V, \chi) \tag{30}$$

$$\dot{\lambda}_R = -\frac{\partial H}{\partial R} \tag{37}$$

$$\dot{\lambda}_\psi = -\frac{\partial H}{\partial \psi} \quad , \quad \lambda_\psi (t_f) = 0 \tag{38}$$

$$\dot{\lambda}_V = -\frac{\partial H}{\partial V} \quad , \quad \lambda_V(t_f) = 0 \tag{39}$$

$$\dot{\lambda}_\chi = -\frac{\partial H}{\partial \chi} \quad , \quad \lambda_\chi(t_f) = 0 \tag{40}$$

Assuming that an optimal control solution exists, the Maximum Principle can be used to express the optimal controls in terms of the state and adjoint variables:

$$\lambda_V > 0 : \eta^* = 1 \ , \qquad \mu^* = \min [\ |\mu_u| , \mu_{max}]\ \mathrm{sign}(\lambda_\chi) \tag{41}$$

$$\lambda_V < 0 : \eta^* = 0 \ , \qquad \mu^* = \mu_{max}\ \mathrm{sign}(\lambda_\chi) \ , \tag{42}$$

where μ_u is the unconstrained optimal bank angle, given by:

$$\mu_u = \arctan \left[\frac{\lambda_\chi}{\lambda_V} \frac{W}{2VD_i} \right] \tag{43}$$

A singular throttle arc, along which $\lambda_V = 0$ over a non-zero time interval, is likely to occur in isolated situations only [30, 32].

Since time does not appear explicitly in the equations and the final time is not specified, one also has:

$$H^* = H|_{t_f} = 0 \tag{44}$$

Substantial simplification of the two-point-boundary-value problem is

The condition of vertical force equilibrium, obtained by substituting in Eq.(5) $\gamma = F_\gamma = 0$, relates the bank angle μ to the load factor n by:

$$n = 1/\cos\mu \ , \tag{31}$$

allowing to eliminate the load factor n from the problem formulation. As a consequence, the equations of motion (4) and (6) reduce to:

$$\dot{V} = (g/W)[\eta T_{max} - D_\circ - (1 + \tan^2\mu)D_i] \triangleq f_V(\eta, V, \mu) \tag{32}$$

$$\dot{\chi} = (g/V)\tan\mu \triangleq f_\chi(V, \mu) \tag{33}$$

In this formulation the interceptor's motion is governed by two independent control variables, the throttle parameter η and the bank angle μ. The maximum admissible value of the bank angle is determined by the load factor constraints given by Eqs.(15)-(16), i.e.:

$$|\mu| \le \mu_{max}, \tag{34}$$

where:

$$\mu_{max} = \sec^{-1}\{ \min [n_L, n_{max}] \} \tag{35}$$

2. OPTIMAL CONTROL FORMULATION

The variational Hamiltonian for the horizontal problem is:

$$H = -1 + \lambda_R f_R + \lambda_\psi f_\psi + \lambda_V f_V + \lambda_\chi f_\chi + \text{constraints} \tag{36}$$

The adjoint differential equations and the corresponding transversality conditions are:

obtained by closed-form integration of the adjoint equations [32]. It is readily verified that the solutions to Eqs. (37)-(40) are:

$$\lambda_R = \frac{\cos(\psi - \psi_f)}{f_R(\psi_f, V_f, \chi_f)} \tag{45}$$

$$\lambda_\psi = \frac{-R\sin(\psi - \psi_f)}{f_R(\psi_f, V_f, \chi_f)} \tag{46}$$

$$\lambda_V f_V(\eta, V, \mu) = -[\lambda_\chi f_\chi(V, \mu) + \frac{V_f\cos(\psi_f - \chi_f) - V\cos(\psi_f - \chi)}{f_R(\psi_f, V_f, \chi_f)}] \tag{47}$$

$$\lambda_\chi = \frac{R\sin(\psi - \psi_f) - V_T\sin\psi_f(t_f - t)}{f_R(\psi_f, V_f, \chi_f)} \tag{48}$$

The special case $\dot{\lambda}_\chi = \lambda_\chi = 0$ is of particular interest. Substitution of Eqs.(45)-(46) into Eq.(40) shows that:

$$\dot{\lambda}_\chi = -\frac{\partial H}{\partial \chi} = \frac{V\sin(\psi_f - \chi)}{f_R(\psi_f, V_f, \chi_f)} \tag{49}$$

From Eqs. (49) and (43) it is then clear that in this case the extremal is a straight line given by:

$$\chi = \psi_f = \text{constant} \tag{50}$$

If $\lambda_V > 0$, this straight-line trajectory is flown with full throttle and zero bank angle. However, the possibility of a zero-throttle bank angle chattering arc arises if $\lambda_V < 0$.

The quantities λ_V and λ_χ in Eqs.(41)-(43) vanish at the final time. Thus the control at t_f depends upon the derivatives of these quantities. Taking the limit in Eq.(43) as $t \to t_f$ resuits in:

$$\lim_{t \to t_f} \mu_u = \text{arc tan}\left[\frac{W}{2D_i} \tan(\psi_f - \chi_f)\right] \tag{51}$$

Due to the dependence on the unspecified terminal quantities ψ_f, V_f, χ_f and t_f, the optimal control law can not be implemented in a feedback form. Extremal trajectories can be generated only by backward integration for an assumed termination of the encounter. For a medium-range scenario, such extremals can be characterized as consisting of a turning phase followed by a phase of acceleration. During the initial phase the interceptor executes a hard turn to the direction of the final line-of-sight, possibly decelerating in the process. Then in the second phase the interceptor accelerates to the final velocity, flying a nearly straight line trajectory at full throttle. It is evident from Eq.(51) that for this type of engagement the final parameters ψ_f and χ_f must be selected such that the absolute value of the difference of the two is very small. Recall from Eq.(50), that if $\psi_f = \chi_f$, a degenerated straight-line extremal is obtained.

In order to obtain an approximation of the optimal control in feedback form for the variable speed horizontal interception problem, the method of forced singular perturbation is applied.

B. SINGULAR PERTURBATION ANALYSIS

1. MODELING CONSIDERATIONS

The success of the singular perturbation approach depends largely on the ability to identify time-scale separations of the state variables. The assessment of the time-scale separations is largely based on an understanding of the system's dynamic behavior, depending on such factors as aerodynamic characteristics, engine performance, vehicle weight, atmospheric conditions, capture conditions and the initial conditions of the encounter. It has to be noted that different assumptions concerning the system's dynamic behavior, may lead to a different ordering of the dynamics. For instance, in [34] velocity is assumed to be the fastest variable of all. This particular ordering implies sufficient control over velocity, i.e. a very high thrust to weight ratio, and is therefore appropriate mostly for rocket propelled vehicles. The time-scale selection employed here is based on the following observations:

- In a medium-range scenario, the initial separation distance is relatively large and therefore the rate of change in the direction of the line-of-sight will be slow compared with the turning rate of the interceptor.
- Longitudinal accelerations of a fighter aircraft are generally much smaller than the lateral accelerations used for turning.
- The equations describing the relative motion of the two aircraft in polar coordinates are highly coupled and should therefore be analyzed on the same time-scale.

By assuming the velocity dynamics to be faster than the relative position dynamics, a rather simple closed-form feedback solution can be obtained [9]. This formulation requires the initial conditions of the encounter to be such as to allow the interceptor to reach maximum speed. Unfortunately, even for high-performance fighters the assumed time-scale separation can not be warranted for all initial conditions of interest. Hence this formulation does not apply to a medium-range scenario as defined in Section II,A.

Based on the above it seems to be appropriate that for the variable speed medium-range horizontal interception problem the state variables R, ψ, V are considered on the same "slow" time-scale, while χ is designated the role of "fast" variable.

The dynamic equations and boundary conditions associated with the singularly perturbed dynamic model selected here, are:

$$\dot{R} = f_R(\Psi, V, \chi) \quad , \quad R(t_o) = R_o \quad , \quad R(t_f) = d \tag{52}$$

$$\dot{\psi} = f_\psi(R, \psi, V, \chi) \quad , \quad \psi(t_o) = \psi_o \tag{53}$$

$$\dot{V} = f_V(\eta, V, \mu) \quad , \quad V(t_o) = V_o \tag{54}$$

$$\varepsilon\dot{\chi} = f_{\chi}(V, \mu) \quad , \qquad \chi(t_o) = \chi_o \tag{55}$$

The system of adjoint equations and corresponding transversality conditions may be written as:

$$\dot{\lambda}_R = -\frac{\partial H}{\partial R} \tag{56}$$

$$\dot{\lambda}_{\psi} = -\frac{\partial H}{\partial \psi} \quad , \qquad \lambda_{\psi}(t_f) = 0 \tag{57}$$

$$\dot{\lambda}_V = -\frac{\partial H}{\partial V} \quad , \qquad \lambda_V(t_f) = 0 \tag{58}$$

$$\varepsilon\dot{\lambda}_{\chi} = -\frac{\partial H}{\partial \chi} \quad , \qquad \lambda_{\chi}(t_f) = 0 \tag{59}$$

where H remains as defined in Eq.(36). The optimality conditions given by Eqs.(41)-(44) also remain unchanged in the singular perturbation formulation.

A detailed analysis leading to the derivation of a feedback guidance law based on the above FSP model is presented in [10] and is briefly reviewed in the next subsection.

2. HORIZONTAL GUIDANCE LAW SYNTHESIS

Taking the limit in Eqs.(55) and (59) as $\varepsilon \to 0$, the following necessary conditions for optimality of the "reduced-order" solution are obtained:

$$\mu^r = 0 \ , \quad \frac{\partial H^r}{\partial \chi} = -\frac{V_r \sin(\psi_f^r - \chi^r)}{f_R(\psi_f^r, V_f^r, \chi_f^r)} = 0 \tag{60}$$

where superscript 'r' is used to denote the "reduced-order" solution. Note from Eq.(60), that in the "reduced-order" problem χ takes on the role of control variable, whereas the original control μ now becomes a constraint. It is clear, that the optimal trajectory of the interceptor in the "reduced-order" problem is a straight line. The optimal heading χ^r is readily found from Eq.(60):

$$\chi^r = \psi_f^r = \text{constant} \tag{61}$$

Though the "reduced-order" system of state equations is three-dimensional, the optimal direction χ^r of the straight-line trajectory can be solved in terms of the initial state with a minimal computational effort, compatible with real-time requirements. To support the development of a geometrical solution for the reduced-order problem (see Figure 2), the following integrals are introduced:

$$I_t(V_o, V_f) \triangleq \int dt = \int_{V_o}^{V_f} \frac{dt}{dV}\, dV = \int_{V_o}^{V_f} \frac{(W/g)}{(T_{max} - D_o - D_i)}\, dV \tag{62}$$

$$I_s(V_o, V_f) \triangleq \int V dt = \int_{V_o}^{V_f} V \frac{dt}{dV}\, dV = \int_{V_o}^{V_f} \frac{V\,(W/g)}{(T_{max} - D_o - D_i)}\, dV \tag{63}$$

Equations (62)-(63) give the time and distance needed by the interceptor to reach the velocity V_f starting at V_o, while flying a straight line trajectory. Based on these expressions, the interception geometry can be solved in Cartesian coordinates:

$$x(t_f) = d\cos\chi^r = R_o\cos\psi_o + I_t(V_o, V_f^r)V_T - I_s(V_o, V_f^r)\cos\chi^r \tag{64}$$

$$y(t_f) = d\sin\chi^r = R_o\sin\psi_o - I_s(V_o, V_f^r)\sin\chi^r \tag{65}$$

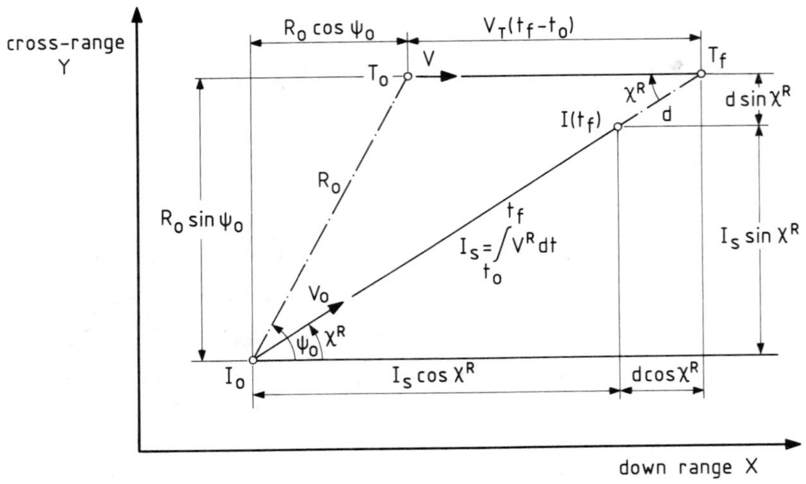

Fig. 2. "Reduced-order" Solution in the Horizontal Plane.

Equations (64)-(65) can be solved by a simple one-dimensional search, yielding both V_f^r and χ^r in terms of the initial conditions (R_o, ψ_o, V_o):

$$V_f^r = \overline{V}_f (R_o, \psi_o, V_o), \qquad \chi^r = \overline{\chi} (R_o, \psi_o, V_o) \tag{66}$$

Based on Eqs.(66) and (61), the "reduced-order" solution can be completed by evaluating the initial values of the "slow" adjoints, using Eqs.(45)-(47):

$$\lambda_R^r(t_o) = \frac{\cos [\psi_o - \overline{\chi}(R_o, \psi_o, V_o)]}{V_T \cos [\overline{\chi}(R_o, \psi_o, V_o)] - \overline{V}_f(R_o, \psi_o, V_o)} \triangleq \overline{\lambda}_R(R_o, \psi_o, V_o) \tag{67}$$

$$\lambda_\psi^r(t_o) = \frac{-R\sin[\psi_o - \bar{\chi}(R_o, \psi_o, V_o)]}{V_T\cos[\bar{\chi}(R_o,\psi_o,V_o)] - \bar{V}_f(R_o, \psi_o, V_o)} \triangleq \bar{\lambda}_\psi(R_o, \psi_o, V_o) \quad (68)$$

$$\lambda_V^r(t_o) = \frac{V_o - \bar{V}_f(R_o,\psi_o,V_o)}{V_T\cos[\bar{\chi}(R_o,\psi_o,V_o)] - \bar{V}_f(R_o, \psi_o, V_o)} \frac{W/g}{(T_{max} - D_o - D_i)_o}$$

$$\triangleq \bar{\lambda}_V(R_o, \psi_o, V_o) \quad (69)$$

The boundary-layer system (denoted by superscript "b") is obtained by introducing the time transformation:

$$\tau = (t - t_o)/\varepsilon \ , \quad (70)$$

into the original system given by Eqs.(52)-(59), yielding:

$$\frac{dR^b}{d\tau} = \varepsilon f_R(\psi, V, \chi) \quad (71)$$

$$\frac{d\psi^b}{d\tau} = \varepsilon f_\psi(R, \psi, V, \chi) \quad (72)$$

$$\frac{dV^b}{d\tau} = \varepsilon f_V(\eta, V, \mu) \quad (73)$$

$$\frac{d\chi^b}{d\tau} = f_\chi(V, \mu) \quad (74)$$

$$\frac{d\lambda_R^b}{d\tau} = -\varepsilon \frac{\partial H}{\partial R} \quad (75)$$

$$\frac{d\lambda_\psi^b}{d\tau} = -\varepsilon \frac{\partial H}{\partial \psi} \quad (76)$$

$$\frac{d\lambda_V^b}{d\tau} = -\varepsilon \frac{\partial H}{\partial V} \quad (77)$$

$$\frac{d\lambda_\chi^b}{d\tau} = -\varepsilon\frac{\partial H}{\partial \chi} \tag{78}$$

In the zeroth order boundary-layer approximation, obtained by setting $\varepsilon = 0$ in Eqs.(71)-(78), all "slow" state variables (R^b, ψ^b, V^b) and corresponding adjoints $(\lambda_R^b, \lambda_\psi^b, \lambda_V^b)$ remain frozen at their initial values at τ=0. Moreover, according to the Matching Principle [25], these values have to be equal to the initial values of the "reduced-order" solution, i.e.:

$$R^b(\tau) = R_o \quad , \quad \psi^b(\tau) = \psi_o \quad , \quad V^b(\tau) = V_o \tag{79}$$

$$\lambda_R^b(\tau) = \lambda_R^r(t_o) \quad , \quad \lambda_\psi^b(\tau) = \lambda_\psi^r(t_o) \quad , \quad \lambda_V^b(\tau) = \lambda_V^r(t_o) \tag{80}$$

Substitution of Eqs.(79), (80) and (67)-(69) into Eq.(44), allows the "fast" heading adjoint in the turning boundary layer $\lambda_\chi^b(\tau)$ to be expressed in terms of the initial "slow" states, the active state $\chi^b(\tau)$ and the control $\mu^b(\tau)$. Assuming "unconstrained" control for τ=0, substitution of this result into the optimal control solution given by Eqs.(41)-(43), yields the following expression for the optimal bank angle in the boundary layer:

$$\tan\mu^b(0) = \tan\mu_{ss}(V_o)\left[\frac{2V_o}{\overline{V}_f(R_o, \Psi_o, V_o) - V_o}\right]^{1/2}\sin\left[\frac{\overline{\chi}(R_o, \Psi_o, V_o)}{2}\right]$$

$$= \tan\overline{\mu}[R_o, \psi_o, V_o, \chi_o] \quad , \tag{81}$$

where:

$$\tan\mu_{ss}(V_o) = \left[\frac{T_{max} - D_o}{D_i} - 1\right]^{1/2} \quad , \tag{82}$$

is the bank angle for a steady-state horizontal turn.

A uniformly valid zeroth-order feedback law can be synthesized by merely replacing the initial state by the current state.

3. DISCUSSION

Analysis of singular perturbation approximate solutions reveals a characteristic behavior very similar to that of open-loop extremals for medium-range scenario's. Both the exact and the approximate control strategies are characterized by a gradually decreasing rate-of-turn as the flight direction asymptotically reaches its reference value. The main difference between the two control strategies is the reference direction. In the singular perturbation solution the final line-of-sight angle, which is the reference direction in the exact solution, is approximated by the instantaneous collision course. The larger the difference between the instantaneous collision course and the final line-of-sight angle, the worse the accuracy of the singular perturbation approximation. Obviously, a large difference is merely a reflection of a lack of "true" time-scale separation.

The guidance law derived in [9] (and also used in [11] and [13]), which is based on treating the velocity dynamics on a separate intermediate time-scale, exhibits great similarity with the guidance law given by Eq.(81). In fact, the guidance laws are the same, except that in [9] the reference velocity $\overline{V}_f(R,\psi,V)$ is replaced by the maximum velocity and the reference direction $\overline{\chi}(R,\psi,V)$ is replaced by the so-called "modified collision course". It is evident that the performance of this guidance law suffers substantially in encounters where the initial separation distance is not sufficiently large to permit the interceptor to attain maximum velocity.

C. NUMERICAL EXAMPLES

In order to evaluate the accuracy of the zeroth-order feedback strategy and the eventual need for first-order corrections, two numerical examples are considered. In both examples the approximate feedback solutions are compared with the respective open-loop extremal solutions. The first example is typical of a medium-range scenario. The boundary conditions, the selected parameters and some characteristic results for this example are summarized in Table 1. The geometric perturbation parameter ε_g, borrowed from the constant-speed interception problem [9,28,31], is introduced in Table 1 to serve as a measure of time-scale separation.

Table 1 Comparison of Exact and Approximate Solutions for a Medium-range Example in the Horizontal Plane

target	$h_T = 12000$ m $V_T = 300$ m/s		
initial state	$R_o = 11192$ m $\psi_o = 64.57°$ $V_o = 349.6$ m/s $\chi_o = -89.61°$	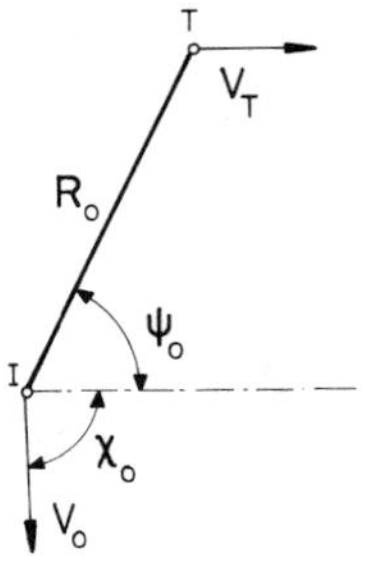	
final state	$R_f = d = 2000$ m		
	exact solution	approx. solution at initial state [Eq. (81), Ref. 10]	approx. solution at initial state [Ref. 13]
"free" final state	$\psi_f = 28.65573°$ $V_f = 455.2$ m/s $\chi_f = 28.65°$	$\bar{V}_f = 436.8$ m/s $\bar{\chi} = 25.45°$	$\bar{V}_f = 605.2$ m/s $\bar{\chi} = 41.69°$
final time	$t_f = 105.0$ s	$\bar{t}_f = 105.2$ s	$\bar{t}_f = 111.2$ s

Interceptor's best turning radius at $V = V_o$: $r_{min} = 2543.5$ m

Geometric perturbation parameter (r_{min}/R_o) : $\varepsilon_g = 0.227$

Although for the presented example the accuracy of the guidance law of Eq.(81) is more than adequate, improvements may be called for in case of relatively short-range engagements. Such improvements can be obtained by incorporating first-order correction terms in the singular perturbation control approximation.

The only difference between the exact optimal solution and the zeroth-order singular perturbation approximation is that in the singular perturbation approach the slow adjoints $(\lambda_R, \lambda_\psi, \lambda_V)$ are approximated by the "reduced-order" feedback expressions $(\bar{\lambda}_R, \bar{\lambda}_\psi, \bar{\lambda}_V)$. Thus, improving the approximate feedback control solution implies improving the estimates of the slow adjoints, while preserving the feedback form.

Improved estimates of the slow adjoints can be obtained by extending the singular perturbation analysis to include first- and higher-order corrections. In [10] the guidance law of Eq.(81) was improved by incorporating first-order corrections , obtained using the method of Matched Asymptotic Expansions (MAE) [22-26], in a feedback form.

In order to demonstrate the improvements that can be obtained by incorporating first-order corrections in the guidance law, an example of a relatively short-range engagement is presented. The conditions and results for this example are summarized in Table 2.

These results show that in this case with a rather large value of ε_g, the payoff error of the zeroth-order approximation is of the order of 3%. The error in the reference (predicted final) values is even larger. The first-order singular perturbation approximation is rather successful in predicting the final values of the state-variables and, as a consequence, it provides an outstanding payoff accuracy even for an engagement starting within the best turning circle of the interceptor.

The improved accuracy comes at the expense of some additional computations. Although in relative terms the first-order algorithm requires about twice as much CPU time as the zeroth-order algorithm, in absolute terms, the overall computational effort is still rather modest.

Table 2 Comparison of Exact and Approximate Solution for a Short-range Example in the Horizontal Plane

target	$h_T = 12000$ m $V_T = 350$ m/s		
initial state	$R_o = 6318.5$ m $\psi_o = 18.87°$ $V_o = 598.2$ m/s $\chi_o = -44.391°$	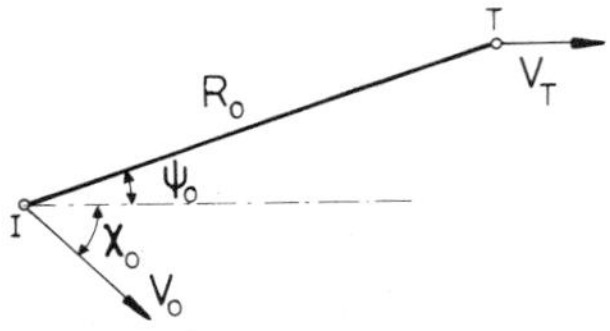	
final state	$R_f = d = 2000$ m		
	exact solution	zeroth-order approx. solution at initial state [Eq. (81), Ref. 10]	first-order approx. solution at initial state [Ref. 10]
"free" final state	$\psi_f = 30.08239°$ $V_f = 543.4$ m/s $\chi_f = 28.65°$	 $\overline{V}_f = 600.1$ m/s $\overline{\chi} = 9.6°$	 $\overline{V}_f = 555.4$ m/s $\overline{\chi} = 26.2°$
final time	$t_f = 27.0$ s	$\overline{t}_f = 27.8$ s	$\overline{t}_f = 27.1$ s
Interceptor's best turning radius at $V = V_o$: $r_{min} = 7447.2$ m Geometric perturbation parameter (r_{min}/R_o) : $\varepsilon_g = 1.18$			

V. INTERCEPTION IN A VERTICAL PLANE

A. FORMULATION OF THE VERTICAL INTERCEPTION PROBLEM

1. EQUATIONS OF MOTION

The equations of motion for an air-to-air interception confined to a vertical plane are obtained by setting in Eqs.(1-6) $y = 0$. This leads via $F_y = 0$ to $\chi = 0$ (or $\chi = \pi$) and consequently to $\mu = 0$ (or $\mu = \pi$). The resulting equations of motion are:

$$\dot{x} = V \cos\gamma - V_T \overset{\Delta}{=} f_x(V,\gamma) \qquad (83)$$

$$\Delta\dot{h} = V \sin\gamma \overset{\Delta}{=} f_h(V,\gamma) \qquad (84)$$

$$\dot{V} = g[(\eta T_{max} - D_o - n^2 D_i)/W - \sin\gamma] \overset{\Delta}{=} f_V(h,V,\gamma,\eta,n) \qquad (85)$$

$$\dot{\gamma} = (g/V)[n - \cos\gamma] \overset{\Delta}{=} f_\gamma(V,\gamma,n) \qquad (86)$$

In these equations $-\pi \le \gamma \le \pi$, which allows both for incoming and outgoing targets (see Fig.3).

In a frequently used alternative formulation the specific energy:

$$E = h + V^2/2g \qquad (87)$$

is used as a state variable replacing V, which merely continues to serve as an abbreviation for:

$$V = \{2g(E - h)\}^{1/2} \qquad (88)$$

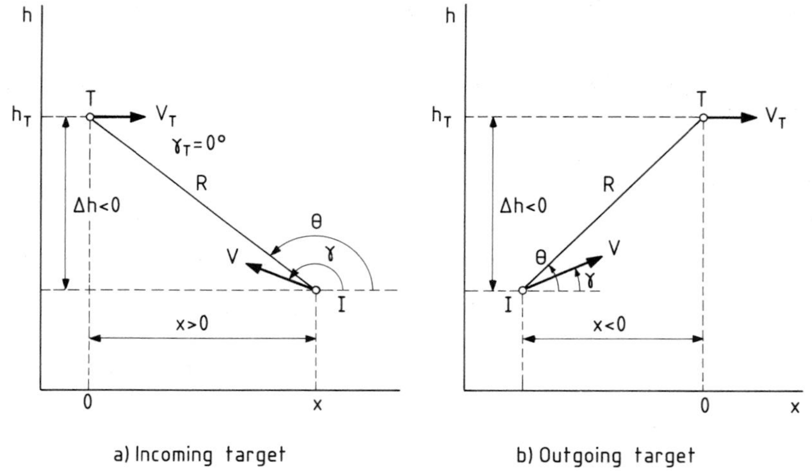

Fig. 3. Vertical Interception Geometries.

In this case Eq. (85) is replaced by:

$$\dot{E} = [\eta T_{max} - D_o - n^2 D_i]V/W \triangleq f_E(h, E, \eta, n) \tag{89}$$

In the sequel of this Chapter this "energy-state" formulation is used.

The terminal manifold of the vertical interception is defined by:

$$\Phi = x(t_f)^2 + [\Delta h(t_f)]^2 - d^2 = 0 \tag{90}$$

2. OPTIMAL CONTROL FORMULATION

The variational Hamiltonian for vertical interception is:

$$H = -1 + \lambda_x f_x + \lambda_h f_h + \lambda_E f_E + \lambda_\gamma f_\gamma + \text{constraints} \tag{91}$$

The necessary conditions for optimality include the adjoint differential equations and transversality conditions:

$$\dot{\lambda}_x = -\frac{\partial H}{\partial x} = 0 \quad , \quad \lambda_x(t_f) = 2\nu x(t_f) \tag{92}$$

$$\dot{\lambda}_h = -\frac{\partial H}{\partial h}\bigg|_E \quad , \quad \lambda_h(t_f) = 2\nu\Delta h(t_f) \tag{93}$$

$$\dot{\lambda}_E = -\frac{\partial H}{\partial E}\bigg|_h \quad , \quad \lambda_E(t_f) = 0 \tag{94}$$

$$\dot{\lambda}_\gamma = -\frac{\partial H}{\partial \gamma} \quad , \quad \lambda_\gamma(t_f) = 0 \quad , \tag{95}$$

where:

$$\frac{\partial H}{\partial h}\bigg|_E = \frac{\partial H}{\partial h}\bigg|_V - \frac{g}{V}\frac{\partial H}{\partial V}\bigg|_h \tag{96}$$

and:

$$\frac{\partial H}{\partial E}\bigg|_h = \frac{g}{V}\frac{\partial H}{\partial V}\bigg|_h \tag{97}$$

Unfortunately, the adjoint equations of the vertical interception cannot be integrated in a analytical form, as it was done for the horizontal case. The main reason for this difficulty is the dependence of the thrust and the aerodynamic forces on the altitude. However, the optimal control solution, if it exists, can be expressed by using the Maximum Principle in terms of the state and adjoint variables:

$$\lambda_E > 0: \ \eta^* = 1, \quad n^* = \min\left[\,|n_u|, n_{max}\right] \operatorname{sign}(\lambda_\gamma) \tag{98}$$

$$\lambda_E < 0: \ \eta^* = 0, \quad n^* = n_{max} \operatorname{sign}(\lambda_\gamma) \ , \tag{99}$$

where n_u is the unconstrained optimal load factor given by:

$$n_u = \frac{\lambda_\gamma}{\lambda_E} \frac{W}{2V^2 D_i} \tag{100}$$

A singular throttle arc, along which $\lambda_E = 0$ over a non-zero time interval, is likely to occur only in isolated situations.

Since time is not explicitly involved and the final time is unspecified:

$$H^* = H\Big|_{t_f} = 0 \ , \tag{101}$$

which allows to determine the value of the multiplier ν in Eqs.(92) and (93):

$$\nu = \{[V(t_f)\cos\gamma(t_f) - V_T]x(t_f) + V(t_f)\sin\gamma(t_f)\Delta h(t_f)\}/2 \tag{102}$$

Since at t_f both λ_E and λ_γ vanish, the control at the final time depends upon the derivatives of these variables. Taking the limit of Eq. (100) as $t \to t_f$ results in:

$$\lim_{t \to t_f} n_u = \frac{W}{2D_i} \tan[\theta(t_f) - \gamma(t_f)] \ , \tag{103}$$

where θ is the line of sight angle in the vertical plane (see Fig 3).

B. SINGULAR PERTURBATION ANALYSIS

1. MODELING CONSIDERATIONS

In order to obtain an approximation of the optimal control solution in a feedback form the approach of forced singular perturbation (FSP), outlined in Section III, is to be applied. As a first step in this direction the following observations, referring to the time-scale separation of the state variables, are made:

- The rate of change of the horizontal range component x is rather gradual, suggesting that it should be considered as a slow variable.
- In a medium-range scenario the maximum speed (specific energy) of the interceptor cannot be reached. Therefore, the specific energy has to be considered on the same time-scale as the horizontal range component.
- In most known "energy-state" models altitude is considered as a fast variable compared to the specific energy.
- The flight path angle γ can change much faster than the relative geometry expressed by x and Δh.

In some previous studies [30,34] considerable attention was paid to the coupling between the altitude and the flight path angle, having its origin in the phugoid mode of an uncontrolled aircraft. Full consideration of this coupling leads to analyze both variables on the same time-scale and requires as a consequence to solve a numerically very sensitive two-point boundary value problem. In [5] it was, however, shown that the error induced by neglecting the coupling and analyzing h and γ on separate time-scales is mainly due to a reduced damping of the resulting second order dynamic response, while the frequency remains almost unchanged. Moreover, it was also reported [5] that this error could be corrected by slightly modifying the cost function. A similar result was achieved by a feedback approximation derived in [11].

Based on the above, the following FSP model is used in the present analysis:

$$\dot{x} = f_x(V,\gamma) \quad , \quad x(t_o) = x_o \tag{104}$$

$$\dot{E} = f_E(h, E, \eta, n) \quad , \quad E(t_o) = E_o \tag{105}$$

$$\varepsilon \Delta \dot{h} = f_h(V,\gamma) \quad , \quad \Delta h(t_o) = h_o - h_T \tag{106}$$

$$\varepsilon^2 \dot{\gamma} = f_\gamma(V,\gamma,n) \quad , \quad \gamma(t_o) = \gamma_o \tag{107}$$

The terminal manifold and the variational Hamiltonian remain the same as in Eqs.(90) and (91). The adjoint equations and the transversality conditions corresponding to the FSP model are:

$$\dot{\lambda}_x = -\frac{\partial H}{\partial x} = 0 \quad , \qquad \lambda_x(t_f) = 2\nu x(t_f) \tag{108}$$

$$\dot{\lambda}_E = -\frac{\partial H}{\partial E}\bigg|_h \quad , \qquad \lambda_E(t_f) = 0 \tag{109}$$

$$\varepsilon\dot{\lambda}_h = -\frac{\partial H}{\partial h}\bigg|_E \quad , \qquad \lambda_h(t_f) = 2\nu\Delta h(t_f) \tag{110}$$

$$\varepsilon^2\dot{\lambda}_\gamma = -\frac{\partial H}{\partial \gamma} \quad , \qquad \lambda_\gamma(t_f) = 0 \quad , \tag{111}$$

2. BASIC GUIDANCE LAW SYNTHESIS

The FSP model, represented by Eqs.(104)-(111), was solved in [11] and is briefly reviewed in the following. By setting $\varepsilon = 0$ in these equations, the "reduced order" problem (the variables are denoted by the superscript "r") with $f_h^r = f_\gamma^r = 0$ is obtained. Consequently, one has:

$$\sin\gamma^r = 0 \tag{112}$$

and:

$$n^r = \cos\gamma^r \tag{113}$$

For an outgoing target ($|\gamma| \leq \pi/2$) this leads to $\gamma^r = 0$ and $n^r = 1$, while for an incoming target ($|\gamma| \geq \pi/2$) $\gamma^r = \pi$ and $n^r = -1$ (see Fig. 3).

Substituting these results into Eq. (91) and using Eqs. (101) and (109) leads to:

$$\lambda_x^r = \frac{1}{V_f^r \cos \gamma^r - V_T} = \frac{1}{\pm V_f^r - V_T} , \tag{114}$$

as well as to:

$$\lambda_E^r = \frac{(V_f^r - V^r) \cos \gamma^r}{V_f^r \cos \gamma^r - V_T} \frac{1}{f_E^r} = \frac{\pm (V_f^r - V^r)}{\pm V^r - V_T} \frac{1}{f_E^r} , \tag{115}$$

where V_f^r, the value of V^r at t_f, is an unknown parameter depending on the initial conditions and the prescribed terminal range "d". The control variables of the "reduced-order" problem are η^r and h^r. For a successful interception V_f^r must be greater than V_T. Inside the flight envelope the aircraft can accelerate (i.e. $V_f^r > V^r$ and $f_E^r > 0$), and therefore $\eta^r = 1$.

Moreover, the optimal altitude is given by:

$$h^r = \arg \{\max_h \frac{f_E (h, E^r, \eta = 1 , n = \pm 1)}{(V_f^r - V^r)}\} \triangleq \tilde{h}^r (E^r, V_f^r) , \tag{116}$$

describing a "reduced-order" optimal flight path in the h-V plane. A family of such flight paths, parameterized by V_f^r is depicted in Fig. 4. The value of V_f^r can be obtained, for a given set of initial and terminal conditions, by a fast converging iterative process, called "range matching" [5,11]. It solves the implicit integral equation:

$$|x_o| - d = \left| \int_{E_o}^{E_f} \frac{V^r \cos \gamma^r - V_T}{f_E [\tilde{h}^r(E, V_f^r), E, \eta = 1, n \pm 1]} dE \right| \triangleq I_x(E_o, V_f^r) , \tag{117}$$

where V^r is, based on Eq.(88), a function of E^r and $\tilde{h}^r(E^r, V_f^r)$. The solution yields therefore $V_f^r = \tilde{V}_f^r(x_o, E_o)$, abbreviated in the sequel as $\tilde{V}_f^r$.

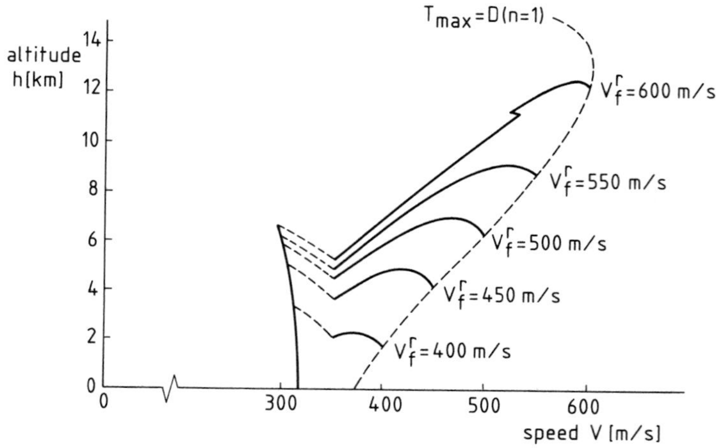

Fig. 4 Family of "Reduced-order" Trajectories in the Vertical Plane

Since the "reduced-order" solution, cannot satisfy the initial conditions of the "fast" variables, Δh and γ, two consecutive boundary layer solutions are needed. Substitution of the stretched time-scale:

$$\tau_1 = (t - t_o)/\varepsilon \quad , \tag{118}$$

into Eqs.(104)-(111) provides the first boundary layer, where the variables are denoted by the superscript "1". By setting $\varepsilon=0$ in these equations the following results are obtained:

$$x^1(\tau_1) = x_o \quad , \quad E^1(\tau_1) = E_o \quad , \quad f^1_\gamma(\tau_1) = 0 \Rightarrow n^1 = \cos\gamma^1 \tag{119}$$

$$\lambda^1_x(\tau_1) = \lambda^r_x \quad , \quad \lambda^1_E(\tau_1) = \lambda^r_E(t_o) \Rightarrow \eta^1 = 1 \quad , \quad \frac{\partial H}{\partial \gamma^1} = 0 \tag{120}$$

As a consequence of Eq.(120), the active control in this boundary layer is the flight path angle γ^1 which can be approximated [11] by the following relatively simple feedback form:

$$\gamma^1 = \arc\cos\left\{\frac{V^1 f^r_E(E_o,\tilde{V}^r_f)}{\tilde{V}^r_f[f^r_E(E_o,\tilde{V}^r_f) - f^1_E(E_o)] + V^r_o f^1_E(E_o)}\right\} \operatorname{sign}[\tilde{h}^r(E_o,\tilde{V}^r_f) - h^1]$$

$$= \tilde{\gamma}^1(E_o,\tilde{V}^r_f,h^1) \ , \qquad (121)$$

where $f^r_E(E_o,\tilde{V}^r_f)$ is an abbreviation for $f_E[\tilde{h}^r(E_o,\tilde{V}^r_f),\ E_o,\eta{=}1,n{=}\pm 1]$, $f^1_E(E_o)$ is an abbreviation for $f_E(h^1,E_o,\eta{=}1,n{=}\cos\gamma^1)$, and V^r_o is the value of V^r expressed by Eq. (88) at E_o and $\tilde{h}^r(E_o,\tilde{V}^r_f)$. The active adjoint variable in this boundary layer is approximated by:

$$\lambda^1_h(\tau_1) = \lambda^r_x \tan\gamma^1 \qquad (122)$$

In the next boundary layer the time-scale is further stretched:

$$\tau_2 = \tau_1/\varepsilon \qquad (123)$$

Substitution of Eq.(123) into the state and adjoint equations and setting again $\varepsilon = 0$, while denoting the variables of this boundary layer by the superscript "2", yields:

$$x^2(\tau_2) = x_o \ , \qquad E^2(\tau_2) = E_o \ , \qquad h^2(\tau_2) = h_o \qquad (124)$$

$$\lambda^2_x(\tau_2) = \lambda^r_x \ , \quad \lambda^2_E(\tau_2) = \lambda^r_E(t_o) \Rightarrow \eta^2 = 1 \ , \quad \lambda^2_h(\tau_2) = \lambda^r_x \tan\tilde{\gamma}^1(E_o,\tilde{V}^r_f,h_o) \qquad (125)$$

These results allow to obtain a feedback expression for the unconstrained load factor at $\tau_2=0$:

$$n_u^2(0) = \cos\gamma_o + A(E_o,\tilde{V}_f^r,h_o)\,\sin\{\frac{\tilde{\gamma}^1(E_o,\tilde{V}_f^r,h_o) - \gamma_o}{2}\} = \tilde{n}_u^2(x_o,E_o,h_o,\gamma_o)\,, \tag{126}$$

where:

$$A(E_o,\tilde{V}_f^r,h_o) = [\frac{2}{\tilde{V}_f^r - V_o^r}\;\frac{W\,f_E\,[\tilde{h}^r(E_o,\tilde{V}^r),\,E_o,\,\eta=1,\,n=\cos\tilde{\gamma}^1]}{D_i(h_o,E_o)\,\cos\tilde{\gamma}^1(E_o,\,\tilde{V}_f^r,\,h_o)}]^{1/2} \tag{127}$$

Based on Eq.(126), a uniformly valid zeroth-order feedback control law can be synthesized by replacing the initial values of the state variables with their current values. Since this guidance law is totally independent of the target trajectory, in [11] an intuitive feedback control was proposed for the terminal phase of the interception.

3. GUIDANCE LAW IMPROVEMENTS

Further analysis [12] revealed several deficiencies and error sources in the guidance algorithm summarized in the previous subsection:

(i) The convergence of the aircraft trajectory guided by Eq. (126) towards the reference trajectory of the "reduced-order" solution is very slow.

(ii) The reference trajectory is discontinuous at the transonic region (the "transonic jump" phenomenon).

(iii) The solution for the terminal phase of the interception (terminal boundary layer) is not satisfactory.

The origin of the first deficiency lies in Eq.(112), obtained for the "reduced-order" solution from $f_h^r=0$. This result is clearly incompatible with the actual flight path angle associated with the "reduced-order" optimal flight path, defined by Eq.(116). Along every smooth segment of this trajectory the flight path angle is given by:

$$\sin\gamma^R = \frac{T - D}{W} \;/\; (1 + \frac{V}{g}\frac{dV^r}{dh^r}) = \sin\gamma^R(E, V_f^r) \neq 0 \tag{128}$$

In order to avoid this problem one can follow the approach introduced in [35] and replace the altitude by another "fast" variable Ω, defined as:

$$\Omega \triangleq \frac{\partial}{\partial h}\bigg|_E \{\frac{f_E}{(V_f - V)\cos\gamma}\} = \Omega(h, V, V_f) \tag{129}$$

This variable reaches an equilibrium on the "reduced-order" optimal trajectory as requested by Eq. (116) and its time derivative is given by:

$$\dot{\Omega} = \frac{\partial\Omega}{\partial h}\dot{h} + \frac{\partial\Omega}{\partial V}\dot{V} = \sin\gamma\,(V\frac{\partial\Omega}{\partial h} - g\frac{\partial\Omega}{\partial V}) + \frac{T - D}{W}g\frac{\partial\Omega}{\partial V} \triangleq f_\Omega(h, V, \gamma) \tag{130}$$

Clearly, by replacing Eq. (106) with:

$$\varepsilon\dot{\Omega} = f_\Omega \tag{131}$$

and setting $\varepsilon = 0$ one obtains Eq.(128) because:

$$-(\frac{\partial\Omega}{\partial h}) / (\frac{\partial\Omega}{\partial V}) = \frac{dV}{dh}\bigg|_\Omega \tag{132}$$

Indeed, by adding γ^R to Eq.(121), i.e. redefining:

$$\gamma^{1*} \triangleq \gamma^1 + \gamma^R \;, \tag{133}$$

to replace γ^1 in Eqs.(126) and (127), a significant improvement is achieved.

The "transonic jump" is a well known characteristic of most supersonic aircraft. It is the result of the "reduced-order" modeling, where altitude is considered as a pseudo-control and as such the corresponding optimal trajectory can be discontinuous. In the open-loop optimal solution of the complete point-mass model such discontinuity does not exist. In terms of rigorous singular perturbation theory the "transonic jump" has to be treated as an "internal" boundary layer. Such a solution was obtained recently [36], but unfortunately only in an open-loop form. In the frame of a feedback algorithm an acceptable solution is to replace the transonic discontinuity by a continuous reference flight path, such as a backward extrapolation of the supersonic subarc of the "reduced-order" optimal trajectory. Since such a flight path is not optimal in the sense of the "reduced-order" modeling, it may happen that the argument of Eq.(121) is greater than 1. In order to avoid such problem, it was proposed [12] to use instead of Eq (127) another feedback control law:

$$n_u(E,x,h,\gamma) = \cos\gamma + \{\omega^2[h^r_{ex}(E,x)-h] + 2V\zeta\omega[\gamma^R_{ex}(E,x)-\gamma]\}/g, \quad (134)$$

where $h^r_{ex}(E,x)$ and $\gamma^R_{ex}(E,x)$ are values obtained from the extrapolated supersonic subarc. In this expression ω and ζ are control parameters associated with the linearized second-order response of the aircraft. For the damping ratio the value of $\zeta = 0.7$ was selected, while ω was obtained by some numerical experimentation for an optimal performance. For several aircraft models the best value of ω was in the range of 0.08 - 0.15.

The two modifications, which are outlined above, created an improved feedback control algorithm for steering the interceptor towards the "reduced-order" flight path asymptotically. This algorithm, being independent of the target altitude, cannot guarantee that the terminal manifold of Eq. (90), which involves a "fast" variable, is reached. This is an inherent limitation of a "reduced order" solution, to which singular perturbation theory provides an answer in the form of a "terminal" boundary layer. Such a "terminal" boundary layer solution imposes two difficulties. It must be stable in a reversed stretched time-scale [37], which means a structural instability in a real-time (forward) implementation. Another difficulty is to determine the point of transition for starting the "terminal" boundary layer solution.

If the target altitude is lower than the final part of the "reduced-order" trajectory, the terminal phase of the time-optimal interception is a dive along the maximum dynamic pressure limit. This type of trajectories were treated extensively in [3-5]. In other investigations [11-14] the attention is focused on interception of high flying targets. In such a situation the classical energy-state approximation [1] calls for a "zoom" climb at a constant specific energy until the terminal manifold is reached.

In [12] a closed form analytical solution for such "zoom" trajectory is derived and proposed as an approximation for the "terminal" boundary layer solution of the minimum-time interception in the vertical plane. The derivation of this closed form solution is summarized in the Appendix. Here the main results are repeated, denoting the variables of the constant specific energy "zoom" by the superscript "z".

The solution is characterized by:

$$V^z/\cos\gamma^z = V(t_f)/\cos\theta(t_f) = A_f \tag{135}$$

and the nominal load factor to generate the trajectory is given by:

$$n^z = 2\cos\gamma^z \tag{136}$$

The "zoom" trajectory can be completely determined by the closed form solution. Thus, the unknown constant A_f, the starting time t_o^z and the corresponding horizontal distance $x(t_o^z)$ can be found. At the same time the entire "zoom" trajectory can be precomputed and stored. Since the closed form solution is only an approximation one must monitor the deviation of the actual trajectory and correct it. For this purpose the following feedback control law, motivated by Eq. (103), has been proposed:

$$n(t \geq t_o^z) = 2\cos\gamma + \frac{W}{2D_i} \tan[\gamma^z(\theta) - \gamma] \tag{137}$$

By incorporating the appropriate corrections to all three deficiencies, mentioned at the beginning of this subsection, a modified feedback control algorithm was synthesized [12]. It was also demonstrated [12] that this new algorithm leads to improvements of the order of 1% in the pay-off in comparison with the uncorrected guidance law.

4. ACCURACY ASSESSMENT

The absolute pay-off accuracy of the improved guidance algorithm was tested by a comparison with the open-loop optimal solution obtained by a very accurate numerical multiple shooting method [14]. In the numerical examples an unclassified F-4 type model was used. This comparison, which is summarized in Table 3, demonstrated a rather satisfactory pay-off accuracy of the "original" FSP solution, in the order of 1% .

Table 3 Comparison of Interception Time in the Exact and Approximate Solutions for Medium-range Examples in the Vertical Plane

initial state	$R_o = 82800$ m $V_o = 250$ m/s $\gamma_o = 180°$ (incoming target) $h_o = 1000$ m			
final state	$R_f = d = 1000$ m			
example No.		1	2	3
target	h_T (m)	4180	7000	12000
	V_T (m/s)	300	300	400
interception time (s)	exact solution	122.78	124.03	113.99
	original FSP sol.	123.56	124.86	115.02
	modified FSP sol.	122.88	124.26	114.54

The comparison revealed, however, a discrepancy between the FSP and the optimal trajectories in the subsonic region. There is a clear indication, as can be seen in Fig.5, that the subsonic subarc of the reduced-order trajectory is far from "optimal". The subsonic subarc in the reduced-order solution features a constant Mach number of $M = 0.92$ up to an altitude of about 5 km, where the "transonic jump" starts. In contrast, the open-loop optimal trajectory is an accelerating climb into the supersonic region. The origin of this nonoptimality is the incompatibility of the assumption of $|\cos\gamma^r|=1$, imbedded in the "reduced-order" optimization, and Eq.(128) which predicts flight path angles of the order of 30-35 degrees in the subsonic region. In the supersonic region, the flight paths angles generally do not exceed 5-6 degrees and the assumption of $|\cos\gamma^r|=1$ remains valid. It is important to note, that even if the approach of the modified "fast" variable Ω is used, the value of $\cos\gamma^r$ which maximizes the Hamiltonian will still be 1.

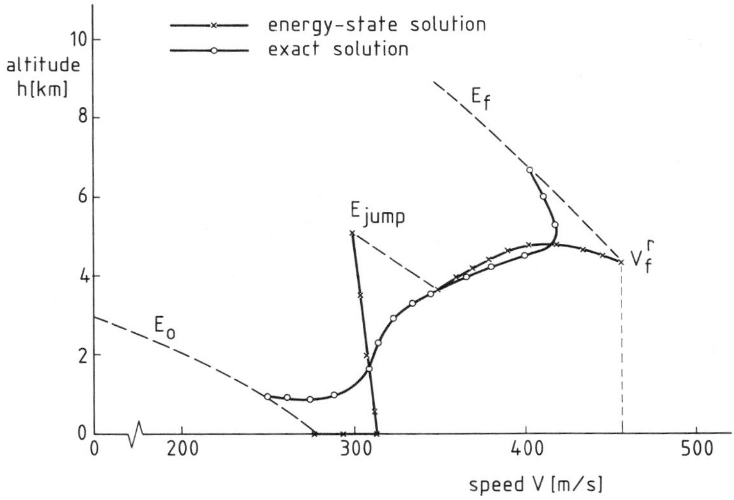

Fig. 5. Non-optimality of the Subsonic Subarc (F-4 model).

The practical conclusion drawn from this comparison has been that the "reduced-order" subsonic subarc should not be used as a reference trajectory for the FSP feedback algorithm. This conclusion is entirely compatible with the previously introduced correction in the transonic region. As an acceptable reference trajectory for the sub and transonic regions a smooth flight path, which starts at zero altitude and blends into the supersonic subarc of the "reduced-order" trajectory,has been proposed [38]. Such a modification reduced the error of the FSP feedback approximation for the medium-range interception in the vertical plane to be less than 0.5%.

VI. THREE-DIMENSIONAL INTERCEPTION

A. MODELING CONSIDERATIONS

Since the FSP models used for the interception in the horizontal and the vertical planes are different, the extension of these feedback algorithms to a three-dimensional scenario is by no means trivial. In [4-6] it was assumed that the azimuth angle χ is a slower variable than the altitude and the flight path angle γ. This assumption, however, has not been confirmed by the results obtained in these very works. It turns out that all three-dimensional interceptions are composed of two phases: a relatively brief initial turning phase followed by an interception in the vertical plane. This observation seems to suggest that χ is a faster variable than h and γ.

In other works [13,14] it was claimed that since horizontal and vertical turning rates are the components of the same physical quantity,- namely the total turning rate of the velocity vector,- both χ and γ have to be analyzed on the same time-scale. This approach is adopted here in the sequel.

The FSP model, proposed for the analysis of the three-dimensional interception, incorporates also the replacement of the altitude by $\Omega(h,V)$ defined by Eq.(129) and it is given as follows:

$$\dot{x} = F_x \quad , \quad x(t_o) = x_o \tag{138}$$

$$\dot{y} = F_y \quad , \quad y(t_o) = y_o \tag{139}$$

$$\dot{E} = F_E = f_E \quad , \quad E(t_o) = h_o + V_o^2/2g \tag{140}$$

$$\varepsilon\dot{\Omega} = F_\Omega = f_\Omega \quad , \quad \Omega(t_o) = \Omega(h_o, V_o) \tag{141}$$

$$\varepsilon^2\dot{\gamma} = F_\gamma \quad , \quad \gamma(t_o) = \gamma_o \tag{142}$$

$$\varepsilon^2\dot{\chi} = F_\chi \quad , \quad \chi(t_o) = \chi_o \ , \tag{143}$$

with the terminal manifold:

$$\Phi = x(t_f)^2 + y(t_f)^2 + [\Delta h(t_f)]^2 - d^2 = 0 \tag{144}$$

The corresponding variational Hamiltonian is:

$$H = -1 + \lambda_x F_x + \lambda_y F_y + \lambda_E F_E + \lambda_\Omega F_\Omega + \lambda_\gamma F_\gamma + \lambda_\chi F_\chi + \text{constraints} \tag{145}$$

The adjoint differential equations and transversality conditions are:

$$\dot{\lambda}_x = -\frac{\partial H}{\partial x} = 0 \quad , \quad \lambda_x(t_f) = 2\nu x(t_f) \tag{146}$$

$$\dot{\lambda}_y = -\frac{\partial H}{\partial y} = 0 \quad , \quad \lambda_y(t_f) = 2\nu y(t_f) \tag{147}$$

$$\dot{\lambda}_E = -\frac{\partial H}{\partial E}\bigg|_\Omega \quad , \quad \lambda_E(t_f) = 0 \tag{148}$$

$$\varepsilon\dot{\lambda}_\Omega = -\frac{\partial H}{\partial \Omega}\bigg|_E \quad , \quad \lambda_\Omega(t_f) = 2\nu[\Delta h(t_f)] / \frac{\partial \Omega}{\partial h}\bigg|_{E(t_f)} \tag{149}$$

$$\varepsilon^2\dot{\lambda}_\gamma = -\frac{\partial H}{\partial \gamma} \quad , \quad \lambda_\gamma(t_f) = 0 \tag{150}$$

$$\varepsilon^2\dot{\lambda}_\chi = -\frac{\partial H}{\partial \chi} \quad , \quad \lambda_\chi(t_f) = 0 \tag{151}$$

B. GUIDANCE LAW SYNTHESIS

Based on the this FSP model the feedback guidance law for three-dimensional interception is synthesized by following the approach outlined in [14], [38] and [39]. The solution of the "reduced-order" problem (obtained by setting $\varepsilon = 0$), provides a "required flight trajectory" (RFT). This trajectory lies in a vertical plane. Using Eqs.(151), (146) and (147), the orientation of this plane can be established:

$$\tan \chi^r = y(t_f)/x(t_f) = \text{constant} \tag{152}$$

This "reduced-order" solution is based on a three-dimensional "range matching" process using Eqs.(64-65), with the integrals I_t and I_s evaluated along the RFT, while the velocity is replaced by the specific energy as the independent variable. As the outcome of this process the values of the constants χ^r and V_f^r as well as the RFT profile, defined by $\tilde{h}^r(E,V_f^r)$, and $\gamma^R(E,V_f^r)$ are obtained. Based on the experience gained in [14], two alternative approaches can be proposed for obtaining the RFT profile for any given V_f^r.

In the first approach the RFT profile is composed of the supersonic subarc of the "energy state approximation", defined by Eq.(116), and a smooth extrapolation back to the initial point of the subsonic subarc at sea level. The second alternative is to compute the open-loop optimal trajectory based on an "exact" point-mass aircraft model in the vertical plane, between the same initial point and the prescribed terminal velocity V_f^r on the maximum speed boundary. The initial and terminal flight path angles have to be selected such as to satisfy (at least approximately) $n = \cos\gamma$.

The use of point-mass extremals to build up a family of reference trajectories was motivated by the perturbation feedback scheme of [40]. Although the guidance law of [40] has not been formally derived using singular perturbation analysis, it makes use of singular perturbation ideas in terms of an assumed hierarchical trajectory-family (i.e. boundary-layer) structure. Considering, for instance, guidance in the vertical plane, the feedback law of [40] provides optimal altitude/path-angle transients that funnel into a single energy-range climb path (RFT). In fact, the structure of the vertical guidance law in [40] is identical to the one given by Eq.(134), except that in [40] the feedback gains (and thus ζ and ω) are specified as a function of specific energy, ensuring optimal transient behavior.

In both cases the practical approach is to precompute and store a sufficiently large family of reference trajectories, parameterized by V_f^r. Though the "range matching" is an iterative process, - a one dimensional search for the appropriate value of V_f^r to satisfy Eqs.(64-65), - it converges rapidly and is compatible with a real time on-line implementation.

The other element of the guidance law synthesis consists of determining the actual aircraft controls in order to reach the RFT and tracking it until the terminal phase of the interception starts. If one assumes that the coupling between horizontal and vertical turning maneuvers can be neglected, both turning rates can be optimized independently. This is an important simplification, which allows to compute the horizontal and vertical components of the required load factor separately by using explicit feedback expressions obtained in the planar analyses. The horizontal load factor component n_h, defined by Eq.(17), is computed using Eq.(81) by replacing the initial conditions with the current value of the state variables as well as $\overline{\chi}$ and $\overline{V}_f$ with χ^r and V_f^r. The vertical load factor component n_v, defined by Eq.(18), is computed by using Eqs.(116), (121), (126) and (133), similarly replacing all initial conditions with the current values of the state variables. The bank angle μ, to be used as the guidance command is given by:

$$\mu = \arc\tan (n_h/n_v) \ , \tag{153}$$

while the resultant load factor is expressed as:

$$n = \min \{n_u, n_{max}, n_L\} \ , \tag{154}$$

where:

$$n_u = [n_v^2 + n_h^2]^{1/2} \ , \tag{155}$$

as shown in Fig. 6.

In order to test the accuracy of the three-dimensional FSP feedback control synthesis, several examples were computed in [14] and [38]. The results of the comparison to the exact open-loop optimal solution were equally encouraging. They confirmed the validity of the assumption for neglecting the coupling between the horizontal and vertical turning maneuvers in a medium-range scenario. They also demonstrated that the three-dimension feedback algorithm has about the same pay-off accuracy of 1% as the one for vertical interception. Moreover, slight modifications of the algorithm, such as continuous updating of the RFT and correction for the nonoptimality of the subsonic subarc, lead to an improved pay-off accuracy, better than 0.5% .

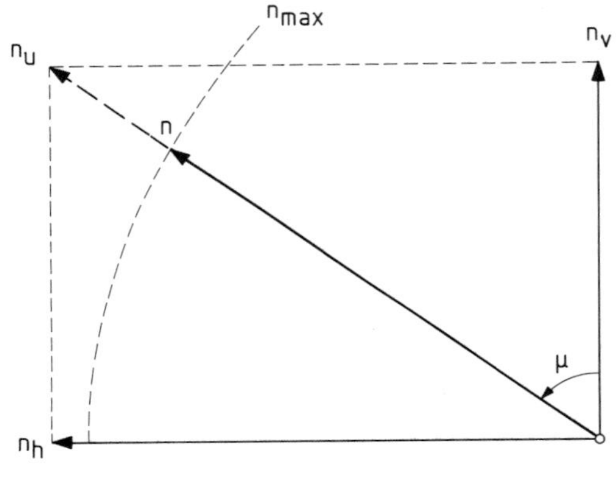

Fig. 6. Three-dimensional Load Factor Synthesis.

C. NUMERICAL EXAMPLE

All numerical examples presented thus far, make use of the same aircraft model, which is representative of a previous generation fighter aircraft (F-4). It seems therefore appropriate to evaluate at the present time the accuracy of the FSP feedback approximation on the basis of a more advanced fighter aircraft, featuring a relatively high thrust-to-weight ratio. For this reason, a new numerical example is presented in this subsection, based on an unclassified approximation of a F-15 aircraft model, taken from [40].

A comparison of the solutions based on an "exact" set of point-mass equations of motion with "reduced-order" model solutions has revealed that the influence of the neglected dynamics in the reduced-order solution is far more significant in the case of the F-15 than for the F-4 aircraft model. Figure 7 presents a typical example of such a comparison of the full and reduced-order solutions for the F-15 model. The results clearly indicate that the reduced-order solution is hardly suitable to serve as a RFT in a feedback guidance scheme.

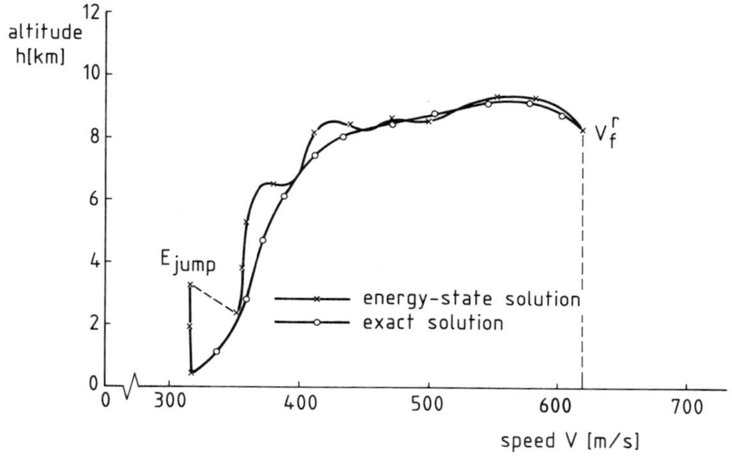

Fig. 7. Full and "Reduced-order" Solutions (F-15 model).

For this reason the approach to the digital simulation of the FSP guidance laws driving the F-15 point-mass system model, has been to store a one-parameter family of precomputed full-order point-mass reference trajectories. The "range matching" feature of the algorithm is then used to select (on-line) the appropriate member of this reference family, to serve as a RFT.

The use of point-mass solutions for serving as RFT, turned out to have the additional advantage that, since no failures of the vertical FSP law could be observed, there was no need to resort to the guidance law of Eq.(134) in the transonic region.

The initial conditions for the example, as well as some important results, are given in Table 4. The example is characterized by a low initial speed and altitude and a relatively large heading error for the interceptor. Figure 8 shows that the initial conditions of the interceptor in the (V,h) plane are close to the corner velocity locus, implying that the interceptor is capable of achieving relatively high turn rates in the initial phase. As a result, the heading error is corrected very rapidly (in fact, 90% of the horizontal turn is completed within 15 seconds !).

Table 4. Comparison of Exact and Approximate Solutions for a Three-dimensional Medium-range Example

target	$h_T = 11278$ m $V_T = 200$ m/s		
initial state	$R_o = 47424$ m $\psi_o = -1.88°$ $\Delta h_o = -7778$ m $V_o = 200$ m/s $\gamma_o = 0°$ $\chi_o = -120°$		
final state	$R_f = d = 3048$ m		
		exact solution	FSP feedback solution
"free" final state	ψ_f (°)	0.0	0.2
	Δh_f (m)	-738.4	-878.1
	V_f (m/s)	609.0	606.0
	γ_f (°)	10.7	14.9
	χ_f (°)	0.0	0.2
final time	t_f (s)	174.0	174.65

The ground track of the initial turn is shown in Fig.9. Note that the projection of the target velocity vector on the horizontal plane is aligned with the X-axis of the coordinate frame. As heading error is decreased, the emphasis shifts towards high energy and range rates and the control actions are such that the flight is directed towards the RFT. In this transient flight, the interceptor executes a dive in order to gain speed, as can be seen in Fig.8, as well as in Figs.10 and 11. Also note in Fig. 11, the rather large rate-of-change of the path-angle in the initial phase, typical of a high-performance fighter. Since the target is flying at a relatively high altitude (some 2000 m above the RFT), a zoom climb is commanded in the terminal phase.

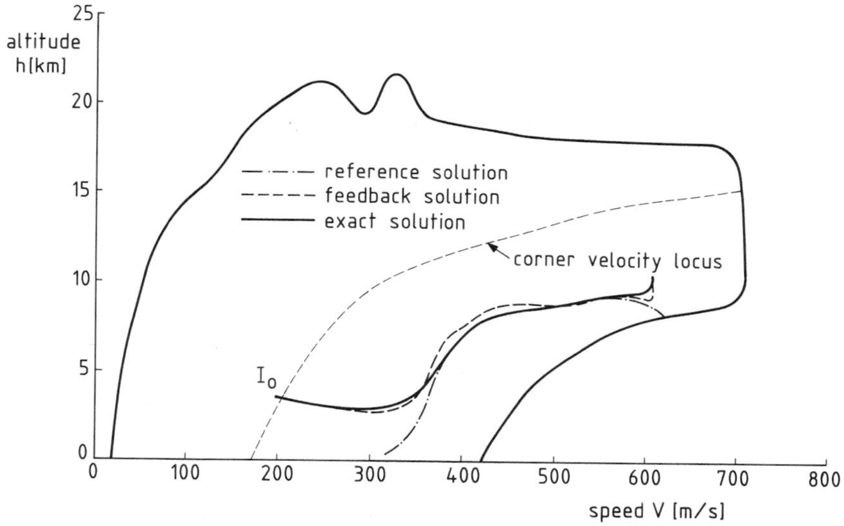

Fig. 8. Comparison of Interceptor Trajectories in the h-V plane.

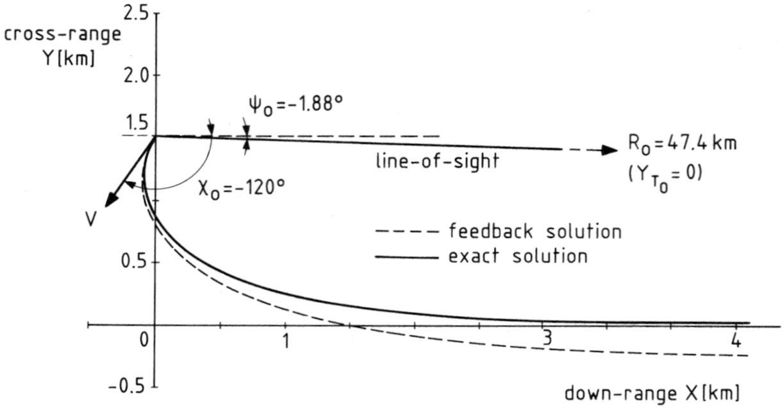

Fig. 9. Comparison of Horizontal Trajectory Projections.

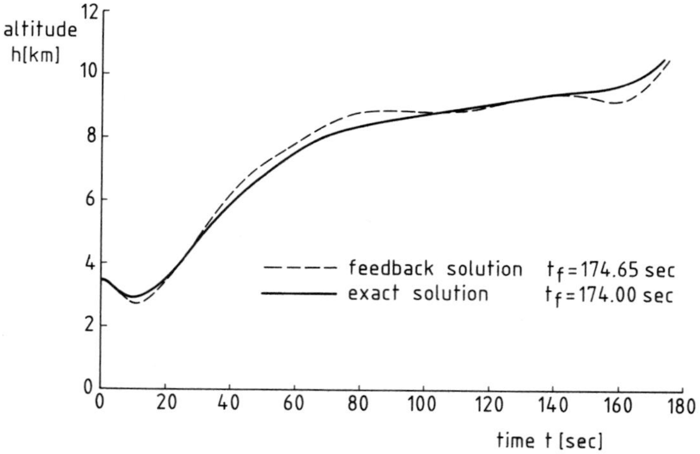

Fig. 10. Comparison of Altitude Time Histories.

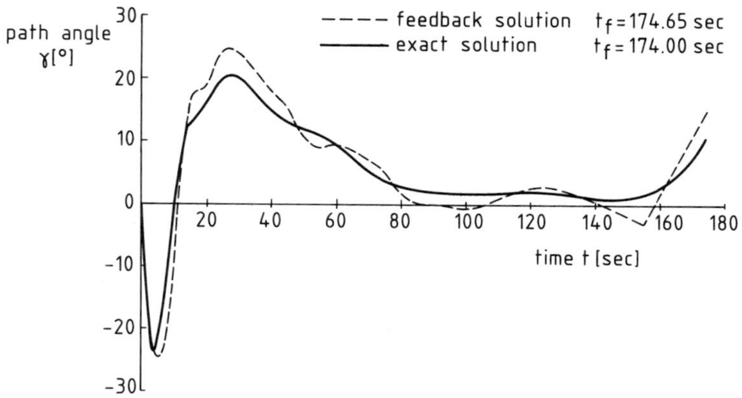

Fig. 11. Comparison of Flight Path Angle Time Histories.

In Figs.8 through 11, a comparison of the simulated feedback solution and the exact open-loop solution is given. The correspondence between the two solutions is remarkably close. As a result, the difference in time-of-capture between the optimal solution and the feedback approximation is only about 0.65 sec (or 0.4%). About 30% of this error can be attributed to the nonoptimality of the terminal zoom maneuver and approximately 60% to the initial turning maneuver. It can be concluded that the pay-off accuracy of the FSP feedback guidance law is very satisfactory, even for highly "dynamic" maneuvers, such as the one demonstrated in the present F-15 example.

VII. CONCLUSIONS

In this Chapter an extensive investigation, oriented towards the synthesis of a feedback guidance law for an interceptor aircraft in a medium-range scenario, is summarized.The guidance algorithm is based on the application of a singular perturbation approach. It is composed of two major elements. A fast converging iterative algorithm provides a three-dimensional "reference flight trajectory" (RFT). It is the solution of a "reduced-order" problem, where the interceptor velocity vector can be instantaneously oriented towards the optimal direction. The RFT is tracked by using two feedback control laws for steering the velocity vector in the horizontal and the vertical planes.

In all phases of the reported multi-year investigation the results, obtained by this zeroth-order feedback approximation of the optimal control, were compared with the "exact" (open-loop) optimal solution. This process allowed to introduce modifications in the suboptimal solution leading to a very satisfactory (better than 0.5%) pay-off accuracy.

The new example incorporated in this Chapter, featuring an F-15 aircraft model, has demonstrated the structural robustness of the feedback guidance law. Though in generating the RFT the computational scheme had to be fitted to the aircraft model characteristics, the feedback guidance formulae have remained unchanged.

The feedback algorithm presented here can be distinguished from similar works reported in the open literature by the following:

(i) It incorporates all phases of an air-to-air interception by providing a closed form approximation to the terminal zoom trajectory.

(ii) It has been validated by using different aircraft models and demonstrated a very satisfactory pay-off accuracy, better than any known result in the open literature.

The hierarchical structure of the algorithm, the explicit feedback form of the control laws and the validated high accuracy make it a very attractive candidate for a real-time implementation on board of a future interceptor aircraft.

ACKNOWLEDGEMENT

The present Chapter was completed during the first author's Sabbatical leave spent as a Visiting Professor at the Faculty of Aerospace Engineering, Delft University of Technology, Delft, The Netherlands.

REFERENCES

1. A.E. Bryson, M.N. Desai and W.C. Hoffman, "Energy-state Approximation in Performance Optimization of Supersonic Aircraft," *J. Aircraft, 6*, 481-488 (1969).
2. R.L. Schultz and N.R. Zagalsky, "Aircraft Performance Optimization," *J. Aircraft, 9*, 108-114 (1972).
3. A.J. Calise and D.D. Moerder, "Singular Perturbation Analysis of Minimum time Long Range Intercept," Drexel University Report (1978).

4. R.K. Mehra, R. Washburn, S. Sajan and J. Caroll, "A Study of Application of Singular Perturbation Theory," NASA CR-3167 (1979).
5. A.J. Calise and D.D. Moerder, "Singular Perturbation Techniques for Real-time Aircraft Trajectory Optimization and Control," NASA CR-3597 (1982).
6. D.B. Price, A.J. Calise and D.D. Moerder, "Piloted Simulation of an On-board Trajectory Optimization Algorithm," *J. Guidance and Control, 7*, 335-360 (1984).
7. F.P. Jones, E.I. Duke and A.J. Calise. " Flight Test Experience from a Three-dimensional Optimal Interception of a Maneuvering Target," presented at the Second International Symposium on Differential Game Applications, Williamsburg, Virginia, August 21-22, 1986.
8. H.T. Huynh and D. Moraigne, "Quasi-optimal On-line Guidance Law for Military Aircraft," AIAA Paper No.85-1977-CP, presented at the AIAA Guidance and Control Conference, Snowmass, Colorado, August 19-22, 1985.
9. J. Shinar and A. Merari, "Aircraft Performance Optimization By Forced Singular Perturbations," *Proc. 12th ICAS Congress*, Munich W.Germany, 758-772 (1980).
10. H.G. Visser and J. Shinar, "A Highly Accurate Feedback Approximation for Horizontal Variable Speed Interceptions," *J. Guidance, Control and Dynamics, 9*, 691-698 (1986).
11. J. Shinar and M. Negrin, "An Explicit Feedback Approximation for Medium-range Interceptions in a Vertical Plane," *Optimal Control Applications & Methods, 4*, 303-323 (1983).
12. J. Shinar and V. Fainstein, "Improved Feedback Algorithms for Optimal Maneuver in a Vertical Plane,"AIAA Paper No.85-1976-CP, presented at the AIAA Guidance and Control Conference, Snowmass, Colorado, August 19-22, 1985.
13. J. Shinar, M. Negrin, K. H. Well and E. Berger, "Comparison Between the Exact and a Approximate Feedback Solution for Medium Range Interception Problems," Proc. of the 1981 Joint Automatic Control Conference, TA-1A (1981).
14. J. Shinar, K.H. Well and B. Jarmark, "Near Optimal Feedback Control for Three-dimensional Interceptions," Proc. of the 15th ICAS Congress, 161-171 (1986).
15. A. Erdelyi, "Singular Perturbations," *Bull. Amer. Math. Soc., 68*, (1962).

16. S. Kaplan, "*in* "Fluid Mechanics and Singular Perturbations," (D. A. Lagerstrom, L.N. Howard and C.S. Lin, eds.), Academic Press, New York, 1967.
17. R.E. O'Malley, "Introduction to Singular Perturbations," Academic Press, New York, 1974.
18. R.E. Meyer and S.V. Parter, eds. "Singular Perturbations and Asymptotics," Academic Press, New York, 1980.
19. J. Keworkian and J. Cole, "Perturbation Methods in Applied Mathematics," Springer Verlag, New York, 1980.
20. W. Wasow, "Asymptotic Expansions for Ordinary Differential Equations," Interscience, New York, 1965.
21. W. Eckhaus, "Matched Asymptotic Expansions and Singular Perturbations," North Holland, Amsterdam, 1973.
22. C.R. Hadlock, " The Existence and Dependence on a Parameter of Solutions of a Nonlinear Two Point Boundary Value Problem," *J. Differential Equations, 14*, (1973).
23. R.R. Wilde and P.V. Kokotovic, "Optimal Open and Closed Loop Control of Singularly Perturbed Linear Systems," *IEEE Trans. Automat. Contr., AC-18*, (1973).
24. M.I. Freedman and J.L. Kaplan, "Singular Perturbations of Two Point Boundary Value Problems Arising in Optimal Control," *J. Optimiz. Theory Appl., 19*, (1976).
25. M.D. Ardema, ed., "Singular Perturbations in Systems and Control," Springer Verlag, New York (1982).
26. P.V. Kokotovic, H.K. Khalil and J. O'Reilley, "Singular Perturbation Methods in Control: Analysis and Design," Academic Press, New York, 1986.
27. A.J. Calise, "Singular Perturbation Methods for Variational Problems in Aircraft Flight," *IEEE Trans. Automat. Contr., AC-21* (1976).
28. J. Shinar, "On Applications of Singular Perturbation Techniques in Nonlinear Optimal Control," *Automatica, 19*, 203-211 (1983).
29. J.V. Breakwell, J. Shinar and H.G. Visser, "Uniformly Valid Feedback Expansions for Optimal Control of Singularly Perturbed Dynamic Systems," *J.Optimiz. Theory Appl., 46*, 441-454 (1985).
30. H.J. Kelley, "*in* "Control and Dynamic Systems," (C.T. Leondes, ed.), Vol.10, p.131-178, Academic Press, New York, 1973.
31. H.G. Visser and J. Shinar, "First-order Corrections in Optimal Feedback Control of Singularly Perturbed Nonlinear Systems," *IEEE Trans. Automat. Contr., AC-31*, 387-393 (1986).

32. N. Rajan, U.R. Prasad and N.J. Rao, "Pursuit-Evasion of Two Aircraft in the Horizontal Plane," *J. Guidance and Control, 3*, 261-267 (1980).
33. A.J. Calise, "Optimal Thrust Control with Proportional Navigation Guidance,". *J. Guidance and Control, 3*, 312-318 (1980).
34. M.D. Ardema, "Solution of Minimum Time to Climb Problem by Matched Asymptotic Expansions," *AIAA J., 14*, 843-850 (1976).
35. H.J. Kelley, E.M. Cliff and A.R. Weston, "Energy State Revisited," *Optimal Control Applications & Methods, 7*, 195-200 (1986).
36. M.D. Ardema and L. Yang, "Interior Transition Layers in Flight Path Optimization," *J. Guidance, Control and Dynamics, 11*, 12-18 (1988).
37. A.B. Vasileva, "Asymptotic Solutions of Two Point Boundary Layer Problems for Singularly Perturbed Conditionally Stable Systems, "*in* "Singular Perturbations: Order Reduction in Control System Design," p. 57-62, ASME, New York, 1972.
38. A. Spitzer, "Development of an Autonomous Guidance for an Aircraft in Air-to-Air Missions," M.Sc. Thesis, Department of Aeronautical Engineering, Technion, Israel Institute of Technology, 1988.
39. J. Shinar, "Concept of Automated Aircraft Guidance System for Air-to-Air Missions," AIAA Paper No.86-2285-CP presented at the 13th Atmospheric Flight Mechanics Conference, Williamsburg, Virginia, August 18-20, 1986.
40. H.G. Visser, H.J. Kelley and E.M. Cliff, " Energy Management of Three Dimensional Minimum Time Intercept," *J. Guidance, Control and Dynamics, 10*, 574-580 (1987).

APPENDIX: TIME-OPTIMAL ZOOM INTERCEPTION AT CONSTANT SPECIFIC ENERGY

This is an idealized version of the terminal phase of a minimum-time interception in a vertical plane obtained by assuming that the specific energy of the interceptor aircraft is constant and the flight path angle γ is a control variable. In order to avoid confusion, the variables in this simplified model are denoted by a superscript "z". Thus, the first assumption is expressed by:

$$h^z + (V^z)^2/2g = E^z = \text{constant} \tag{156}$$

As a consequence, the velocity V^z is not an independent variable, but merely an abbreviation for:

$$V^z = [2g(E^z - h^z)]^{1/2} \tag{157}$$

By considering the flight path angle γ as a control variable, the following "reduced-order " dynamic system, starting to operate at t_o^z is obtained:

$$\dot{x}^z = V^z \cos\gamma^z - V_T \quad , \qquad x(t_o^z) = x_o^z \tag{158}$$

$$\dot{h}^z = V^z \sin\gamma^z \quad , \qquad h(t_o^z) = h_o^z \tag{159}$$

The terminal conditions of the interception are derived from Eq.(90) in polar coordinates (being identical to the terminal conditions of the full order problem they are not indexed by a superscript):

$$x^z(t_f) = - d \cos\theta_f \tag{160}$$

$$h^z(t_f) = h_T - d \sin\theta_f \ , \tag{161}$$

where θ_f, the direction of the line of sight at t_f, is an unspecified parameter.

The variational Hamiltonian of the time-optimal interception for this simplified model is:

$$H^z = -1 + \lambda_x^z(V^z \cos\gamma^z - V_T) + \lambda_h^z(V \sin\gamma^z) \tag{162}$$

The adjoint equations and the corresponding transversality condition are:

$$\dot{\lambda}_x^z = -\frac{\partial H^z}{\partial x^z} = 0 \quad , \qquad \lambda_x^z = \lambda_x^z(t_f) = 2\nu\, x^z(t_f) = -2\nu\, d \cos\theta_f \tag{163}$$

$$\dot{\lambda}_h^z = -\frac{\partial H^z}{\partial x^z}\Big|_{E^z} = -\frac{g}{V^z}(\lambda_x^z \cos\gamma^z - \lambda_h^z \sin\gamma^z) \quad ,$$

$$\lambda_h^z(t_f) = 2\nu\, [h^z(t_f) - h_T] = -2\nu\, d \sin\theta_f \tag{164}$$

The optimal flight path angle for this "zoom" maneuver, which maximizes the Hamiltonian is:

$$\gamma^z = \text{arc tan}\ \{\lambda_h^z / \lambda_x^z\} \tag{165}$$

Since time is not explicitly involved and the final time is unspecified one also has:

$$H^z = 0 \tag{166}$$

and consequently, since due to Eq. (165) $\gamma^z(t_f) = \theta_f$:

$$\nu = -1/2d\,[V^z(t_f) - V_T \cos\theta_f] \tag{167}$$

By substituting Eqs.(163),(166) and (167) into Eq. (162) one obtains for all $t \geq t_o^z$:

$$V^z(\lambda_x^z \cos\gamma^z + \lambda_h^z \sin\gamma^z) = V^z(t_f)/[V^z(t_f) - V_T \cos\theta_f] \tag{168}$$

Since Eq.(165) implies that:

$$\lambda_x^z = G(t)\cos\gamma^z \tag{169}$$

and:

$$\lambda_h^z = G(t)\sin\gamma^z \tag{170}$$

Substitution of these expressions into Eq.(168) yields:

$$G(t) = V^z(t_f)/[V^z(V^z(t_f) - V_T \cos\theta_f)] \tag{171}$$

as well as:

$$G(t_f) = 1/[V^z(t_f) - V_T \cos\theta_f] \tag{172}$$

Keeping in mind that according to Eq.(163) λ_x^z is a constant, Eqs.(170)-(172) lead to a fundamental relationship which characterizes the minimum-time "zoom" interception at a constant specific energy:

$$V^z/\cos\gamma^z = V^z(t_f)/\cos\theta_f = \text{constant} \triangleq A_f \tag{173}$$

and provide a feedback expression for the optimal control γ^z for any given A_f:

$$\gamma^z = \arccos\,(V^z/A_f) \quad , \tag{174}$$

where V^z represents E^z and h^z via Eq.(156).

From Eq.(174) the rate of change of γ^z and the corresponding load factor n^z can be computed:

$$\dot{\gamma}^z \sin\gamma^z = -(1/A_f)\, \dot{V}\Big|_{E^z} = (g/A_f V^z)\, \dot{h}^z = g \sin\gamma^z / A_f \qquad (175)$$

Consequently:

$$\dot{\gamma}^z = g/A_f = \text{constant} \qquad (176)$$

By comparing Eq.(86) to Eq.(176) and using Eq.(174) one can directly conclude that:

$$n^z = \cos\gamma^z + (V^z/A_f) = 2\cos\gamma^z \quad , \qquad (177)$$

as quoted in Eq.(136).

The closed form solution allows to express the unspecified parameter θ_f as a function of the initial conditions of the "zoom" maneuver E^z, $h(t_o^z)$ and $\gamma(t_o^z)$. Applying Eq.(156) for both the initial and the terminal state and substituting Eq.(161) leads to the following quadratic equation for $\sin\theta_f$:

$$[E^z - h(t_o^z)]\sin^2\theta_f + d\cos^2\gamma(t_o^z)\sin\theta_f + h(t_o^z) - h_T\cos^2\gamma(t_o^z)$$

$$- E^z\sin^2\gamma(t_o^z) = 0 \qquad (178)$$

It is easy to see that this equation always has a real positive root (smaller than 1) for any feasible "zoom-climb" trajectory.

Once θ_f is computed from Eq.(178) the terminal altitude can be directly obtained from Eq. (161).

The time required to complete the "zoom" maneuver is derived by using Eq.(176):

$$(t_f - t_o^z) = \frac{1}{g} \frac{V(t_o^z)}{\cos\gamma\,(t_o^z)} [\,\theta_f - \gamma(t_o^z)] \tag{179}$$

The corresponding initial horizontal range is obtained by integrating Eq.(158) after the substitution of Eqs.(173)-(176) and using Eqs.(160) and (179):

$$x(t_o^z) = -\,d\cos\theta_f - \frac{V_T}{g} \frac{V(t_o^z)}{\cos\gamma(t_o^z)} [\theta_f - \gamma(t_o^z)]$$

$$+ \frac{1}{2g}\left[\frac{V(t_o^z)}{\cos\gamma(t_o^z)}\right]^2 \{[\theta_f - \gamma(t_o^z)] + \frac{1}{2}[\sin(2\theta_f) - \sin(2\gamma(t_o^z))]\} \tag{180}$$

Moreover, all the variables along the entire "zoom" trajectory can be expressed as explicit functions of the optimal flight path angle γ^z and the initial conditions. By matching the monotonically varying actual horizontal range $x(t)$ with the result obtained in Eq.(180), the starting time of the "zoom" maneuver t_o^z can be selected and then the entire trajectory can be precomputed and stored. Comparison of the actual and the stored trajectories allows to implement a feedback guidance law as proposed in Eq. (137).

MULTISTEP MATRIX INTEGRATORS FOR REAL-TIME SIMULATION

J.A. De ABREU-GARCIA and T.T. HARTLEY
Department of Electrical Engineering
The University of Akron
Akron, OH 44325-3904

I. INTRODUCTION

The real-time simulation of physical systems is a cost effective method for the analysis, design, and testing of today's increasingly complex control systems. In particular, the application of real-time simulation in the evaluation of industrial processes, hardware testing, and personnel training is of special interest to engineers and scientists. However, the hardware constraints associated with real-time simulations can sometimes restrict the simulation timestep to be larger than is allowable for the simulation to be stable. On the other hand, the use of smaller timesteps would guarantee stability but would not allow enough time for all of the required computations to be performed. Hence, real-time results are often difficult to obtain.

A large number of important practical problems is governed by differential equations which exhibit stiffness. A stable system is said to be stiff if the ratio of its largest to its smallest eigenvalue is large. Since, to guarantee stability, most explicit integration techniques restrict the product of an eigenvalue with the timestep to be small ($\approx|\lambda T|\leq 1$), stiff systems require special attention, [1]. The problem is that, although the transients of the most negative eigenvalues decay very fast, they result in large λT products. Therefore, in closed-loop numerical integration, smaller stepsizes must be used to maintain stability. On the other hand, the least negative eigenvalues decay very slowly and yield small λT products. This implies that the real-time simulation of stiff systems requires extremely small timesteps to maintain stability while requiring large computing times to determine the slower and more important responses. Clearly, the wide spread of the eigenvalues of stiff systems makes the real-time simulation of these systems a very difficult problem. There are, however, two possible solutions to this problem. One is to use a faster computer. The other is to use an algorithm that allows a larger timestep to be used while maintaining stability. The latter solution is the one considered in this paper.

To date, several integration techniques for the real-time simulation of stiff systems have been proposed, see for example [1]-[37], and references therein. However, most of these techniques are of the implicit nature. Implicit integration techniques are usually too slow to be used for real-time applications. This stems from the fact that the current output is a function of itself, and past outputs and function evaluations. Thus, implicit integration requires iteration of the solution at each timestep. In contrast, explicit methods require smaller timesteps due to their small stability regions. Moreover, for an explicit method, the actual output is calculated only from previous outputs and function evaluations. Hence, explicit methods are computationally fast, thereby making them excellent candidates for real-time applications. Thus, explicit methods are chosen as the prototypical methods for real-time simulation. In this paper an explicit linear multistep integration technique for vector systems is presented.

Historically, most methods that have been used for numerical integration are scalar methods, although they can be used for vector systems of differential equations. Essentially, the same integrator is used for each component of the vector system. This paper presents a method with regression coefficients that are matrices whose dimension is compatible with that of the system they are integrating. This technique uses the Stability Region Placement (SRP) approach originally proposed in [2] to allow the simulation timestep to be chosen independent from the system eigenvalues. Since this method is essentially the matrix version of that of [2], it will be referred to here as the Matrix Stability Region Placement method (MSRP).

It is important to keep in mind that when using a p-step method in real-time digital simulation, the original n-dimensional set of first order differential equations is replaced by an n-dimensional set of pth-order difference equations. Thus, $n\times(p-1)$ additional roots are introduced into the resulting closed-loop discrete-time system. These additional roots are usually referred to as spurious roots, while the n roots corresponding to those of the original system are the principal roots.

Recall that the location of the roots of the closed-loop discrete transfer function determines the stability of an integration method for a given linear system. Further, recall that a linear discrete-time system is stable provided that all its roots lie inside the unit circle in the z-plane. Thus, it will be shown that for a linear system of differential equations, it is possible to use the MSRP method to place the principal roots of the resulting closed-loop discrete-time system at the exact mappings of the s-plane poles $z=e^{sT}$, while placing all the spurious roots at the origin in the z-plane. In addition, for linear time-invariant systems, the resulting discrete-time system gives exact time responses and has zero steady-state error in the simulation.

For nonlinear systems, MSRP can also be used. To accomplish this, the nonlinear system must first be linearized about the desired steady state operating point. Once the system and input Jacobian matrices have been found for the operating point of interest, the MSRP method can again be used to compute the coefficients of the matrix integrators. The resulting simulations are reasonably accurate as long as the system states are close enough to steady state. However, the larger the transients are, the larger the expected transient error is. This problem can be alleviated by adaptively tracking the system Jacobian matrix as the system evolves. This allows the integrator matrices to track the system Jacobian as it changes, while keeping the timestep constant.

This paper is organized as follows. First, the two-step MSRP matrix integrator (MSRP2) is presented in section II. Some consideration is given to the eigen-structure of the resulting second order closed-loop system. The understanding of matrix integrators gained in this section is then utilized in section III as a vehicle to develop the more general multistep MSRP method. The resulting closed-loop system is rewritten as a linear system using both block diagram and state space representations. These representations are shown to be very useful in determining important system properties such as controllability, observability, and stability. This section is concluded with a single-sweep algebraic algorithm for developing the equations that govern the matrix coefficients of the integrator, and a closed-form solution of these equations. Section IV is devoted to the case when the system matrix is singular. In this section mathematical expressions for determining the integrator coefficients are derived for the two-step method, with extensions to the p-step method being rather obvious. The adaptive integration approach referred to earlier in connection with integrators for nonlinear systems is developed in section V. Here the integrator is adapted to the given system at each point in time based on a system Jacobian estimation rather than changing the timestep as do most common adaptive integrators. Section VI gives a detailed discussion of some numerical and implementational aspects of the techniques developed here. Included in the developments mentioned above are several illustrative examples of practical interest. Finally, section VII gives some concluding remarks concerning the work being developed here.

II. TWO-STEP MATRIX INTEGRATORS

The method presented here is essentially that of [2] with the only difference being that a vector system of equations is considered rather than a scalar system. The material presented in this section also borrows heavily from the MSRP2 integration method proposed in [3]. However, here the results are presented in a more compact form, and derived using different techniques. Also, it should be pointed out that this method is a generalization of the usual zero-order-hold approach.

When performing numerical integration it is important to maintain the overall system transient response while the integration operator maintains the character of an integrator. This is normally done by deriving integrators that reproduce the transient response better as the timestep goes to zero [2]. However, for a specific system, the SRP method derives an integrator that improves the transient response for a given nonzero timestep.

To pave the way to the development of the general p-step MSRP method presented in section III, an MSRP2 method is considered in what follows. In [2] it was shown that MSRP2 has the minimum number of coefficients necessary to maintain both the transient response and the integration property for a scalar system, while also allowing arbitrary choice of timestep. It is shown here that this is also true for vector systems.

The linear MSRP2 integrator can be written as:

$$x_{k+2}=-A_1 x_{k+1}-A_0 x_k+T(B_1 \dot{x}_{k+1}+B_0 \dot{x}_k) \qquad (1)$$

where T is the integration timestep, kT is the time, x is the n-dimensional state vector at time t, and A_i and B_i : i=0,1 are the n×n regression coefficient matrices.

Usually, the system of interest is described by a set of linear differential equations. However, it should be realized that this method applies equally well to nonlinear systems provided that they are first linearized about some desired steady state operating point (this is the topic of part V). Therefore, without loss of generality, linearity of the system being used will be assumed. That is:

$$\dot{x}_K = Jx_K + Gu_K \tag{2}$$

where, u is an m-vector of input functions, and J and G are the system and input matrices, respectively.

In actual implementation, the above steps form a closed-loop discrete-time linear system which can be derived by substituting eq. (2) into eq. (1), viz.

$$x_{K+2} = -A_1 x_{K+1} - A_0 x_K + T(B_1 J x_{K+1} + B_0 J x_K) + T(B_1 G u_{K+1} + B_0 G u_K) \tag{3}$$

It is worthwhile noting that the resulting system, eq. (3), is 2n-dimensional. Therefore, this system has 2n roots associated with it. The n principal roots are those corresponding to the mappings of the eigenvalues of J. The remaining n roots are due to the fact that the first order differential equation, eq. (2), is being replaced by a second order difference equation, eq. (1). These roots are spurious roots and act as noise sources in the simulation.

The stability of the integration process is governed by the stability of eq. (3). Therefore, the location of the 2n eigenvalues of the resulting closed-loop system determines whether the integration process is stable. It is well known that the classical z-transform technique provides a vehicle for stability analyses of discrete-time closed-loop systems. Thus, taking the z-transform of eq. (3) yields:

$$\left[z^2 I + (A_1 - TB_1 J)z + (A_0 - TB_0 J)\right] X(z) = \left[B_1 z + B_0\right] TGU(z) \tag{4}$$

Clearly, eq. (4) represents a linear system with transfer function matrix (TFM) H(z) given as:

$$H(z) = \left[z^2 I + (A_1 - TB_1 J)z + (A_0 - TB_0 J)\right]^{-1} \left[B_1 z + B_0\right] TG \tag{5}$$

Recall that the regression coefficients, namely A_i

and B_i : i=0,1; are n-dimensional square matrices. Therefore, there are $4n^2$ unknowns to be determined which place the 2n eigenvalues of eq. (5). It is clear that the poles of the resulting closed-loop system can be arbitrarily placed in an infinite number of locations. The problem is how to determine the regression coefficients A_i and B_i : i=0,1 so that the system transient response is maintained while still having eq. (1) perform as an integrator.

To solve the first problem, it would be reasonable to place n of the eigenvalues (principal roots) at the exact z-plane mapping of the eigenvalues of J, that is, the usual discrete-time system matrix e^{JT}. To guarantee the accuracy of the method, the n spurious roots should be placed as close to the origin in the z-plane as possible [1]-[2]. From the foregoing, it can readily be seen that the desired pole location can be achieved provided that A_i and B_i : i=0,1 satisfy the following matrix equations:

$$A_1 - TB_1 J = -e^{JT} \tag{6a}$$

$$A_0 - TB_0 J = 0, \quad \text{where 0 is the } n \times n \text{ null matrix} \tag{6b}$$

Note that here there are $2n^2$ equations and $4n^2$ unknowns. Therefore, more constraints are required. The remaining $2n^2$ constraints are obtained by forcing eq. (1) to act as an integrator. In [1]-[2] it is shown that for a linear multistep method, and thus MSRP2, to be convergent it is necessary and sufficient that the method be consistent and zero-stable. Equivalently, the steady state gain of the discrete-time system should be that of the continuous-time system, and an open-loop integrator pole should be at Z=+1, respectively. Algebraically, these constraints can be written, again respectively, as (see [1]):

$$A_1 - B_0 - B_1 = -2I, \quad \text{where I is the } n \times n \text{ identity matrix} \tag{7a}$$

$$I + A_0 + A_1 = 0 \tag{7b}$$

It is shown in [2] that these constraints imply

that the integrator is first order accurate.

Equations (6a)-(7b) then give $4n^2$ equations to be solved for the $4n^2$ unknowns in the A_i's and B_i's. Provided that J is invertible, this set of equations can be solved using linear algebra techniques. That is:

$$\begin{bmatrix} I & \vdots & I \\ \cdots & & \cdots \\ -(JT)^{-1} & \vdots & I-(JT)^{-1} \end{bmatrix} \begin{bmatrix} A_0 \\ \cdot\cdot \\ A_1 \end{bmatrix} = \begin{bmatrix} -I \\ \cdots \\ e^{JT}(JT)^{-1}-2I \end{bmatrix} = \begin{bmatrix} \Phi_1 \\ \cdot\cdot \\ \Phi_2 \end{bmatrix} \tag{8a}$$

Post-multiplying the first row in eq. (8a) by $(JT)^{-1}$ and adding the result to the second row yields:

$$\begin{bmatrix} I & \vdots & I \\ \cdot & \vdots & \cdot \\ 0 & \vdots & I \end{bmatrix} \begin{bmatrix} A_0 \\ \cdot\cdot \\ A_1 \end{bmatrix} = \begin{bmatrix} \Phi_1 \\ \cdots \\ \Phi_2+\Phi_1(JT)^{-1} \end{bmatrix} = \begin{bmatrix} \hat{\Phi}^1 \\ \cdot\cdot \\ \hat{\Phi}_2 \end{bmatrix} \tag{8b}$$

From which A_1 and A_2 follow directly as:

$$A_1=\hat{\Phi}_2=\Phi_2+\Phi_1(JT)^{-1}=e^{JT}(JT)^{-1}-(JT)^{-1}-2I \tag{9a}$$

$$A_0=\hat{\Phi}_1-A_1=\Phi_1-A_1=-e^{JT}(JT)^{-1}+(JT)^{-1}+I \tag{9b}$$

The B_i's follow directly from the A_i's as:

$$B_1=[A_1+e^{JT}](JT)^{-1} \tag{9c}$$

$$B_0=A_0(JT)^{-1} \tag{9d}$$

The expressions just given for the A_i's and B_i's are easily computed using a software package such as PC-Matlab [38] by inputting T and J.

Regarding the above development, it should be mentioned that the $\hat{\Phi}_i$ notation was introduced to facilitate the transition from the two-step method to the p-step method presented in section III.

Referring back to eq. (3), and using system theoretical concepts, important properties of MSRP integrators can be found. Careful consideration of eqs. (3) and (9a)-(9d) reveals that the eigenstructure of both eq. (3) and the A_i's and B_i's is entirely determined by that of the original system matrix J. This can easily be seen once it is realized that the regression coefficients are a function of J only. It is a well known fact that functions of a matrix share the same eigenvectors [39]. Consequently, A_i, B_i, J, and e^{JT} commute as they all have the same eigenvectors. This implies that any changes experienced by J will be also experienced by the A_i's, B_i's, and e^{JT}. This is emphasized further in what follows.

Before proceeding any further it is expedient to reformulate the 2n-dimensional closed-loop system of eq. (3) into a block diagram representation. One such representation is given in fig. 1 below.

Notice that the relationship between the states of the actual closed-loop system and the state variables of fig. 1 is $x_k = x_2(k)$. It should also be noticed that with the pole assignment given by eqs. (6a)-(6b) the block diagram representation of eq. (3) has only one feedback loop; namely, the feeback associated with the term $A_1 - TB_1 J = -e^{JT}$. This fact can be exploited in order to provide a 2n-dimensional state space representation of the closed-loop system of eq. (3). Of special interest, from the control viewpoint, are the well known canonical form realizations. One such representation of eq. (3) is given in eq. (10) below.

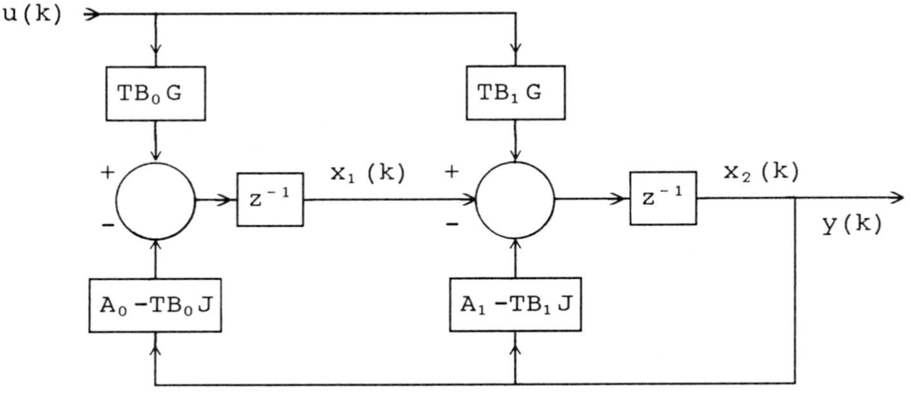

Figure 1. Block diagram implementation of MSRP2

The state space realization of fig. 1 is:

$$x(k+1)=\begin{bmatrix} 0 & \vdots & 0 \\ \cdots & \vdots & \cdots \\ I & \vdots & e^{J\dot{T}} \end{bmatrix} x(k)+\begin{bmatrix} TB_0G \\ \cdots \\ TB_1G \end{bmatrix} U(k) \quad \Rightarrow \quad x(k+1)=Ax(k)+Bu(k) \tag{10}$$

$$y(k)=\begin{bmatrix} 0 & \vdots & I \end{bmatrix} x(k)=x_2(k)=x_K \quad \Rightarrow \quad y(k)=Cx(k)=x_2(k)=x_K$$

It should be realized that with the assignments made in eqs. (6a)-(6b), the closed-loop integration is reduced to the model that is normally derived using the standard zero-order-hold transformation provided that there are no inputs to the system [40]. It can be shown, however, that MSRP2 emulates a first-order-hold, when inputs are considered.

The eigenstructure of the closed-loop system of eq. (10) is now given in the following theorem.

Theorem: If P is the matrix whose columns are the eigenvectors of J (P is the modal matrix of J), that is, P is such that $P^{-1}JP=\Lambda$, with Λ being diagonal, then the matrix Q whose columns are the eigenvectors of A is related to P as follows:

$$Q = \begin{bmatrix} P & \vdots & 0 \\ \cdots & \vdots & \cdots \\ -Pe^{-\Lambda\dot{T}} & \vdots & P \end{bmatrix} \tag{11}$$

Proof: Suppose that Q has the structure given in eq. (11). Further, suppose that Q has the eigenvectors of A in eq. (10) as its columns. Then, it must follow that $Q^{-1}AQ$ is a diagonal matrix. Thus, to proof the theorem, it is sufficient to show that Q is the matrix which diagonalizes A, that is:

$$Q^{-1}AQ=\begin{bmatrix} P^{-1} & \vdots & 0 \\ \cdots & \vdots & \cdots \\ e^{-\Lambda\dot{T}}P^{-1} & \vdots & P^{-1} \end{bmatrix}\begin{bmatrix} 0 & \vdots & 0 \\ \cdots & \vdots & \cdots \\ I & \vdots & e^{J\dot{T}} \end{bmatrix}\begin{bmatrix} P & \vdots & 0 \\ \cdots & \vdots & \cdots \\ -Pe^{-\Lambda\dot{T}} & \vdots & P \end{bmatrix} \tag{12a}$$

$$Q^{-1}AQ=\left[\begin{array}{c:c} 0 & 0 \\ \hdashline P^{-1} & P^{-1}e^{JT} \end{array}\right]\left[\begin{array}{c:c} P & 0 \\ \hdashline -Pe^{-\Lambda T} & P \end{array}\right] \tag{12b}$$

$$=\left[\begin{array}{c:c} 0 & 0 \\ \hdashline I-P^{-1}e^{JT}Pe^{-\Lambda T} & P^{-1}e^{JT}P \end{array}\right] \tag{12c}$$

However, $P^{-1}e^{JT}P=e^{P^{-1}JTP}=e^{\Lambda T}$. Hence, substituting this result into eq. (12c) yields:

$$Q^{-1}AQ=\left[\begin{array}{c:c} 0 & 0 \\ \hdashline I-e^{\Lambda T}e^{-\Lambda T} & e^{\Lambda T} \end{array}\right]=\left[\begin{array}{c:c} 0 & 0 \\ \hdashline 0 & e^{\Lambda T} \end{array}\right] \tag{12d}$$

From the foregoing it is easy to see that Q, as specified in the theorem, is the modal matrix of A.

Thus, the proof is complete.

Q.E.D.

Clearly, the n-eigenvectors of this representation (eq. (10)) are the n-eigenvectors of e^{JT}, which are the eigenvectors of J. Hence, using the MSRP2 design presented here the system mode shapes are preserved.

To demonstrate the utility of the integration method presented in this section, the third order Brennan and Leake turbojet engine model of [41] is considered. The nonlinear model is described by the following set of nonlinear differential equations grouped into algebraic and dynamic relations as:

-algebraic relations:

$$T_3=0.64212+0.35788N^2$$

$$\dot{W}_3=1.3009N-0.13982\left[P_4^2-\mathrm{sqrt}(P_4+0.41688N^2-0.0899P_4N)\right]$$

-dynamic relations:

$$\dot{P}_4=\dot{W}_f\left[0.93586\frac{P_4}{P_b}+31.486\right]+21.435\dot{W}_3T_3-53.86\frac{P_4^2}{P_b}$$

$$\dot{N}=\frac{1.258}{N}\left[\frac{P_4^2}{\dot{P}_b}-\dot{W}_3N^2\right]$$

$$\dot{P}_b=37.78\dot{W}_3-38.448P_4+0.66849\dot{W}_f$$

where, $\dot{W}_f$ is the fuel input mass rate
P_4 is the combustor pressure
P_b is the combustor density
N is the rotor speed
W_3 is the compressor discharge mass flow
T_3 is the compressor discharge temperature.

This model, linearized and normalized about $\dot{W}_f=1$ with initial conditions $(x_1,x_2,x_3)=(0.53831,1.77504,0.54589)$, where $x_1=P_4$, $x_2=P_b$, and $x_3=N$, has system and input Jacobians, J and G respectively, given as:

$$J=\begin{bmatrix}-112.266 & 52.925 & 42.259\\ -48.1501 & 0 & 47.443\\ 2.8377 & -1.258 & -4.096\end{bmatrix}\qquad G=\begin{bmatrix}32.422\\ 0.668\\ 0\end{bmatrix}$$

The eigenvalues of the linearized model are -3.15, -32.33, and -82.24. Thus, this is a stiff system and a timestep of approximately -1/max(eigenvalue), or T=0.01 is the largest value that could possibly be used for most linear multistep integrators. Using PC-Matlab [38], eqs. (9a)-(9d), and T=0.02 , the following values are obtained for the A_i's and B_i's. Notice that this timestep is twice as large as is normally possible.

$$A_0=\begin{bmatrix}0.6573 & -0.2626 & -0.3021\\ 0.2356 & 0.1028 & -0.3574\\ -0.0161 & 0.0057 & 0.0369\end{bmatrix}\qquad A_1=\begin{bmatrix}-1.6573 & 0.2626 & 0.3021\\ -0.2356 & -1.1028 & 0.3574\\ 0.0161 & -0.0057 & -1.0369\end{bmatrix}$$

$$B_0 = \begin{bmatrix} -0.2507 & -0.1044 & -0.1091 \\ 0.0941 & -0.4715 & -0.1275 \\ -0.0062 & 0.0023 & -0.4874 \end{bmatrix} \qquad B_1 = \begin{bmatrix} 0.5934 & 0.3671 & 0.4111 \\ -0.3297 & 1.3687 & 0.485 \\ 0.0223 & -0.0081 & 1.4505 \end{bmatrix}$$

Figure 2 shows the natural response for the same initial conditions and can be compared with the actual response of the nonlinear system simulated with AB-2 (two-step Adams-Bashforth) for T=0.002 shown in fig. 3. Additional comparisons can be observed in fig. 4 for which a timestep of T=0.01 was used. Notice that the responses are effectively identical to those of fig. 3.

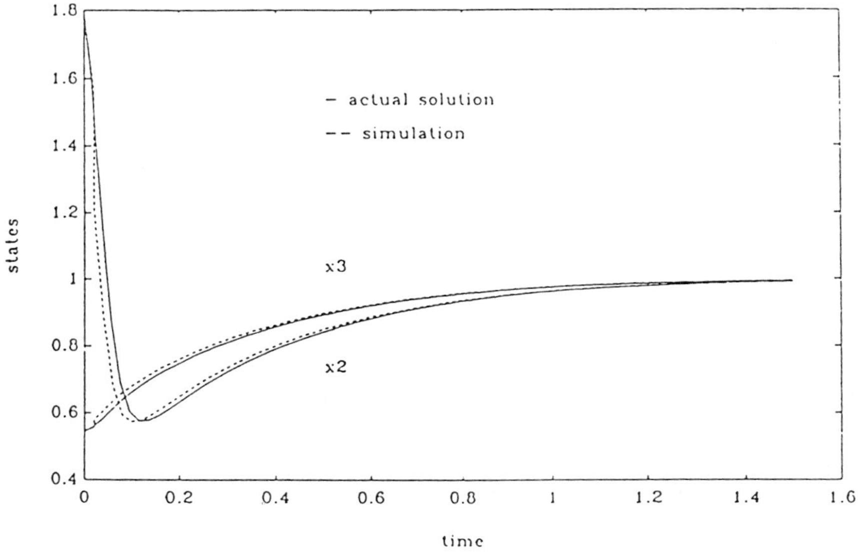

Figure 2. MSRP2 simulation of linearized Brennan and Leake turbojet engine model with T=0.02

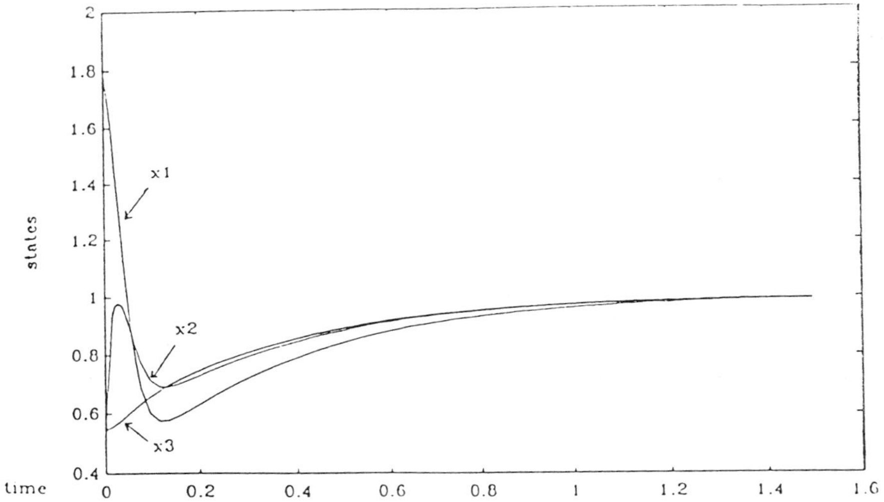

Figure 3. AB-2 simulation of the nonlinear Brennan and Leake turbojet engine model with T=0.002

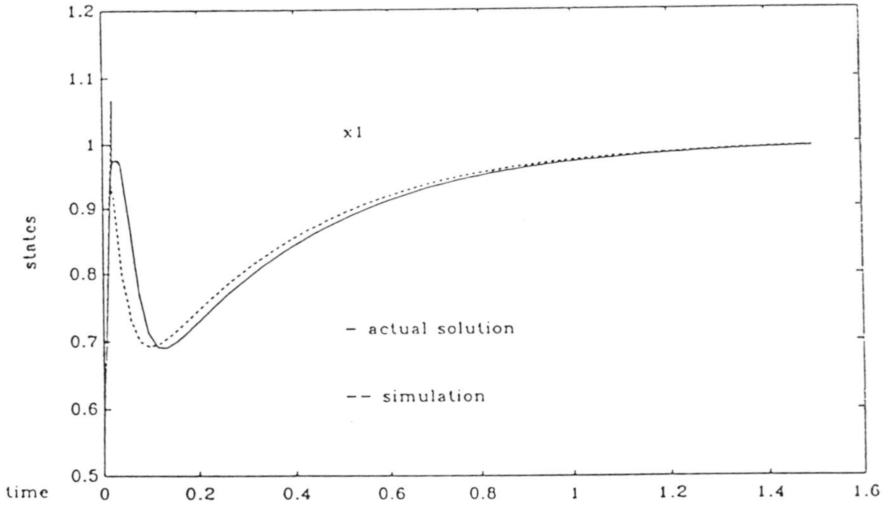

Figure 4. MSRP2 simulation of linearized Brennan and Leake turbojet engine model with T=0.01

This completes the development of the two-step matrix integration method being presented here. The case when the method has more than two steps is the topic of the next section.

III. P-STEP MATRIX INTEGRATORS FOR LINEAR SYSTEMS

In this section, the two-step method just presented in section II is generalized to the case when p>2 [17]. Again, it should be realized that this is also a generalization of the zero-order-hold method, as it generates (p-1)-order-hold representations. As in the case when p=2, the explicit p-step MSRP integrator can also be described by a difference equation similar to eq. (1), viz.

$$x_{k+p}=\sum_{i=0}^{p-1}-A_i x_{k+i}+T\sum_{i=0}^{p-1}B_i \dot{x}_{k+i} \tag{13}$$

where $\dot{x}_k$ is specified in eq. (2).

Using eq. (2) into eq. (13) yields:

$$x_{k+p}=\sum_{i=0}^{p-1}(TB_i J-A_i)x_{k+i}+T\sum_{i=0}^{p-1}B_i G u_{k+i} \tag{14}$$

The resulting closed-loop linear discrete-time system can be determined by taking the z-transform of eq. (14) and grouping similar terms on either side of the equality sign to give:

$$\left[z^p I+\sum_{i=0}^{p-1}(A_i-TB_i J)z^i\right]X(z)=\left[T\sum_{i=0}^{p-1}B_i G z^i\right]U(z) \tag{15}$$

which can be rewritten in the usual input-output form with TFM H(z) given by:

$$H(z)=\left[z^p I+\sum_{i=0}^{p-1}(A_i-TB_i J)z^i\right]^{-1}\left[T\sum_{i=0}^{p-1}B_i G z^i\right] \tag{16}$$

The system of eq. (15) can be represented in a block diagram form as shown in fig. 5, with the actual state being $x_{p-1}(k)=x_k$.

It is worthwhile noting that fig. 5 is precisely the block diagram representation corresponding to the well known observer canonical form for multivariable linear systems. This implies that the resulting closed-loop system is guaranteed to be observable, which in turn guarantees that the integrated state can be completely recovered from output observations. The state space representation of the closed-loop system of eqs. (15)-(16) can also be written as indicated below:

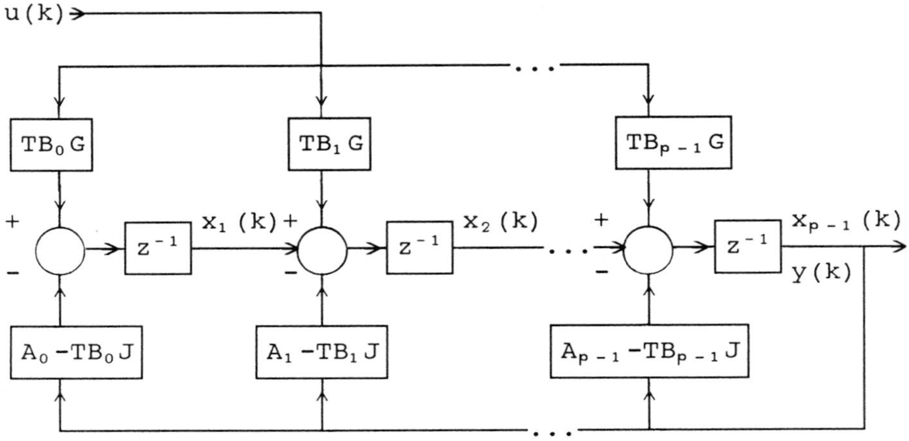

Figure 5. Block diagram implementation of a p-step MSRP integrator

$$A=\begin{bmatrix} 0 & 0 & 0 & . & . & . & 0 & -(A_0-TB_0J) \\ I & 0 & 0 & . & . & . & 0 & -(A_1-TB_1J) \\ 0 & I & 0 & . & . & . & 0 & -(A_2-TB_2J) \\ . & . & . & . & & & . & . \\ . & . & . & & . & & . & . \\ . & . & . & & & . & . & . \\ 0 & 0 & 0 & . & . & . & 0 & -(A_{p-2}-TB_{p-2}J) \\ 0 & 0 & 0 & . & . & . & I & -(A_{p-1}-TB_{p-1}J) \end{bmatrix} \quad B=\begin{bmatrix} TB_0G \\ TB_1G \\ TB_2G \\ . \\ . \\ . \\ TB_{p-2}G \\ TB_{p-1}G \end{bmatrix} \qquad (17a)$$

$$C=\begin{bmatrix} 0 & 0 & 0 & . & . & . & 0 & I \end{bmatrix}$$

Equivalently, this equation can be rewritten as:

$$x(k+1)=Ax(k)+Bu(k), \quad y(k)=Cx(k)=x_{p-1}(k)=x_k \qquad (17b)$$

It is clear that the closed-loop stability of the

integrator is given by the closed-loop stability of the system described by eqs. (17a)-(17b). Therefore, the same philosophy used in the two-step method can also be applied here to place the poles at the desired locations. From eq. (16) this is seen to be equivalent to forcing the A_i and B_i coefficients to satisfy the following set of matrix equations:

$$A_{p-1} - TB_{p-1}J = -e^{JT} \tag{18a}$$

$$A_i = TB_iJ \; : \; i=0,1,2,\ldots,p-2 \tag{18b}$$

To guarantee that eq. (13) maintains its character of an integrator, the matrix version of the algebraic equations given in [1] regarding consistency and zero-stability must be satisfied. Hence, as in the two-step case, the method is consistent and zero-stable provided that the following equations are satisfied [2]:

$$I + \sum_{i=0}^{p-1} A_i = 0 \quad \text{and} \quad \sum_{i=1}^{p-1} iA_i - \sum_{i=0}^{p-1} B_i = -pI \tag{19}$$

Notice that the pole placement constraints imposed by eqs. (18a)-(18b) yield pn^2 equations, while eq. (19) gives only $2n^2$ additional equations. However, in total there are $2pn^2$ unknowns. Therefore, $(p-2)n^2$ more equations are needed to determine the A_i and B_i coefficients uniquely. These extra equations can be found by increasing the order of accuracy of the integrator using Taylor series expansion of x and its derivative as suggested in [1], that is:

$$\frac{1}{q!}\left[\sum_{i=1}^{p-1} i^q A_i\right] - \frac{1}{(q-1)!}\left[\sum_{i=1}^{p-1} i^{q-1} B_i\right] = -\frac{p^q}{q!} I, \quad q=2,3,\ldots,p-1 \tag{20}$$

Again, notice that the principal roots of the closed-loop system lie at the exact z-plane mappings of the s-plane poles, while all the spurious roots lie at the origin. This can clearly be seen by rewriting the

A matrix of the closed-loop system as follows:

$$A=\begin{bmatrix} 0 & 0 & 0 & . & . & . & 0 & 0 \\ I & 0 & 0 & . & . & . & 0 & 0 \\ 0 & I & 0 & . & . & . & 0 & 0 \\ . & . & . & . & & & . & . \\ . & . & . & & . & & . & . \\ . & . & . & & & . & . & . \\ 0 & 0 & 0 & . & . & . & 0 & 0 \\ 0 & 0 & 0 & . & . & . & I & e^{JT} \end{bmatrix} \qquad (21)$$

Equation (21) guarantees the stability of the integration process.

Associated with some of the control aspects of MSRP integrators are the fundamental concepts of controllability and observability. It has already been mentioned that the closed-loop system resulting from implementing the integrator is observable (see fig. 5). This can easily be seen from the observability matrix Ξ_q of the system, viz.

$$\Xi_q=\begin{bmatrix} 0 & 0 & 0 & . & . & .0 & 0 & I \\ 0 & 0 & 0 & . & . & .0 & I & e^{JT} \\ 0 & 0 & 0 & . & . & .I & e^{JT} & e^{2JT} \\ . & . & . & & & . & . & . \\ . & . & . & & . & . & . & . \\ . & . & . & . & & . & . & . \\ 0 & 0 & I & . & . & .e^{(np-5)JT} & e^{(np-4)JT} & e^{(np-3)JT} \\ 0 & I & e^{JT} & . & . & .e^{(np-4)JT} & e^{(np-3)JT} & e^{(np-2)JT} \\ I & e^{JT} & e^{2JT} & . & . & .e^{(np-3)JT} & e^{(np-2)JT} & e^{(np-1)JT} \end{bmatrix} \qquad (22)$$

Regarding system controllability, it is expedient to first transform the system into Jordan co-ordinates. It is interesting to note that p-step MSRP integrators have the property that the A matrix of the resulting closed-loop system always has 2n linearly independent eigenvectors and (p-2)n generalized eigenvectors. This implies that the system of eqs. (17a)-(17b) cannot be diagonalized for $p\geq 3$. Rather, a block lower Jordan form results when using the modal matrix of A to similarly transform this system. In general, the modal matrix of

A and the resulting block lower Jordan form of A, Q and A_{j0}, respectively, can be written as:

$$Q=\begin{bmatrix} P & 0 & 0 \cdots & 0 & 0 & 0 \\ -Pe^{-\Lambda T} & P & 0 \cdots & 0 & 0 & 0 \\ 0 & -Pe^{-\Lambda T} & P \cdots & 0 & 0 & 0 \\ \vdots & \vdots & \vdots & \vdots & \vdots & \vdots \\ 0 & 0 & 0 \cdots & P & 0 & 0 \\ 0 & 0 & 0 \cdots & -Pe^{-\Lambda T} & P & 0 \\ 0 & 0 & 0 \cdots & 0 & -Pe^{-\Lambda T} & P \end{bmatrix} \tag{23a}$$

$$A_{j0}=\begin{bmatrix} 0 & 0 & 0 & \cdots & 0 & 0 & 0 \\ I & 0 & 0 & \cdots & 0 & 0 & 0 \\ 0 & I & 0 & \cdots & 0 & 0 & 0 \\ \vdots & \vdots & \vdots & & \vdots & \vdots & \vdots \\ 0 & 0 & 0 & \cdots & 0 & 0 & 0 \\ 0 & 0 & 0 & \cdots & I & 0 & 0 \\ 0 & 0 & 0 & \cdots & 0 & 0 & e^{\Lambda T} \end{bmatrix} \tag{23b}$$

Once the Jordan form of A has been determined, it is easy to show that the system of eq. (17a) is always controllable. This can be seen by considering the controllability matrix Ξ_p of the system, viz.

$$\Xi_p=\begin{bmatrix} TB_0G & 0 & \cdots 0 & 0 & 0 & 0 \cdots 0 \\ TB_1G & TB_0G & \cdots 0 & 0 & 0 & 0 \cdots 0 \\ TB_2G & TB_1G & \cdots 0 & 0 & 0 & 0 \cdots 0 \\ \vdots & \vdots & \vdots & \vdots & \vdots & \vdots \\ TB_{p-3}G & TB_{p-4}G & \cdots TB_0G & 0 & 0 & 0 \cdots 0 \\ TB_{p-2}G & TB_{p-3}G & \cdots TB_1G & TB_0G & 0 & 0 \cdots 0 \\ \mu_0 & \mu_1 & \cdots \mu_{p-3} & \mu_{p-2} & \mu_{p-1} & \mu_p \quad \mu_{np-p-1} \end{bmatrix} \tag{24}$$

where $\mu_i=e^{i\Lambda T}(TB_{p-1}G)$, $i=0,1,2,\ldots\ldots,np-p-1$

From the foregoing, the system of eq. (17a) is controllable provided that both TB_0G and μ_{p-1} are nonsingular. It can be shown that B_0 and B_{p-1} are both nonsingular (see section IV). Since G is full column rank and $T \neq 0$, it follows that Ξ_p is full column rank.

To implement the p-step MSRP integrator, it is required that eqs. (18a)-(20) be solved for the A_i's and B_i's. However, careful consideration of these equations reveals that the B_i coefficients can be readily obtained once the A_i coefficients have been determined. Therefore, the problem of interest at this point is to solve the above mentioned set of equations for the A_i's. Then the B_i's can be computed directly from the A_i's. In what follows, eqs. (18a)-(20) are solved for the A_i regression coefficients.

The technique used here to solve eqs. (18a)-(20) for the A_i's mimics the technique used earlier in the two-step method to solve eqs. (6a)-(7b) for the A_i's. This technique involves only the well known matrix column and row operations. The first step is to rewrite eqs. (18a)-(20) as a function of the A_i's only. This step results in a matrix equation of the same form as that of eq. (8a). The second step involves carrying out a series of elementary row operations on this matrix equation to obtain the upper triangular form given in eq. (8b). These two steps yield a linear system of matrix equations written here as:

$$M\alpha = R\Phi \tag{25a}$$

The parameters of eq. (25a) are as follows:

i) M is a $pn \times pn$ matrix, $M=[m_{ij} I]$, with I being the $n \times n$ identity matrix and the m_{ij}'s being given as:

$$m_{ij} = \begin{cases} 1, & i=1 \quad j=1,2,\ldots,p \\ 0, & i=2,3,\ldots,p \quad j=i-1,i-2,\ldots,1 \\ (j-i+1)m_{i-1\,j}, & i=2,3,\ldots,p \quad j=i,i+1,\ldots,p \end{cases} \tag{25b}$$

ii) α is a $pn \times n$ vector whose entries are the A_i's, viz.

$$\alpha' = \left[A_0',\ A_1',\ A_2',\ \ldots\ldots,\ A_{p-1}' \right] \tag{25c}$$

where α' is the transpose of α.

iii) R is a pn×pn matrix, $R=[r_{ij}I]$, with the r_{ij}'s given as:

$$r_{ij}=\begin{cases} 0, & i=2,3,\ldots,p \quad j=1,i+1,i+2,\ldots,p-1 \\ 1, & i=1,2,\ldots,p \quad j=i \\ (-1)^{i+j}\{(i-2)|r_{i-1j}| + |r_{i-1j-1}|\}, & i=3,4,\ldots,p \quad j=i-1,i-2,\ldots,2 \end{cases} \quad (25d)$$

iv) Φ is a pn×n vector whose entries are specified as:

$$\Phi'=\left[\hat{\Phi}_1', \ \hat{\Phi}_2', \ \hat{\Phi}_3', \ \ldots\ldots, \ \hat{\Phi}_p'\right] \quad (25e)$$

$$\hat{\Phi}_k=\Phi_k+(k-1)\hat{\Phi}_{k-1}(JT)^{-1} \quad , \quad k=1,2,\ldots,p$$

$$\Phi_k=(k-1)(p-1)^{k-2}e^{JT}(JT)^{-1}-p^{k-1}I$$

Notice that eq. (25a) still does not give a direct solution for the A_i's as the vector containing the A_i's is being pre-multiplied by M. Therefore, to determine the A_i's it is necessary to find the inverse of M. Although this might seem like a formidable task, it turns out that the upper triangular nature of M not only guarantees the existence of its inverse, but also it can be exploited to obtain a closed-form inverse of M. In what follows, an algorithm to determine this inverse, element by element, is given. This algorithm does not require the use of any special software packages. Rather, it utilizes only the usual arithmetic operations supported by most hardware.

The entries w_{ij} of W, the inverse of the matrix M in eq. (25a) above, are given by:

$$w_{ij}=\begin{cases} 0, & i=2,3,\ldots,p \quad j=i-1,i-2,\ldots,1 \\ (i!)^{-1}, & i=0,1,\ldots,p-1 \quad j=i \\ (-1)^{i+j}(j-i)^{-1}w_{i-1j}, & i=1,2,\ldots,p \quad j=i+1,i+2,\ldots,p \end{cases} \quad (25f)$$

Therefore, eq. (25a) can now be rewritten as:

$$\alpha=M^{-1}R\Phi=WR\Phi \quad (25g)$$

This completes the development of the matrix equation required to obtain the A_i's when the method step p is greater than or equal to 2.

As an illustrative example, consider the case when a seven-step method is to be used for the simulation; that is, p equals 7. Then, following the process outlined in eqs. (25a)-(25e) matrices M and R, and vectors α and Φ are found as indicated below:

$$\begin{bmatrix} I & I & I & I & I & I & I \\ 0 & I & 2I & 3I & 4I & 5I & 6I \\ 0 & 0 & 2I & 6I & 12I & 20I & 30I \\ 0 & 0 & 0 & 6I & 24I & 60I & 120I \\ 0 & 0 & 0 & 0 & 24I & 120I & 360I \\ 0 & 0 & 0 & 0 & 0 & 120I & 720I \\ 0 & 0 & 0 & 0 & 0 & 0 & 720I \end{bmatrix} \begin{bmatrix} A_0 \\ A_1 \\ A_2 \\ A_3 \\ A_4 \\ A_5 \\ A_6 \end{bmatrix} =$$

$$\begin{bmatrix} I & 0 & 0 & 0 & 0 & 0 & 0 \\ 0 & I & 0 & 0 & 0 & 0 & 0 \\ 0 & -I & I & 0 & 0 & 0 & 0 \\ 0 & 2I & -3I & I & 0 & 0 & 0 \\ 0 & -6I & 11I & -6I & I & 0 & 0 \\ 0 & 24I & -50I & 35I & -10I & I & 0 \\ 0 & -120I & 274I & -225I & 85I & -15I & I \end{bmatrix} \begin{bmatrix} \hat{\Phi}_1 \\ \Phi_2 \\ \Phi_3 \\ \Phi_4 \\ \Phi_5 \\ \Phi_6 \\ \Phi_7 \end{bmatrix} \qquad (26)$$

Similarly, from eq. (25f) W, the inverse of M, is found to be:

$$\begin{bmatrix} I & -I & 1/2I & -1/6I & 1/24I & -1/120I & 1/720I \\ 0 & I & -I & 1/2I & -1/6I & 1/24I & -1/120I \\ 0 & 0 & 1/2I & -1/2I & 1/4I & -1/12I & 1/48I \\ 0 & 0 & 0 & 1/6I & -1/6I & 1/12I & -1/36I \\ 0 & 0 & 0 & 0 & 1/24I & -1/24I & 1/48I \\ 0 & 0 & 0 & 0 & 0 & 1/120I & -1/120I \\ 0 & 0 & 0 & 0 & 0 & 0 & 1/720I \end{bmatrix} \qquad (27)$$

The results just given above can be easily ascertained by keeping careful record of the operations required to reduce the original matrix system (same structure as that of eq. (8a) for p=2) to an upper triangular form (equivalent to eq. (8b) for p=2).

As a more complete example, consider simulating the Brennan and Leake turbojet engine of [41] using a three-step MSRP method. Thus, letting p=3 in eq. (13), the integrator equation is:

$$x_{k+3}=-A_2 x_{k+2}-A_1 x_{k+1}-A_0 x_k+T(B_2 \dot{x}_{k+2}+B_1 \dot{x}_{k+1}+B_0 \dot{x}_k)$$

Clearly, when starting the simulation x_3 requires knowledge of x_2 and x_1. To find x_1, Euler's method or preferably a zero-order-hold can be used with x_0 as initial data. Once this computation has been completed, x_2 can be easily determined using MSRP2. Three-step MSRP can then be used to find x_{k+3}. The first two steps just outlined have already been carried out in the illustrative example of section II. Therefore, only those computations pertaining to three-step MSRP are included here.

To implement the three-step integrator it is necessary to compute the integrator coefficients, namely A_i and B_i : i=0,1,2. The A_i's can be obtained by solving eq. (25g). Replacing the numerical values of the A_i's in eq. (18b) allows the B_i's to be determined. For p=3, eq. (25g) is as follows:

$$\begin{bmatrix} A_0 \\ A_1 \\ A_2 \end{bmatrix} = \begin{bmatrix} I & -I & 1/2I \\ 0 & I & -I \\ 0 & 0 & 1/2I \end{bmatrix} \begin{bmatrix} I & 0 & 0 \\ 0 & I & 0 \\ 0 & -I & I \end{bmatrix} \begin{bmatrix} \hat{\Phi}_1 \\ \hat{\Phi}_2 \\ \hat{\Phi}_3 \end{bmatrix}$$

To obtain equivalent expressions for the A_i coefficients the matrix product indicated above can be carried out. This product yields three equations, one per regression coefficient. These equations, having only the A_i's as unknowns, are:

$$A_0=\hat{\Phi}_1-3/2\hat{\Phi}_2+1/2\hat{\Phi}_3$$

$$A_1=2\hat{\Phi}_2-\hat{\Phi}_3$$

$$A_2=-1/2\hat{\Phi}_2+1/2\hat{\Phi}_3$$

From eq. (25e) the $\hat{\Phi}_i$'s can be found as:

$$\hat{\Phi}_1=\Phi_1=-I$$

$$\hat{\Phi}_2=\Phi_2+\hat{\Phi}_1(JT)^{-1}=e^{JT}(JT)^{-1}-3I-(JT)^{-1}$$

$$\hat{\Phi}_3=\Phi_3+2\hat{\Phi}_2(JT)^{-1}=4e^{JT}(JT)^{-1}-9I+2e^{JT}(JT)^{-2}-6(JT)^{-1}-2(JT)^{-2}$$

Substituting these expressions for the $\hat{\Phi}_i$'s into those for the A_i's yields:

$$A_0=e^{JT}(JT)^{-2}-(JT)^{-2}+1/2e^{JT}(JT)^{-1}-3/2(JT)^{-1}-I$$

$$A_1=-2e^{JT}(JT)^{-2}+2(JT)^{-2}-2e^{JT}(JT)^{-1}+4(JT)^{-1}+3I$$

$$A_2=e^{JT}(JT)^{-2}-(JT)^{-2}+3/2e^{JT}(JT)^{-1}-5/2(JT)^{-1}-3I$$

As with MSRP2, three-step MSRP was implemented using a timestep T=0.02. The simulation results are shown in figs. 6-8, where three-step MSRP is compared with both AB-2 and MSRP2. Notice that for three-step MSRP the response shows some noise at the beginning. This is due to the fact that as the method step increases, the region of stability decreases; thereby, making the integration process more sensitive to changes in the parameters of J and to startup errors. For this timestep the integrator coefficients were determined using the above expressions for A_i and eq. (18) for B_i, as:

$$A_0=\begin{bmatrix}-0.5779 & 0.2357 & 0.2601\\ -0.2118 & -0.0798 & 0.3062\\ 0.0142 & -0.0052 & -0.0310\end{bmatrix} \qquad B_0=\begin{bmatrix}0.2246 & 0.0811 & 0.0831\\ -0.0731 & 0.3963 & 0.0970\\ 0.0047 & -0.0018 & 0.4071\end{bmatrix}$$

$$A_1=\begin{bmatrix}1.8132 & -0.7341 & -0.8222\\ 0.6593 & 0.2625 & -0.9699\\ -0.0445 & 0.0161 & 0.0990\end{bmatrix} \qquad B_1=\begin{bmatrix}-0.7001 & -0.2667 & -0.2753\\ 0.2404 & -1.2641 & -0.3215\\ -0.0156 & 0.0060 & -1.3017\end{bmatrix}$$

$$A_2=\begin{bmatrix}-2.2353 & 0.4983 & 0.5621\\ -0.4474 & -1.1827 & 0.6637\\ 0.0302 & -0.0109 & -1.0680\end{bmatrix} \qquad B_2=\begin{bmatrix}0.8180 & 0.4482 & 0.4942\\ -0.4028 & 1.7650 & 0.5819\\ 0.0270 & -0.0099 & 1.8576\end{bmatrix}$$

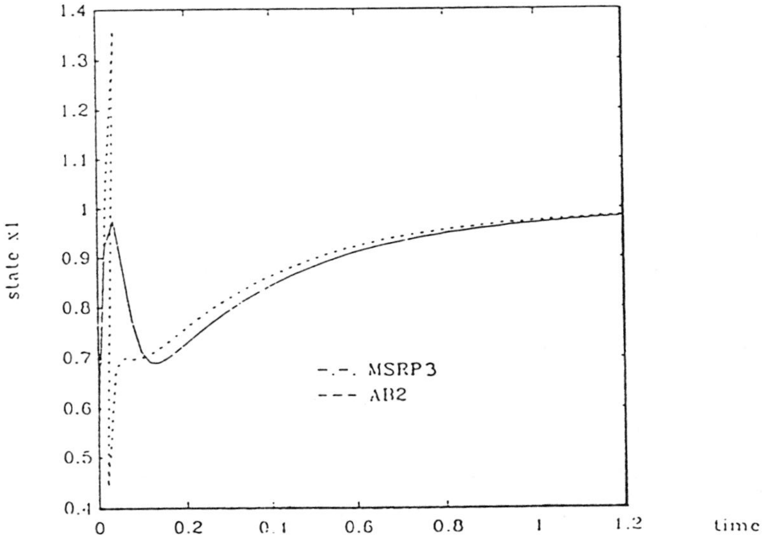

Figure 6. AB-2 and three-step MSRP simulation of x_1 of the Brennan and Leake turbojet engine model with T=0.02

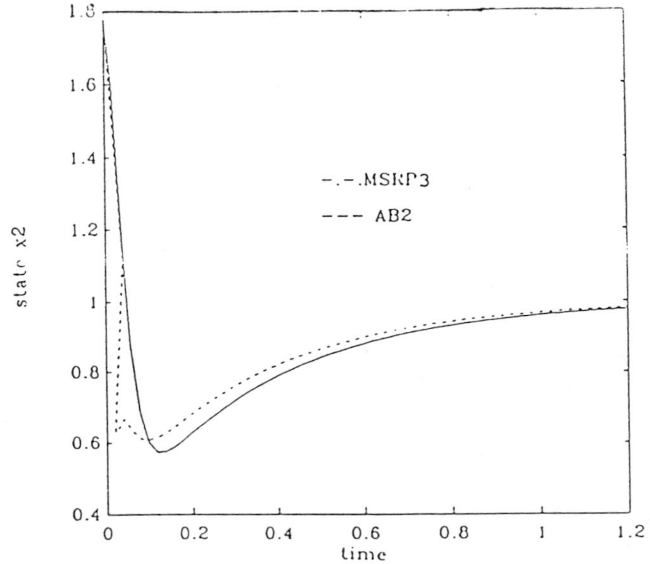

Figure 7. AB-2 and three-step MSRP simulation of x_2 of the Brennan and Leake turbojet engine model with T=0.02

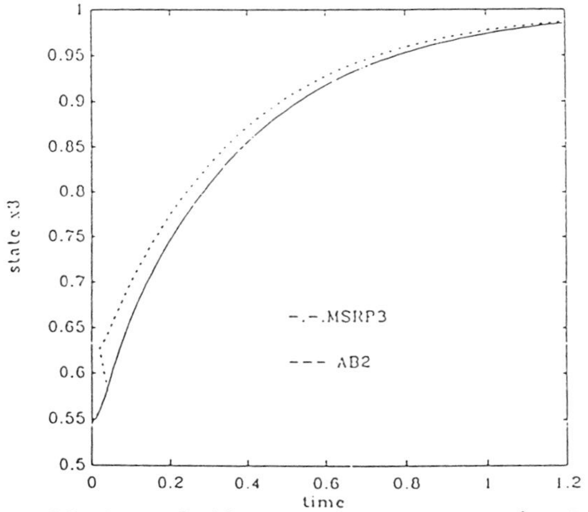

Figure 8. AB-2 and three-step MSRP simulation of x_3 of the Brennan and Leake turbojet engine model with T=0.02

This completes the development of the multistep MSRP method integrator for vector systems.

IV. MULTI-STEP MATRIX INTEGRATORS FOR LINEAR SYSTEMS: J SINGULAR

The development of MSRP just presented in sections II and III implicitly assumes that the inverse of the system Jacobian J exists. This assumption is evident in most of the equations encountered in these two sections (eqs. (9a-9d) and eq. (25e)). However, there may very well be situations in which the system matrix J has eigenvalues at the origin. This could particularly be the case in adaptive integration in which the system Jacobian evolves with time and thus its eigenvalues are constantly changing. In a situation such as this the results of sections II and III are inapplicable. Therefore, it is necessary to develop a method which removes the restrictions imposed by the nonexistence of the inverse of J on the integration process. It is shown in this section that it is always possible to use the matrix stability region placement technique of sections II and III even in situations where the system matrix J has multiple eigenvalues at the origin. This is done next in what follows for MSRP2.

Suppose first, that J is singular. Then eqs. (6a)-(7b) can no longer be solved for the A_i's and B_i's directly. However, consider post-multiplying eq. (7a) by JT, viz.

$$A_1JT-B_0JT-B_1JT=-2JT \tag{28}$$

Now eq. (6a) implies that $B_1JT=A_1+e^{JT}$ and eq. (6b) implies that $B_0JT=A_0$. Since the integrator coefficients A_i and B_i, the Jacobian J, and e^{JT} share the same eigenstructure [39], they commute with each other. Therefore, substituting B_1JT and B_0JT in eq. (28), and exploiting the commutativity of A_1 with JT ($A_1JT=JTA_1$) eq. (28) can be rewritten as:

$$[JT-I]A_1-A_0=e^{JT}-2JT \tag{29}$$

Equations (29) and (7b) form a system of two matrix equations in two unknowns; namely A_0 and A_1. This system of equations is:

$$\left[\begin{array}{c:c} I & I \\ \hdashline -I & JT-I \end{array}\right]\left[\begin{array}{c} A_0 \\ \hdashline A_1 \end{array}\right]=\left[\begin{array}{c} -I \\ \hdashline e^{JT}-2JT \end{array}\right] \tag{30}$$

Adding the first row to the second row in eq. (30) yields:

$$\left[\begin{array}{c:c} I & I \\ \hdashline 0 & JT \end{array}\right]\left[\begin{array}{c} A_0 \\ \hdashline A_1 \end{array}\right]=\left[\begin{array}{c} -I \\ \hdashline e^{JT}-2JT-I \end{array}\right] \tag{31}$$

Since J is considered singular, the inverse of the left hand side of eq. (31) does not exist. Therefore, this equation cannot be solved for the A_i's using the techniques that were used in sections II and III.

However, in what follows it is shown that it is still possible to solve eq. (31) for the A_i's without resorting to matrix inversion provided the system is first transformed into a suitable set of co-ordinates. Any transformations performed on the original system can always be done off-line and the results stored until the actual run. In this section only transformations that put the system into either a Jordan or a diagonal form will be considered. Recall that, generally speaking, a Jordan form results when the system has repeated eigenvalues, while a diagonal form results whenever the system has either distinct eigenvalues or the system Jacobian has a full set of eigenvectors. Although computing the Jordan canonical form of a matrix is usually not a numerically reliable process, in most practical applications repeated eigenvalues do not occur. If this is of concern to the design engineer, stable upper triangular decompositions such as upper real Schur or Hessenberg forms can always be used. Thus, there is a price to pay for the solvability of eq. (31) when the system has zero eigenvalues. A detailed account of the computational cost associated with this process is given in section VI.

The foregoing suggests that the problem of solving eq. (31) for the integrator coefficients when the system being integrated has eigenvalues at the origin can be resolved by transforming the original system into Jordan co-ordinates. Having done this, the integration process can be performed on the new system using one integrator per system state and then transforming the resulting state vector back to the original set of co-ordinates.

Recall that in Jordan co-ordinates J and e^{JT} are both block diagonal; that is,

$$J=\text{Diag}\left[J_1, J_2, \ldots, J_r\right] \qquad e^{JT}=\text{Diag}\left[e^{J_1 T}, e^{J_2 T}, \ldots, e^{J_r T}\right] \tag{33}$$

where the J_i blocks of J, and the $e^{J_i T}$ blocks of e^{JT}, for i=1,2,...,r have, respectively, the following structure:

$$J_i=\begin{bmatrix} \lambda_i & 1 & 0 & . & . & .0 & 0 \\ 0 & \lambda_i & 1 & . & . & .0 & 0 \\ 0 & 0 & \lambda_i & . & . & .0 & 0 \\ . & . & . & . & & . & . \\ . & . & . & & . & . & . \\ . & . & . & & & .. & . \\ 0 & 0 & 0 & . & . & .\lambda_i & 1 \\ 0 & 0 & 0 & . & . & .0 & \lambda_i \end{bmatrix} \qquad \begin{array}{l} J_i \in \mathbb{C}^{n_i \times n_i} \; : \; i=1,2,\ldots,r \\ n=\sum_{i=1}^{r} n_i \\ \quad =\text{order of minimal realization} \end{array} \tag{34}$$

$$e^{J_i T}=\begin{bmatrix} e^{T\lambda_i} & Te^{T\lambda_i} & T^2/2!e^{T\lambda_i} & . & . & .T^{n_i-1}/(n_i-1)!e^{T\lambda_i} \\ 0 & e^{T\lambda_i} & Te^{T\lambda_i} & . & . & .T^{n_i-2}/(n_i-2)!e^{T\lambda_i} \\ 0 & 0 & e^{T\lambda_i} & . & . & .T^{n_i-3}/(n_i-3)!e^{T\lambda_i} \\ . & . & . & & . & . \\ . & . & . & & . & . \\ . & . & . & & & .. \\ 0 & 0 & 0 & . & . & .Te^{T\lambda_i} \\ 0 & 0 & 0 & . & . & .e^{T\lambda_i} \end{bmatrix} \tag{35}$$

Now consider the second equation in eq. (31) involving A_1, viz.

$$JTA_1=e^{JT}-2JT-I \tag{36a}$$

Then, as J is in Jordan co-ordinates, and as e^{JT} is block diagonal with each block being upper triangular, it follows that both A_0 and A_1 are also block diagonal with each block being upper triangular. Moreover, careful consideration of eq. (36a) reveals that each block of A_1 is upper Toeplitz; that is, the blocks of A_1 are constant along the diagonals. This is easily seen once it is realized that the blocks of J and e^{JT} are also constant along the diagonals. Hence, A_1 can be determined by specifying the elements of either the top row or the right column of each one of its blocks. Thus, considering the equation associated with the last column of each block A_{1i} of A_1 in eq. (36a) gives:

$$JT\left[A_{1i}\right]_n=\left[e^{JT}\right]_n-\left[2JT\right]_n-\left[I\right]_n \quad : \ i=1,2,\ldots,r \tag{36b}$$

where the notation $[.]_n$ is used here to denote the nth column of the matrix inside the square bracket.

Alternatively, eq. (36b) can be written out in detail so as to illustrate the structure of the matrices involved. This gives:

$$\begin{bmatrix} T\lambda_i & T & \cdots & 0 & 0 \\ 0 & T\lambda_i & \cdots & 0 & 0 \\ 0 & 0 & \cdots & 0 & 0 \\ \vdots & \vdots & & \vdots & \vdots \\ 0 & 0 & \cdots & T\lambda_i & T \\ 0 & 0 & \cdots & 0 & T\lambda_i \end{bmatrix} \begin{bmatrix} a_{1\,n_i} \\ a_{1\,n_i-1} \\ a_{1\,n_i-2} \\ \vdots \\ a_{12} \\ a_{11} \end{bmatrix} = \begin{bmatrix} T^{n_i-1}/(n_i-1)!e^{T\lambda_i} \\ T^{n_i-2}/(n_i-2)!e^{T\lambda_i} \\ T^{n_i-3}/(n_i-3)!e^{T\lambda_i} \\ \vdots \\ Te^{T\lambda_i}-2T \\ e^{T\lambda_i}-2T\lambda_i-1 \end{bmatrix}$$

This matrix equation can be easily solved to obtain the following recursive expressions for the entries of each block A_{1i} of A_1; namely the scalars a_{1j} for $j=1,2,..,n_i$:

$$a_{1j}=\begin{cases} 1/T\lambda_i\left[e^{T\lambda_i}-2T\lambda_i-1\right], & j=1 \\ 1/T\lambda_i\left[Te^{T\lambda_i}-2T-Ta_{11}\right], & j=2 \\ 1/T\lambda_i\left[T^{j-1}/(j-1)!e^{T\lambda_i}-Ta_{1\,j-1}\right], & j=3,4,\ldots,n_i \end{cases} \tag{37a}$$

However, when $\lambda_i=0$, the entries of A_1 can be computed sequentially starting with a_{11} (from the second last row of eq. (36b)). Once a_{11} has been determined, a_{12} can be obtained by simply backsubstituting the value of a_{11} just found into the appropriate row of the above equation involving a_{12}.

In the same fashion, the remaining a_{ij}'s can also be determined once the value of the $a_{i\,j-1}$'s have been computed. Mathematically this process can be written as:

$$a_{1j} = \begin{cases} -1, & j=1 \\ T^{n_i-1-k}/(n_i-k)!, & j=n_i-k,\ k=1,2,\ldots,n_i-2 \\ \text{arbitrary}, & j=n_i \end{cases} \tag{37b}$$

The upper Toeplitz structure of the blocks A_{1i} of A_1 can be easily ascertained, for $\lambda_i=0$, from eq. (37b) by writing A_{1i} in matrix form:

$$A_{1i} = \begin{bmatrix} -1 & T/2! & T^2/3! \ldots & T^{n_i-2}/(n_i-1)! & a_{1n_i} \\ 0 & -1 & T/2! \ldots & T^{n_i-3}/(n_i-2)! & T^{n_i-2}/(n_i-1)! \\ 0 & 0 & -1 \quad \ldots & T^{n_i-4}/(n_i-3)! & T^{n_i-3}/(n_i-2)! \\ \vdots & \vdots & \vdots & \vdots & \vdots \\ 0 & 0 & 0 \quad \ldots & -1 & T/2! \\ 0 & 0 & 0 \quad \ldots & 0 & -1 \end{bmatrix} \tag{38}$$

with i running from 1 to r.

Having determined A_1, the coefficient A_0 can be found by substituting the value of A_1 just computed into the first equation in eq. (31); that is, $A_0 = -A_1 - I$. From this expression it is clear that A_0 has the same structure of A_1. Moreover, A_0 is only a function of A_1. Therefore, the entries a_{0j} of the blocks A_{0i} of A_0 can be computed directly from the a_{1j}'s for $\lambda_i \neq 0$ as:

$$a_{0j} = \begin{cases} -1/T\lambda_i \left[e^{T\lambda_i}-1\right]+1, & j=1 \\ -a_{1j}, & j=2,3,\ldots,n_i \end{cases} \tag{39a}$$

and for $\lambda_i=0$ as:

$$a_{0j}=\begin{cases} 0, & j=1 \\ -a_{1j}, & j=2,3,\ldots,n_i \end{cases} \tag{39b}$$

Thus, the blocks of A_0 when $\lambda_i=0$ can be specified as follows:

$$A_{0i}=\begin{bmatrix} 0 & -T/2! & -T^2/3! \ldots & -T^{n_i-2}/(n_i-1)! & -a_{1n_i} \\ 0 & 0 & -T/2! \ldots & -T^{n_i-3}/(n_i-2)! & -T^{n_i-2}/(n_i-1)! \\ 0 & 0 & 0 \quad \ldots & -T^{n_i-4}/(n_i-3)! & -T^{n_i-3}/(n_i-2)! \\ \vdots & \vdots & \vdots & \vdots & \vdots \\ 0 & 0 & 0 \quad \ldots & 0 & -T/2! \\ 0 & 0 & 0 \quad \ldots & 0 & 0 \end{bmatrix} \tag{40}$$

with i running from 1 to r.

From the foregoing it is seen that the blocks of A_1 and A_0 associated with the zero eigenvalues of the system have n_i repeated eigenvalues each located at -1 and 0, respectively.

Returning to the problem of interest, it still remains to determine the B_i coefficients. As pointed out earlier, the commutativity of JT with B_0 can be used in conjunction with eq. (6a) to give $JTB_0=A_0$. Thus, as B_0 is a function of only J and A_0, and as J and A_0 are both block upper triangular matrices with constant elements along the diagonal of each block, B_0 is also upper Toeplitz. Therefore, B_0 can be determined using the same technique that was used for determining the A_i coefficients. More specifically, B_0 can be found by solving the following equation:

$$\begin{bmatrix} T\lambda_i & T & 0 & \cdots & 0 & 0 \\ 0 & T\lambda_i & T & \cdots & 0 & 0 \\ 0 & 0 & T\lambda_i & \cdots & 0 & 0 \\ \vdots & \vdots & \vdots & & \vdots & \vdots \\ 0 & 0 & 0 & \cdots & T\lambda_i & T \\ 0 & 0 & 0 & \cdots & & T\lambda_i \end{bmatrix} \begin{bmatrix} b_{0n_i} \\ b_{0n_i-1} \\ b_{0n_i-2} \\ \vdots \\ b_{02} \\ b_{01} \end{bmatrix} = \begin{bmatrix} a_{0n_i} \\ a_{0n_i-1} \\ a_{0n_i-2} \\ \vdots \\ a_{02} \\ a_{01} \end{bmatrix} \tag{41}$$

with i running from 1 to r.

When $\lambda_i \neq 0$ the entries of the blocks of B_0 are found as:

$$b_{0j} = \begin{cases} a_{0j}/T\lambda_i, & j=1 \\ 1/T\lambda_i\,(a_{0j} - Tb_{0j-1}), & j=2,3,\ldots,n_i \end{cases} \tag{42a}$$

While when λ_i is a zero eigenvalue the b_{0j}'s are found as:

$$b_{0j} = \begin{cases} (a_{0j+1})/T, & j \neq n_i \\ \text{arbitrary}, & j=n_i \end{cases} \tag{42b}$$

Similarly, from eq. (7a), that is $B_1 = A_1 - B_0 + 2I$, the entries in the partitions of B_1 can be found as:

$$b_{1j} = \begin{cases} a_{1j} - b_{0j} + 2, & j=1 \\ a_{1j} - b_{0j}, & j=2,3,\ldots,n_i \end{cases} \tag{43}$$

Notice that the blocks of B_1 and B_0 associated with the zero eigenvalues of the system being integrated have their n_i repeated eigenvalues located at 1.5 and -0.5, respectively.

It is worthwhile noting that the diagonal form of MSRP2, when implemented on systems with only zero eigenvalues, reduces to AB-2 for scalar systems. This can be easily seen by comparing these two integrators. The two-step Adams-Bashforth integrator is:

$$x_{k+2}=x_{k+1}+T(1.5\dot{x}_{k+1}-0.5\dot{x}_k) \tag{44}$$

It follows that when this integrator is implemented in diagonal form, that is, one integrator per system state, the resulting matrices will be diagonal with eigenvalues located, respectively, at 1, 0, 1.5, and -0.5. By comparison, consideration of eq. (1) with the integrator coefficients in diagonal form yields the integrator of eq. (44) whenever the system eigenvalues lie at the origin.

To illustrate the utility of MSRP2 for the simulation of systems with eigenvalues at the origin consider the third order model of a DC motor [42]:

$$\dot{z}=Az+Bu, \quad y=Cz, \quad \text{and}$$

$$A=\begin{bmatrix} -1040.0 & -48500.0 & 0.0 \\ 1.0 & 0.0 & 0.0 \\ 0.0 & 1.0 & 0.0 \end{bmatrix} \qquad B=\begin{bmatrix} 1.0 \\ 0.0 \\ 0.0 \end{bmatrix}$$

$$C=\begin{bmatrix} 0.0 & 0.0 & 400000.0 \end{bmatrix}$$

The state variables correspond to armature current, rotor angular velocity, and rotor angular displacement, respectively. Transforming the system into Jordan form (in this case diagonal form) gives:

$$\dot{x}=Jx+Gu, \quad y=Hx, \quad \text{and}$$

$$J=\text{Diag}\begin{bmatrix} 0.0, -48.9374, -991.0626 \end{bmatrix}$$

$$G'=\begin{bmatrix} 0.00002 & 0.052 & 1.0519 \end{bmatrix} \quad H=\begin{bmatrix} 400000.0 & -166.9891 & 0.4072 \end{bmatrix}$$

Using eqs. (38), (40), (42b), and (43) and an

integration timestep of T=0.01 the integrator coefficients are computed as:

$$A_0 = \begin{bmatrix} 0.0 & 0.0 & 0.0 \\ 0.0 & 0.2092 & 0.0 \\ 0.0 & 0.0 & 0.8991 \end{bmatrix} \quad A_1 = \begin{bmatrix} -1.0 & 0.0 & 0.0 \\ 0.0 & -1.2092 & 0.0 \\ 0.0 & 0.0 & -1.8991 \end{bmatrix}$$

$$B_0 = \begin{bmatrix} \beta & 0.0 & 0.0 \\ 0.0 & -0.4275 & 0.0 \\ 0.0 & 0.0 & -0.0907 \end{bmatrix} \quad B_1 = \begin{bmatrix} 1-\beta & 0.0 & 0.0 \\ 0.0 & 1.2183 & 0.0 \\ 0.0 & 0.0 & 0.1916 \end{bmatrix}$$

Although β is an arbitrary parameter, its optimal value equals -.5 as this value yields an AB-2 integrator (for that particular state). The exact impulse response of the system and the impulse response using MSRP2 are shown in Fig. 9. Notice that an integration timestep of T=0.01 has been used to simulate this system with MSRP2. Interestingly enough, this timestep is ten times that of the largest timestep allowed by most of the explicit linear multistep methods available in the literature; that is, $T_{max}=1/\max(\lambda)\approx 1/991\approx 0.001$.

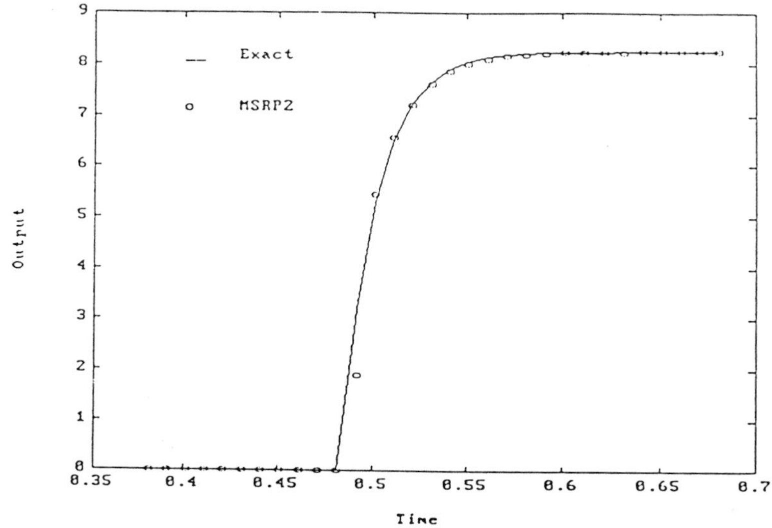

Figure 9. Impulse response of the DC motor

As a final remark, it should be realized that the development just presented in this section not only solves the problem of singular system matrix J, but it also gives the solution to the more general MSRP problem without explicitly computing matrix inverses. In addition, this development can be easily extended to the multistep case; that is, the case when the method step is higher than two.

V. MULTI-STEP MATRIX INTEGRATORS FOR NONLINEAR SYSTEMS

In this section the integration technique just developed for linear systems is extended to nonlinear systems. The main problem with extending this technique to nonlinear systems, however, is that the system matrix, J, is not readily available. Moreover, J is also effectively changing with both time and space. Therefore, some modifications must be made to the general linear theory to compensate for these problems. To this end, three different techniques for accomplishing the aforementioned extension are considered here, namely, single linearization, repeated linearization, and recursive identification of J, see references [3] and [14]-[15].

The first, and most straightforward, approach is to linearize the nonlinear system about its operating point by finding its Jacobian matrix, J, and its input Jacobian, G. Then these Jacobian matrices can be used in eqs. (9a)-(9d), (18b), and (25g) to determine the integrator A_i and B_i coefficients. Although this approach works reasonably well, it requires that G be modified to correct the steady-state error [43]. Another major drawback of this approach is that the accuracy of the resulting simulation decreases when the system moves away from its operating point. However, it is important to point out that preliminary investigations indicate that the closed-loop pole locations of the linear integration are not extremely sensitive to changes in J [14]. The main advantage of this design approach is that both transient response and closed-loop stability can be reasonably maintained for an arbitrary stable system using any timestep desired. The utility for real-time simulations is that it is possible to choose the timestep so that all the required computations can be performed in the interval.

The second approach is somewhat adaptive in nature and is similar to gain scheduling in a control system. In essence, this approach requires that the system Jacobians be computed analytically before programming the simulation. These Jacobians are then updated, or re-evaluated, at each point in space as the system evolves. Then at each timestep the integrator A_i and B_i coefficients are re-computed via eqs. (9a)-(9d), (18b), and (25g) based on the new values of J and G before the next integration step. From the foregoing, it is clear that this approach represents a tremendous computational burden to the simulation process, and perhaps to the user as well. As a result, this method is useful in real-time simulation provided that some kind of distributed processing is used to perform the required computations. To avoid steady-state errors, the derivative values from the actual nonlinear equations should be used. By doing this, the steady-state error is eliminated and the number of computations is reduced, since it is not needed to evaluate the input Jacobian matrix G. Furthermore, some problems with maintaining global stability are possible, especially for systems in which the eigenvalues are changing rapidly.

The third approach is completely adaptive integration. Here, the simulation package uses a recursive identification algorithm to directly estimate the system Jacobian and the input Jacobian, J and G respectively. The simulation package only has access to derivative evaluations at each timestep and provides as an output the updated system states. The only information that the simulation package needs is the order of the system, the desired timestep and initial guess of Jacobians J and G for the identification process. This would allow the system to be simulated without the user doing anything but providing an external derivative evaluation file. Then the algorithm tracks any changes in the Jacobians J and G of the system, and changes the integration coefficients accordingly in order to keep the simulation as accurate and as stable as possible. This idea is more clearly described, for any system, using a block diagram as illustrated in fig. 10 below:

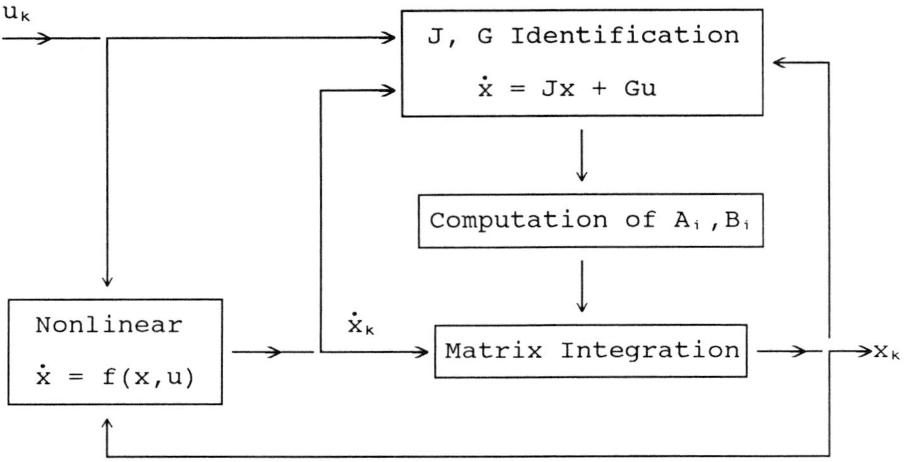

Figure 10. Block diagram of adaptive integration of nonlinear systems

The identification process plays a major role in the adaptive integration. The accuracy of the Jacobian estimation of the system has a direct effect upon the accuracy of the simulation. This is easily seen once it is realized that the integration coefficients are a function of the Jacobian J and the simulation stepsize. A good initial guess of the Jacobians J and G, and of the covariance matrix is somewhat important to speed up the convergence of the estimation which guarantees less noise at the beginning of the simulation prior to the algorithm learning the correct parameters. If the rate of convergence is fast, then on-line integration is possible and there is no need to wait for convergence of Jacobians before the integration is performed. However, this approach does impose a considerable computational burden and multiprocessing would probably be necessary for a real-time application.

To realize the full potential of on-line adaptive integration it is highly desirable that the identification algorithm be simple and easy to implement. Both the orthogonal projection algorithm and the recursive least squares algorithm with selective data weighting and covariance resetting are considered to have these two properties [44]. Moreover, these on-line algorithms are very attractive in practice because the recursive identification is such that the current parameter estimate, $\hat{\Theta}(k)$, is computed in terms of the previous estimates, $\hat{\Theta}(k-1)$. These on-line adaptive integration techniques are presented in what follows. First, consider the already familiar nonlinear system:

$$\dot{x}=f(x,u) \tag{45}$$

where x is the n-dimensional state vector, u is the m-dimensional input vector, and f is a nonlinear vector function of x and u.

Next, let the linear model for the nonlinear system of eq. (45), the parameter matrix, and the regression vector be defined by eqs. (46), (47), and (48), respectively, as:

$$\dot{x}_k = Jx_k + Gu_k \tag{46}$$

$$\Theta_k = \begin{bmatrix} J_k' \\ \cdots \\ G_k' \end{bmatrix} \tag{47}$$

$$\phi_k' = \begin{bmatrix} x_k' & \vdots & u_k' \end{bmatrix} \tag{48}$$

Then the estimated model can be defined as:

$$\hat{\dot{x}}_k' = \phi_k' \hat{\Theta}_k \tag{49}$$

The algorithm used here for adaptive integration

is the orthogonal projection algorithm [44] defined as:

$$\hat{\Theta}_K=\hat{\Theta}_{K-1}+\frac{P_{K-2}\phi_{K-1}}{c+\phi'_{K-1}P_{K-2}\phi_{K-1}}\left[\dot{x}'_K-\phi'_{K-1}\hat{\Theta}_{K-1}\right] \quad (50)$$

$$P_{K-1}=P_{K-2}-\frac{P_{K-2}\phi_{K-1}\phi'_{K-1}P_{K-2}}{c+\phi'_{K-1}P_{K-2}\phi_{K-1}} \quad (51)$$

In eqs. (50)-(51), $\hat{\Theta}_1$ is a given initial estimate, and $P_0=I$. The positive constant c, normally referred to as a forgetting factor, is introduced here to avoid the necessity of checking $\phi'_{K-1}P_{K-2}\phi_{K-1}$ for zero at each step of the algorithm. Effectively, this algorithm gives a way of sequentially solving a set of linear equations for the unknown matrix Θ [44].

When c in eqs. (50)-(51) is one the well-known least squares algorithm results. Of special interest here is a variant of this algorithm, namely the recursive least squares algorithm with selective data weighting and covariance resetting [44], defined as:

$$\hat{\Theta}_K=\hat{\Theta}_{K-1}+\frac{a_{K-1}P_{K-2}\phi_{K-1}}{1+a_{K-1}\phi'_{K-1}P_{K-2}\phi_{K-1}}\left[\dot{x}'_K-\phi'_{K-1}\hat{\Theta}_{K-1}\right] \quad (52)$$

$$P_{K-1}=P_{K-2}-\frac{a_{K-1}P_{K-2}\phi_{K-1}\phi'_{K-1}P_{K-2}}{1+a_{K-1}\phi'_{K-1}P_{K-2}\phi_{K-1}} \quad (53)$$

In eqs. (52)-(53), $\hat{\Theta}_1$ is a given initial estimate, and $P_{-1}=P_0>0$. P_0 is a measure of confidence in the initial estimate of $\hat{\Theta}_0$. If there are measurement errors in $\dot{x}_K$, then a_{K-1} might be chosen as the inverse of the expected mean square error E. This particular choice has the effect of attaching smaller weight to those terms in which $\dot{x}_K$ is expected to have larger errors. A numerically robust form of the algorithm can be obtained by incorporating all data points, weighted differently. This is easily accomplished by choosing the selection criterion as follows:

$$a_{K-1} = \begin{cases} k_1, & \phi'_{K-1} P_{K-2} \phi_{K-1} \geq E \\ k_2, & \phi'_{K-1} P_{K-2} \phi_{K-1} < E \end{cases} \qquad k_1 >> k_2 > 0 \tag{54}$$

As a numerical example, the nonlinear model of section II is considered again here. The orthogonal projection algorithm is applied to this system using a stepsize T=0.01. The initial guesses of J, G, and the initial value of P(-1) given below were used [14], viz.

$$J = \begin{bmatrix} 1 & 1 & 0 \\ 0 & 0 & 0 \\ 0 & 0 & 1 \end{bmatrix} \qquad G = \begin{bmatrix} 1 \\ 0 \\ 0 \end{bmatrix}$$

$P(-1) = P_0 = 100I$, where I is the $n \times n$ identity matrix.

The results obtained using the aforementioned on-line adaptive integration method are shown in fig. 11. Notice that these results are comparable with those of fig. 3. The Jacobians J and G converged to the following matrices:

$$J = \begin{bmatrix} -52.7214 & 23.4486 & 21.0545 \\ -47.2648 & -0.2340 & 46.9170 \\ 2.2266 & -0.9686 & -3.8325 \end{bmatrix} \qquad G = \begin{bmatrix} 8.2966 \\ 0.5421 \\ 2.5738 \end{bmatrix}$$

It is interesting to note that, although this system has eigenvalues at -27+j21.0381, -27-j21.0381, and -2.7879, the response obtained by simulating this system using a stepsize T = 0.01 is more accurate than that obtained by simulating the linearized version of the original system using the same stepsize, see figs. 12 and 4, respectively. Finally, the tracking property of this technique is illustrated in fig. 13 which shows the variation of the eigenvalues as the system evolves. Figure 14 shows the results of using on-line adaptive integration with stepsize T=0.02. For further details regarding this example see [14].

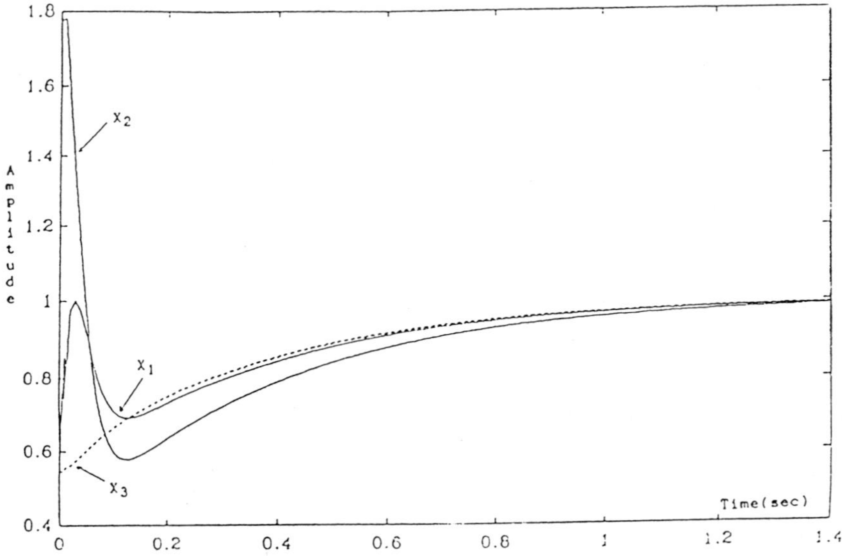

Figure 11. On-line adaptive integration of the Brennan and Leake turbojet engine model using MSRP2 with T=0.01

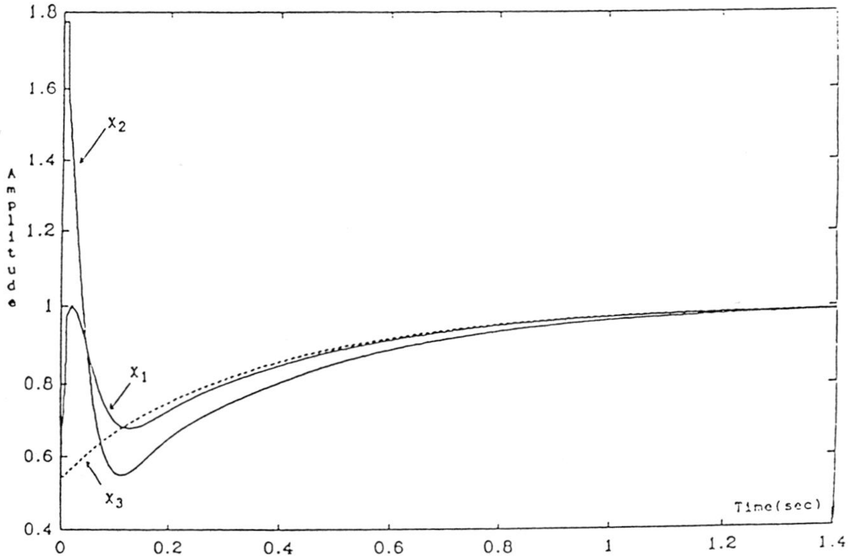

Figure 12. MSRP2 simulation of the Brennan and Leake turbojet engine model using converged linearized estimated model for T=0.01

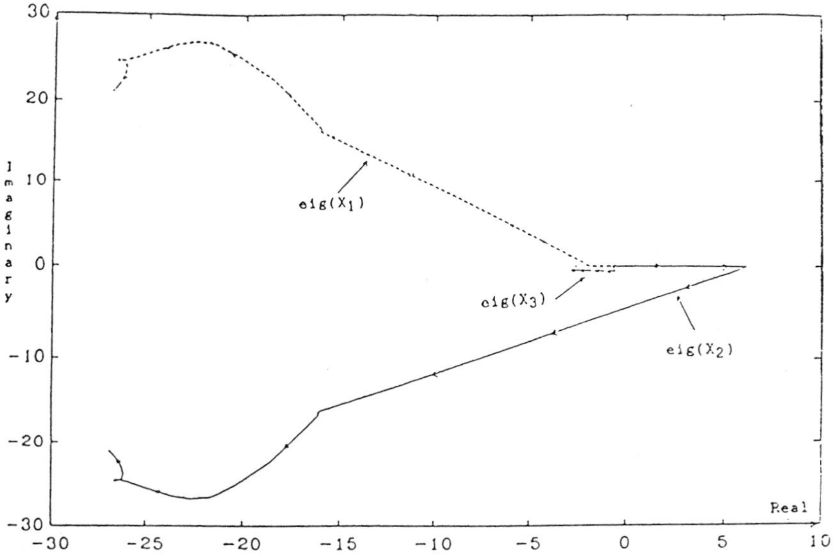

Figure 13. Plot of the eigenvalues of the estimated linearized model of the Brennan and Leake turbojet engine model with T=0.01

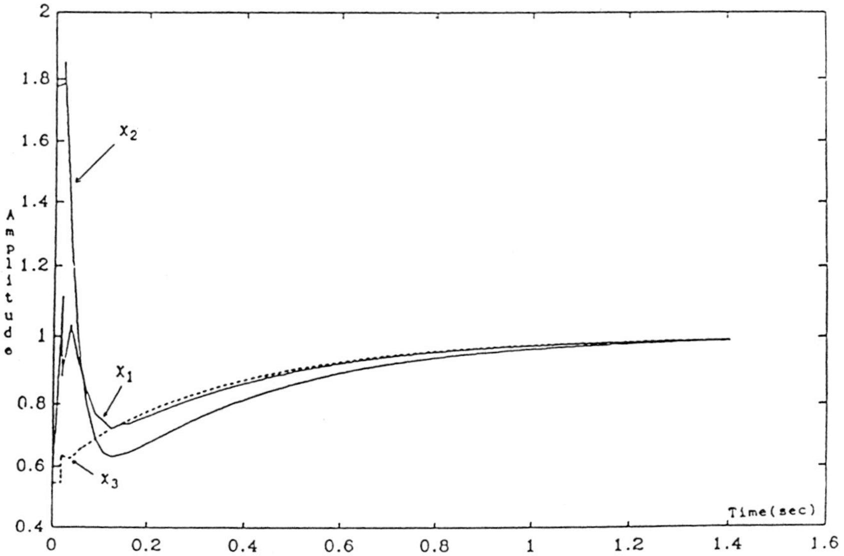

Figure 14. On-line adaptive integration of the Brennan and Leake turbojet engine model using MSRP2 with T=0.02

VI. COMPUTATIONAL ASPECTS OF MATRIX INTEGRATORS

A major disadvantage of the matrix integration technique proposed in the previous sections lies in the fact that the integrators are no longer localized, that is, one per system state. As a consequence, the computational burden associated with matrix integration increases rapidly with the number of states (approximately 4n-squared multiplies in addition to the derivative function evaluation). Therefore, it is necessary that techniques for reducing the number of computations be considered. Also, a detailed analysis of the computational burden associated with MSRP integrators is in order. These are the topics of this section.

Before proceeding further it should be mentioned that comparisons among integrators and numerical details of the different algorithms being used will be done using the number of floating point operations (flops) of each algorithm. A flop is approximately the amount of work involved in a floating point add, a floating point multiply, and the required subscripting [45]. In mathematical terms, this is equivalent to the amount of work associated with the following statement:

$$s := s + a_{ik} b_{kj} \tag{55}$$

which in terms of the well-known Fortran computer language can be written as:

```
S=S+A(I,K)*B(K,J)                                            (56)
```

A. TWO-STEP MSRP AND TWO-STEP ADAMS-BASHFORTH INTEGRATORS

In the usual situation, integration methods such as the Adams-Bashforth methods are used in real-time simulations. These methods then are the most appropriate candidates for establishing a comparison of the computational cost involved in MSRP. For the sake of simplicity, and without loss of generality, only the computational cost of AB-2 and MSRP2 is considered here.

Regarding the real-time simulation method considered here, the bulk of the computations results from products of the type Ax, where A can be either an $n\times n$ or an $n\times m$ matrix and x is either an n-dimensional or an m-dimensional vector, accordingly. It is not hard to see that when A is an $n\times n$ matrix, the product Ax requires n^2 flops. On the other hand, when A is $n\times m$ this product can be performed in nm flops. The amount of work associated with multiplying a given vector by a scalar quantity requires approximately n flops. At this time, it is important to point out that these flop counts are simply rough approximations that are used by computer theorists in an effort to acknowledge the countless operations that take place during program execution (paging, subscripting, etc.). The approximate number of flops required by AB-2 and MSRP2 for every timestep is given in table I below.

Table I. Computational cost associated with AB-2 and MSRP2

Required Operations (real-time)	Flops
$\dot{x}=Jx+Gu$	n^2+nm
AB-2 $x_{k+2}=x_{k+1}+T(1.5\dot{x}_{k+1}-0.5\dot{x}_k)$	2n
MSRP $x_{k+2}=-A_1x_{k+1}-A_0x_k+T(B_1\dot{x}_{k+1}+B_0\dot{x}_k)$	$4n^2$
Total # of flops for AB-2	$n^2+n(2+m)$
Total # of flops for MSRP	$5n^2+nm$

It is clear from this table that MSRP2 requires approximately 5 times more computations than AB-2.

Therefore, to break even, the MSRP2 timestep must be at least 5 times that of the AB-2 timestep (normal timestep). Equivalently,

$$\text{speedup} \approx \frac{T_{MSRP}}{5T_{NORMAL}} \tag{57}$$

Hence, MSRP2 can be made more numerically efficient than AB-2 provided that the timestep for MSRP2 is chosen appropriately. This is clearly illustrated in [15] where the nonlinear seventh order Brennan and Leake turbojet engine model of [41] was integrated using both AB-2 and MSRP2. This system is very stiff since its eigenvalues range from -47117 to $-1\pm j23$. Therefore, the largest possible value that can be used for most linear multistep methods is $T=-1/\max(\text{eig.})$; that is, T=0.000021. However, in [15] a sampling time T=0.005 was used for MSRP2. The resulting simulation compared well with the simulation results obtained using AB-2 with a sampling time T=0.00001.

B. IMPLEMENTATION OF TWO-STEP MSRP INTEGRATORS

In table I above the approximate flop count for MSRP2 is given assuming that the integration is performed using the original system. However, in an effort to reduce the number of computations in real-time, it is always possible to first apply a co-ordinate transformation to the original system. This transformation can be done off-line and the results stored. Then the integration process can be carried out on the resulting system. Of special interest here are transformations which yield diagonal or Jordan, Schur, and Hessenberg forms of the original system matrix J. The diagonal form gives essentially the classical linear multistep method, that is, one integrator per system state. It is important to keep in mind that, although this may seem appropriate, computing the diagonal form of a matrix is, in general, not a numerically reliable process. This is especially the case when the matrix being diagonalized has repeated eigenvalues [45]. In contrast, The well-known Schur and Hessenberg decompositions of a matrix are easily obtained via orthogonal transformations. Since orthogonal matrices are perfectly conditioned, these decompositions are considered to be very stable and numerically robust [45]. Furthermore, both Schur and Hessenberg decompositions of a matrix result in quasi-triangular forms. Thus, it is possible to reduce the number of on-line computations during the integration process by using these matrix decompositions. These approaches are considered next in what follows.

The first approach consists of transforming the initial system to either diagonal or Jordan co-ordinates, then performing the integration process. As mentioned in section IV, in general a diagonal form results when the system poles are all distinct, while a Jordan form results when the system has multiple poles. As the regression coefficients of the MSRP integrator are a function of $(JT)^{-1}$ and e^{JT}, and as these functions of J have a triangular structure whenever J is in Jordan form, and finally, as the Schur and Hessenberg forms of J are also triangular matrices, the case when J is in Jordan form is considered a part of the approach in which the A_i's and B_i's are triangular matrices.

The foregoing indicates that localized integrators can only be obtained provided J is diagonalizable. Recall that diagonalizing a matrix involves determining its eigenstructure. Further, recall that J is, in general, an unsymmetric matrix. In the usual situation, the standard procedure to diagonalize an unsymmetric matrix involves three steps. First, the matrix is reduced to upper Hessenberg form using Householder transformations. Then the Q-R algorithm is used to produce the upper real Schur form of the matrix resulting from step one to yield $H=Q'JQ$ (Q' is the transpose of Q). These two steps require about $15n^3$ flops [45] (this includes the computation of both Q and H). Finally, to obtain the diagonal form of J, a block diagonalization method requiring approximately n^3 extra flops is used. Therefore, the diagonalization of J can be accomplished in about $16n^3$ flops.

Having determined the eigenstructure of J, a transformation matrix whose columns are the eigenvectors of J can be formed and its inverse computed off-line and stored. The latter computation is usually done via the singular value decomposition method, thus requiring about $7n^3$ flops [45]-[46]. Then this transformation matrix is used to transform the original system (2) to diagonal co-ordinates. This process is given in what follows, assuming that the original system is that of eq. (2); that is,

$$\dot{x}_k = Jx_k + Gu_k \tag{58}$$

First, let P be the matrix whose columns are the eigenvectors of J. Then it follows that $P^{-1}JP=\Lambda$, where Λ is a diagonal matrix with the eigenvalues of J along its main diagonal. Next, use P as a co-ordinate transformation matrix to obtain localized integrators, viz.

$$x_k = Pz_k \Rightarrow \begin{bmatrix} z_k = P^{-1}x_k \\ \dot{z}_k = P^{-1}\dot{x}_k \end{bmatrix} \tag{59}$$

Then carry out the integration process, using one integrator per system state, as follows:

$$z_{i\,k+2} = -a_{i1}z_{i\,k+1} - a_{i0}z_{ik} + T\left[b_{i1}\dot{z}_{i\,k+1} + b_{i0}\dot{z}_{ik}\right] \tag{60}$$

for i=1,2,3,...,n

Finally, use P to determine the state vector, x, in the original set of co-ordinates by transforming the state vector z obtained from the integration process, viz.

$$x_{k+2} = Pz_{k+2} \tag{61}$$

This completes the first approach of diagonalization and integration. It is worth mentioning at this point that determining the a_i's and b_i's requires approximately 4n flops. However, these computations are done prior to the actual run and the results are stored. Therefore, the total number of off-line computations is $23n^3+4n$ flops. Also, notice that, when J is diagonal, the total number of flops required per timestep is $4n^2+(4+m)n$. Of this total, the function evaluations of eq. (58) constitute the bulk of the computations, that is, n^2+nm flops. The co-ordinate transformations of eq. (59) require n^2 flops each for z and its derivative. Finally, the computations of eq. (60) represent a total of 4n flops, while the transformation of eq. (61) requires n^2 extra flops.

The second approach, or triangular approach, consists on reducing the original system matrix J to an upper quasi-triangular form. There are several methods to do this. One of these methods involves using a sequence of Householder transformations to reduce the system matrix J to its upper Hessenberg form. Equivalently, determine U and H such that $H=U'JU$, where U (orthogonal) is a product of Householder matrices and H is upper Hessenberg, that is, $h_{ij}=0$ whenever $i>j+1$. This process requires $(7/3)n^3$ flops [45]. A second method is to compute the Schur decomposition of J, that is, determine an orthogonal matrix U such that $H=U'JU$, where each H_{ii} is either a scalar or a 2×2 matrix having complex conjugate eigenvalues. This decomposition can be done using the Q-R algorithm in approximately $15n^3$ flops [45]. At this point it is important to mention that these two processes are very numerically stable.

Finally, as a third method, the Jordan reduction of J is considered. As mentioned before, the Jordan form of a matrix can be obtained in approximately $16n^3$ flops. However, it should be emphasized that this reduction is, in general, ill-conditioned. A common aspect among these three methods is that they all yield upper quasi-triangular integrator coefficients. Thus, the total number of on-line computations decreases as compared to the number of computations required for the unreduced system. However, there might be an increase in the total number of off-line computations. This is seen by considering the amount of work involved in determining the A_i's and B_i's. A detailed description of this process is given next as follows.

As just mentioned, the latter three methods being considered here for reducing the number of computations of the integration process all result in upper quasi-triangular matrices. Therefore, without loss of generality, the number of flops required to compute each integrator coefficient is given next assuming that these coefficients are strictly upper triangular matrices.

For the sake of simplicity, the expressions for the A_i's and B_i's for MSRP2 are repeated here as a function of H, the corresponding form of J, viz.

$$
\begin{aligned}
A_1 &= e^{HT}(HT)^{-1} - (HT)^{-1} - 2I \\
A_0 &= -e^{HT}(HT)^{-1} + (HT)^{-1} + I \\
B_1 &= e^{HT}(HT)^{-1} + A_1(HT)^{-1} \\
B_0 &= A_0(HT)^{-1}
\end{aligned}
\tag{62}
$$

1. **Computation of e^{HT}:** The matrix exponential is usually computed using Padé approximations. Although the number of flops for this algorithm is a function of $||H||_\infty$, a representative number is somewhere between $8n^3$ and $10n^3$ flops. However, since H is upper triangular, only about $6n^3$ flops are required [45].

2. **Computation of $(HT)^{-1}$:** This matrix inversion can be easily done via the singular value decomposition of H. From [46], this algorithm takes about $5n^3$ flops for an upper triangular matrix.

3. **Computation of matrix products:** All the required matrix products involve only upper triangular matrices. Therefore, these products require only $(1/6)n^3$ flops each [45].

From the foregoing it is seen that once that e^{HT} and $(HT)^{-1}$ have been determined and the results stored, the integrator coefficients can be computed in approximately $(1/2)n^3$ flops. Note that in this process the product $e^{HT}(HT)^{-1}$ is computed only once and then stored.

The approximate flop counts just given are only for those operations which can be done off-line. Therefore, it still remains to consider the number of on-line computations required per timestep. These computations include the function evaluations of eq. (58), the state vector co-ordinate transformations of eqs. (59)-(60), and the integration process of eq. (3) with the coefficients replaced by the appropriate matrices, that is, eq. (62). As in the case when J is diagonalizable, the function evaluations and co-ordinate transformations require $4n^2+nm$ flops. Each of the products in the integration process requires $(1/6)n^2$ flops, for a total of $(2/3)n^2$ flops. Therefore, when the integrator coefficients are upper triangular, the total number of on-line computations can be approximated to $(14/3)n^2+nm$ flops. Notice that this flop count compares very favorably with that of the case when J is diagonalized. At the same time, however, this flop count is very close to the one obtained when the original system is used in the integration. That is, $5n^2+nm$ flops. For ease of comparison, the results just given in the paragraphs above are compiled in tables II and III below. The quantities inside the round brackets in tables II and III above indicate the number of times a particular operation is done. From these tables it is clear that transforming the original system to upper Hessenberg form prior to performing the integration process requires the least number of off-line computations, while the number of on-line computations compares well with that of the case when the original J matrix is used. This, coupled with the fact that the Hessenberg decomposition of a matrix can be obtained via orthogonal transformations, makes this approach highly desirable. Moreover, the excellent numerical properties of this decomposition guarantees the reliability of the computations.

Table II Off-line computational cost associated with a two-step MSRP integration method

Required Operations	System Matrix J (Computational Cost in Flops)				
	Original	Diagonal	Jordan	Hessenberg	Schur
Co-ordinate X-formations	0	$16n^3$	$16n^3$	$7n^3/3$	$15n^3$
Matrix Inversion	$7n^3$	$7n^3$	(2) $12n^3$	$5n^3$	$5n^3$
Matrix Exponential	$10n^3$	0	$6n^3$	$6n^3$	$6n^3$
Matrix Products	$3n^3$	$4n$	$1n^3/2$	$1n^3/2$	$1n^3/2$
Off-line computations	$20n^3$	$23n^3+4n$	$69n^3/2$	$83n^3/2$	$53n^3/2$

Table III On-line computational cost associated with a two-step MSRP integration method

Required Operations	System Matrix J (Computational Cost in Flops)				
	Original	Diagonal	Jordan	Hessenberg	Schur
Function Evaluations	$nm+n^2$	$nm+n^2$	$nm+n^2$	$nm+n^2$	$nm+n^2$
Co-ordinate X-formations	0	(3) $3n^2$	(3) $3n^2$	(3) $3n^2$	(3) $3n^2$
Integration Process	$4n^2$	$4n$	$2n^2/3$	$2n^2/3$	$2n^2/3$
On-line computations	$nm+5n^2$	$4n+nm+4n^2$	$nm+14n^2/3$	$nm+14n^2/3$	$nm+14n^2/3$

Regarding tables II and III above, it is expedient

to mention that although diagonalizing J yields the least expensive integration algorithm (from the viewpoint of on-line computations), it should be kept in mind that, in general, not all matrices are diagonalizable and that this process can be numerically ill-conditioned.

C. IMPLEMENTATION OF ADAPTIVE INTEGRATION

The next problem to be considered is adaptive matrix integration. In adaptive MSRP2, and in adaptive MSRP in general, it is still necessary to do the function evaluations of eq. (58) and the integration process of the matrix version of eq. (60), per iteration. As determined before, these computations can be done in approximately $5n^2+nm$ flops. In addition to these computations, it is also necessary to consider the number of flops required in the Jacobian identification and the determination of the integrator coefficients. This is done in the following.

1. **Jacobian identification:** In general, this process involves solving eqs. (50)-(51) to find $\hat{\Theta}_K$ and P_{K-1}, respectively. This requires the following matrix products:

i) $P_{K-2}\phi_{K-1} \Rightarrow [(n+m)\times(n+m)]\times[(n+m)\times 1]$
$\Rightarrow (n+m)^2 = n^2+2nm+m^2$ flops

ii) $\phi'_{K-1}P_{K-2}\phi_{K-1} \Rightarrow [1\times(n+m)]\times[(n+m)\times 1] \Rightarrow n+m$ flops

iii) $\phi'_{K-1}\hat{\Theta}_{K-1} \Rightarrow [1\times(n+m)]\times[(n+m)\times n]$
$\Rightarrow (n+m)n = n^2+nm$ flops

iv) $P_{K-2}\phi_{K-1}\phi'_{K-1}P_{K-2} \Rightarrow [(n+m)\times 1]\times[1\times(n+m)]$
$\Rightarrow (n+m)^2 = n^2+2nm+m^2$ flops

v) $P_{K-2}\phi_{K-1}\left[\dot{x}'_K-\phi'_{K-1}\hat{\Theta}_{K-1}\right] \Rightarrow [(n+m)\times 1]\times[1\times n]$
$\Rightarrow (n+m)n = n^2+nm$ flops

The foregoing reveals that the Jacobian identification requires approximately $4n^2+2m^2+6nm+n+m$

flops per iteration.

Notice that in the above multiplications the order in which the products are carried out is important to avoid operations of order n^3 and higher. This can be seen by considering the product in iv) above. If this product is computed by first multiplying ϕ_{K-1} by its transpose, an $(n+m)\times(n+m)$ matrix would result. Thus, determining this product would then involve a triple matrix product requiring operations of order higher than n^2.

2. **Computation of the integration coefficients:** this involves solving eqs. (9a)-(9d) for the A_i's and B_i's, which in turn requires computing and updating both $(JT)^{-1}$ and e^{JT}. The initial computation of $(JT)^{-1}$ can be done in approximately $7n^3$ flops, while e^{JT} can be obtained in about $10n^3$ flops. To update the inverse of JT, Broyden's procedure can be used [47]. It could also be used as an alternative to the Jacobian identification above. This procedure is outlined next using the typical system definition, that is:

$$\dot{x}=f(x), \text{ with corresponding system Jacobian } J=\frac{\partial f(x)}{\partial x} \tag{63}$$

Then Broyden's updates for J and J^{-1} are defined as:

$$J_{i+1}=J_i-\frac{(J_i\Delta x_i-\Delta f_i)\Delta x_i'}{\Delta x_i'\Delta x_i} \tag{64}$$

$$J_{i+1}^{-1}=J_i^{-1}-\frac{(J_i^{-1}\Delta f_i-\Delta x_i)\Delta x_i' J_i^{-1}}{\Delta x_i' J_i^{-1}\Delta f_i} \tag{65}$$

where, $\Delta f_i \doteq f_{i+1}-f_i$, and $\Delta x_i \doteq x_{i+1}-x_i$

Clearly, updating J requires about $3n^2$ flops, while updating J^{-1} requires $4n^2$ flops.

Unfortunately, the authors are not aware of any methods for updating the matrix exponential (some work is currently being done in this area). Thus, the matrix exponential has to be computed at every iteration.

Once the above computations have been completed,

determining the integrator coefficients require only three matrix products, that is, $3n^2$ flops.

To summarize, the matrix integrator coefficients in adaptive matrix integration can be obtained in $17n^3+3n^2$ flops per iteration.

Associated with specific hardware and software implementations there are several aspects of MSRP that should be emphasized. As MSRP only involves adds and multiplies, and as most hardware and software packages support these two basic arithmetic operations, MSRP allows real-time simulation on a wide variety of computer systems and software packages. Therefore, the usual constraints associated with the real-time simulation of physical systems are no longer encountered when using MSRP integrators. It should be realized, however, that although MSRP allows timesteps much larger than would be normally possible, the hardware which is attached to the simulator may restrict the stepsize. Hence, the only constraining factors in real-time simulation using MSRP are those due to hardware constraints.

Finally, to implement MSRP, three steps are required. The first step consists of obtaining the computation time for the given system. The second step involves choosing the desired timestep T. In the final step the MSRP integrator is designed by solving for the integrator coefficients using the appropriate formulas.

VII. CONCLUSIONS

In this paper an explicit linear multistep matrix integration technique for vector systems of ordinary differential equations has been presented. This technique utilizes the stability region placement approach to allow the timestep to be chosen independently from the system eigenvalues. Closed-form solutions for the general p-step method and the case when the system matrix has zero eigenvalues have been provided. Both block diagram and state space representations of the resulting closed-loop system have been given. These representations have been used to show that the system mode shapes are preserved over the integration process. It has also been shown that this technique can still be applied to systems with eigenvalues at the origin, without the need for computing any matrix inversion. The applicability of this technique in the real-time simulation of stiff linear and nonlinear systems has been demonstrated. To this end, extensions of this approach to nonlinear systems utilizing three different techniques have been considered. The third of these techniques has been shown to be completely adaptive and capable of dealing with the spatially varying parameters of a given system. It has also been shown that the adaptive nature of this method allows its use for time-invariant as well as for time-varying systems. This algorithm also tracks the local eigenvalues of nonlinear systems as they evolve. For smaller stepsizes this method was found to give better results than the other two suggested methods. This was seen to be especially true when a good initial guess of the parameters is available. For a given situation, suggestions for the most appropriate implementation technique have been implicitly given by considering and comparing the number of computations required in each instance. Techniques for implementing adaptive integration and its associated numerical cost have also been included. Finally, several numerical examples have been given to illustrate the work being proposed here.

It should be remembered that the integration method being proposed here generalizes the standard zero-order-hold approach. The details of the generalization of higher order holds using this method will be the subject of another paper.

ACKNOWLEDGMENT

This work was partially supported by the Advanced Control Technology Branch of NASA Lewis Research Center under Grant NAG 3-778.

REFERENCES

1. J.D. Lambert, "Computational Methods in Ordinary Differential Equations," Wiley, NY, 1973.

2. T.T. Hartley and G.O. Beale, "Integration Operator Design for Real-Time Digital Simulation," IEEE Trans. Ind. Elect., Vol. IE-32, No. 4, pp. 393-398, Nov. 1985.

3. T.T. Hartley and G.O. Beale, "Matrix Integrators for Real-Time Simulation," Proc. of the Thirteenth Annual IEEE Industrial Electronics Society Conference (IECON'87), SPIE Vol. 853, Boston, MA, Nov. 1987.

4. G.O. Beale, "Optimal Aircraft Simulator Development by Adaptive Random Search Optimization," Ph.D. Dissertation, University of Virginia, May 1977.

5. G.O. Beale, "Optimal Digital Simulation of Aircraft via Random Search Technique," J. Guidance and Control, Vol. 1, pp. 237-241, July-Aug. 1978.

6. G.O. Beale and T.T. Hartley, "Stability Consideration:Numerical Methods and Control Theory Equivalences," IEEE Trans. Ind. Elec., Vol. IE-34, No. 2, pp. 180-187, May 1987.

7. G.O. Beale and T.T. Hartley, "Optimization of Discrete Transfer Functions for Real-Time Simulation," presented at IECON'84, Tokyo, Japan, Oct. 1984.

8. C.W. Gear, "Numerical Initial Value Problems in Ordinary Differential Equations," Prentice Hall, Englewood Cliffs, NJ,1971.

9. C.W. Gear, "The Automatic Integration of Stiff Ordinary Differential Equations," Proc. of IFIPS Conf., North Holland, Amsterdam, pp. 187-193, 1968.

10. T.T. Hartley, "Numerical Methods for Real Time Simulation," Area Paper, Vanderbilt University, Nov. 1983.

11. T.T. Hartley, "Parallel Methods for the Real Time Simulation of Stiff Nonlinear Systems," Ph.D. Dissertation, Vanderbilt University, Dec. 1984.

12. T.T. Hartley, G.O. Beale, and G. Cook, "Multirate Input Sampling for Real-Time Runge-Kutta Simulation," IEEE Trans. Ind. Elect., Vol. IE-34, No. 3, pp. 387-391, August 1987.

13. T.T. Hartley, "Applications of Control Theory to the Simulation of Physical Systems," Proc. of the Sixteenth Annual Pittsburgh Conference on Modeling and Simulation, April 1985.

14. A. Rahrooh and T.T. Hartley, "Adaptive Matrix Integration for Real-Time Simulation," IEEE Trans. Ind. Elect., Vol. IE-36, No. 1, pp. 18-24, February 1989.

15. A. Rahrooh, T.T. Hartley, and J.A. De Abreu-Garcia, "Applications of Adaptive Matrix Integration," Proc. of the Nineteenth Annual Pittsburgh Conf. on Modeling and Simulation, Pittsburgh, May 1988.

16. T.T. Hartley, "Heuristic Methods for Determining Simulation Stability," Proc. of the Thirty-first Midwest Symposium on Circuits and Systems, St. Louis, August 1988.

17. J.A. De Abreu-Garcia, T.T. Hartley, and G.O. Beale, "Multistep Matrix Integrators," Proc. of the Nineteenth Annual Pittsburgh Conference on Modeling and Simulation, Pittsburgh, May 1988.

18. S.P. Chicatelli and T.T. Hartley, "Software for Determining Accuracy and Stability of Simulation," Proc. of the Nineteenth Annual Pittsburgh Conference on Modeling and Simulation, Pittsburgh, May 1988.

19. W. Liniger and R.A. Willoughby, "Efficient Integration Methods for Stiff Systems of Ordinary Differential Equations," SIAM J. Numer. Anal., Vol. 7, pp. 47-66, Mar. 1970.

20. T.D. Bui, "Solving Stiff Differential Equations in the Simulation of Physical Systems," Simulation, pp. 37-46, Aug. 1980.

21. R.E. Crosbie and S. Javey, "Methods for Solving Stiff Differential Equations," Simulation, pp. 140-142, Apr. 1982.

22. Z. Vukic and G.O. Beale, "Adaptive Stability Region Placement Method," Vanderbilt University Electrical and Biomedical Engineering Technical Report EE-BME-85-04.

23. K. Yen, "Partitioning Methods for Numerical Integration of High Order Systems," Ph.D. Dissertation, Vanderbilt University, 1985.

24. T.T. Hartley, "Chaos and Noise in the Digital Simulation of Nonlinear Systems," IEEE Trans. on Circuits and Systems, (under review).

25. T.T. Hartley, ed., "Simulating Nonlinear Systems Using Large Timesteps: A case Study," University of Akron Technical Report EECS-87-01.

26. T.T. Hartley, "Chaos in the Simulation of Physical Systems," Proc. of the Twelveth Annual IEEE Ind. Elect. Society Conference, Milwaukee, WI, Sept. 1986.

27. T.T. Hartley and S.P. Androulakakis, "Chaos in the Simulation of Systems with Discontinuous Derivatives," Proc. of the 1986 IEEE Conference on Decision and Control, Athens, Greece, Dec. 1986.

28. A. Rahrooh and T.T. Hartley, "Chaos in the Simulation of Systems Using Large Timesteps," Proc. of the Thirtieth Annual Midwest Symposium on Circuits and Systems, Syracuse, NY, Aug. 1987.

29. A. Rahrooh and T.T. Hartley, "Anomalies in the Simulation of Systems with Discontinuous Derivatives," The University of Akron Technical Report EECS-86-02, Sept. 1986.

30. P. Henrici, "Discrete Variable Methods in Ordinary Differential Equations," Wiley, NY, 1964.

31. Keck, D. and Hartley, T.T., "A Comprehensive Study of Linear Two-Step Methods," Proc. of the Seventeenth Annual Pittsburgh Conference on Modeling and Simulation, April 1986.

32. J.D. Lambert and S.T. Sigurdsson, "Multistep Methods with Variable Matrix Coefficients," SIAM J. Num. Anal., Vol. 9, No. 4, pp. 715, Dec. 1972.

33. J.S. Rosko, "Digital Simulation of Physical Systems," Addison-Wesley, Reading, MA, 1972.

34. A.P. Sage and S.L. Smith, "Real-Time Digital Simulation for System Control," Proc. IEEE, Vol. 54, No. 12, Dec. 1966.

35. A.P. Sage, "A Technique for the Real-Time Digital Simulation of Nonlinear Control Processes," Proc. Region III IEEE Conference, April 1966.

36. D.J. Evans, ed., "Parallel Processing Systems," Cambridge Univ. Press, 1982.

37. S. Skelboe, "Time domain steady-state analysis of nonlinear electric systems," Proc. IEEE, Vol. 70, pp. 1210-1228, Oct. 1982.

38. PC-MATLAB, The Mathworks Inc., Sherborn, MA (617-653-1415).

39. M.C. Pease, "Methods of Matrix Algebra," Academic Press, New York, 1965.

40. R.G. Jacquot, "Modern Digital Control Systems," Marcel Dekker, NY, 1981.

41. T.C. Brennan and R.J. Leake, "Simplified Simulation Models for Control Studies of Turbojet Engines," NASA Technical Report EE-757, Nov. 1975.

42. B.C. Kuo, "Automatic Control Systems," Prentice-Hall, NJ, 1987.

43. C.J. Daniele and S.M. Krosel, "Generation of Linear Dynamic Models From a Digital Nonlinear Simulation," NASA Technical Paper 1388, Feb. 1979.

44. G.C. Goodwin and K.S. Sin, "Adaptive Filtering Prediction and Control," Prentice-Hall, NJ, 1984.

45. G.H. Golub and C.F. Van Loan, "Matrix Computations," Johns Hopkins Univ. Press, MD, 1983.

46. C.L. Lawson and R.J. Hanson, "Solving Least Squares Problems," Prentice-Hall Inc., Englewood Cliffs, NJ, 1974.

47. P. Rabinowitz, "Numerical Methods for Nonlinear Algebraic Equations," Gordon and Breach Science Publishers, London, 1970.

THE ROLE OF IMAGE INTERPRETATION IN TRACKING AND GUIDANCE

D.D. Sworder

Department of AMES
University of California, San Diego
La Jolla, CA 92093

P.F. Singer

Hughes Aircraft Co.
El Segundo, CA 90245

I. INTRODUCTION

Design of high performance tracking and guidance systems requires a careful blending of techniques from modern control theory and signal processing. While generalized studies of these systems have a long history, it is only recently that sophisticated sensors and data processors have become available, thus permitting the actual implementation of "smart" trackers. Both control and estimation performance have been enhanced through these expedients, but the most pronounced changes have occurred in signal interpretation. This chapter focuses on changes in tracker architecture which exploit the capabilities of a new generation of sensors and processors.

To understand the revolution in tracking algorithms, some perspective on traditional design practice is useful. Conventional electro-optical (EO) sensors; e.g., radar, provide information on center-of-reflection states of a target. Such sensors treat the target as a point in space, and do not resolve features of an object that is actually distributed over some spatial neighborhood. The associated target motion models are compatible with this phrasing of sensor data. Because the measured target has a local structure, the target state is phrased in terms of point-motion attributes; e.g., position, velocity, etc. There are different ways in which the motion model can be delineated. In what follows the analytical models will be phrased in terms of stochastic differential equations as described by Elliott in [1]. A common parametric model is a Gauss-Markov process:

$$dx_t = Fx_t dt + Gdw_t \tag{1}$$

where $\{x_t\}$ is the target state, and $\{w_t\}$ is a Brownian motion process selected to introduce unpredictability into the target's path. Frequently Eq. (1) is written so that the (formal) derivative of $\{w_t\}$ is target acceleration. In this case, the equation describes a point acted upon by a wide-band (white noise), exogenous process. If the observation $\{z_t\}$ is given by

$$dz_t = Hx_t dt + dn_t \tag{2}$$

where $\{n_t\}$ is an independent Brownian motion, it is well known that the conditional mean of x_t is generated by a Kalman filter tuned to the dynamic hypotheses quantified by Eqs. (1)-(2). The error variance, which is computed as an intermediate step in the evaluation of the filter gains, provides a measure of tracking accuracy.

The motion model Eq. (1) displays the incremental change in the target state as the sum of a deterministic drift ($Fx_t dt$), and a random perturbation (Gdw_t). This form is equivalent to the conventional ordinary differential equation model with a stochastic forcing term; e.g., Eq. (1) could be written

$$(d/dt)x_t = Fx_t + G(d/dt)w_t$$

Despite their less orthodox appearance, the incremental forms as given in Eq. (1) will be the basis for the analytical development in this chapter. This choice is made to permit a unified treatment of both continuous and discontinuous forcing terms. Although discontinuous processes may be supposed to have (formal) derivatives in certain circumstances, the manipulation of these pseudo-derivatives can be difficult to justify without a consistent calculus. Such a calculus does exist for stochastic differential equations, and for this reason the indicated notation will be used.

Equations (1)-(2) illustrate the interplay of the sensor structure---noisy measurements of the point-location target states---and a motion model which gives the equation of evolution of these measured states. Upon these equations are based algorithms for estimation and prediction of target location. These derivative processes are in turn inputs to guidance and control loops. System performance is strongly dependent upon correct appraisal of target motion from sensor data. In the Kalman filter-based approaches alluded to above, tracker performance can be expressed in terms of quantities computed in the design process. Such linear-Gaussian estimation structures are clearly discussed in textbooks like that of Maybeck.[2]

Equation (2) delineates the character of a sensor-preprocessor. The process $\{z_t\}$ is a pseudo-observation insofar as it is a processed form of the raw sensor signal. Although some control can be exercised over this initial processing, this issue will not be explored here. Rather the sensor-processor will be dealt with as an undifferentiated unit which receives and partially interprets relevant electro-optical signals. The observation is the sum of a linear function of the target state $\{Hx_t\}$, and an exogenous term $\{n_t\}$ which represents transmission and clutter disturbances. In the common circumstance in which the sensor measures such quantities as range and bearing, the observation is not a linear function of the target motion state $\{x_t\}$, and the estimation algorithm must be suitably modified. An oft used procedure linearizes the measurement about the estimated target state and an extended Kalman filter (EKF) results. Issues arising in a nonlinear environment will be discussed in the sequel, but no distinction will be made between linear and nonlinear operation in this section.

While the above class of estimators is simply implemented, performance is frequently not acceptable during volatile encounters. It has been noted that "tracking (of an agile target) could be maintained only for those maneuvers remaining within the envelope of the white noise; even then, performance was poor due to erroneous assumption of uncorrelated acceleration."[3] To lessen the observed performance inadequacies, investigators have modified the white-noise acceleration model with the intent of more closely portraying the observed motion of representative vehicles. For example, Singer [4] proposed a correlated Gaussian acceleration process. This choice provides more flexibility in describing target motions while at the same time preserving the fundamental simplicity of the Kalman filter-predictor, albeit with an increase in state dimension. Unfortunately, some artificial acceleration parameters are introduced thereby.

The Linear-Gauss-Markov estimation/prediction paradigm has proven so attractive that it has continued to be employed even when salient features of the encounter are clearly not represented in the model. One premise which underlies the indicated procedures is that the target accelerations are well modeled by a wide-band process, or a process derived therefrom by linear filtering. This hypothesis is, unfortunately, not compatible with reasonably detailed descriptions of target motion. When a target is under attack, it will maneuver so as to avoid being tracked or having its future location predicted. As pointed out by General R.D. Russ, USAF, "a pilot uses the combination of speed and maneuverability to create as much 'miss distance' as possible and defeat the effects of missiles or bullets."[5]

The dynamic assumptions inherent in Eq. (1) are plainly not well founded when that target is maneuvering. In this environment Moose, et al. propose to quantify the agility of the target by introducing a discontinuous random process $\{a_t\}$ to augment the acceleration.[6] This supplementary process produces a target trajectory which contains the "jinking" behavior actually observed. The introduction of a maneuver acceleration into Eq. (1) makes more difficult the design of effective algorithms for state estimation. As pointed out by Bolger, "the maneuver must first be detected. Second the Kalman filter state estimate is corrected for the previous maneuver. Third, after detection and correction, the Kalman filter parameters are correspondingly adjusted in anticipation of future target maneuvers."[7] When the measurements are restricted to point-location properties of the target, a significant time may elapse before this procedure is complete. A change in acceleration must be integrated twice before it is reflected in a change in position. The pervasive nature of exogenous disturbances precludes expeditious and accurate determination of even rate from position measurements. Hence, path changes are a delayed indicator of a maneuver, and the target may have begun another phase of its maneuver pattern before the filter coefficients have stabilized.[8] Indeed, if the maneuver detection is accomplished using a level detector and a trigger process derived from the tracking error residuals, a reset of the trigger process may inhibit the detection of the onset of another maneuver mode [9].

As previously indicated, the form of the motion model is strongly influenced by the structure of the sensor measurements; e.g., if the target is unresolved, the motion model describes the path of a point in space. In truth, a target has spatial extent, and new EO sensors more properly distinguish spatial inhomogeneities within their field of view. At appropriate ranges both the target and the clutter have relevant structure. In this circumstance, Eq. (2) is an incomplete description of tracking data. Though some have sought to overcome these difficulties by shaping the spectrum of the sensor noise, such ad hoc artifices have been only marginally successful. More fundamentally, if the peculiarities of an encounter are to be captured, the form of both motion and observation equations must be changed to more realistically represent the behavior---both actual and sensed---of an agile target.

The manner in which an EO imaging sensor changes the dynamic modeling paradigm is illustrated by the analytical design of a tracker incorporating a forward-looking infrared (FLIR) sensor. A FLIR generates a matrix of time-concurrent signals, one from each pixel. This multiplicity of measurements can be processed, and interpreted in different ways. The most conventional of these locates the center of the ostensible target pixels, and marks the target at this point. The rest of the tracking or guidance system can then be synthesized using traditional methods because the output of the FLIR processor has an orthodox form. In this implementation, the FLIR plays a nonimaging role since there is an immediate mapping of the output of the pixel array to a single point. In such situations, Eqs. (1)-(2) provide a natural model for the encounter even though an imaging sensor is used.

As an alternative to the many-to-one map used above, the dimension of Eq. (2) can be augmented to accommodate the parallel observation paths. This many-to-many map retains the measurement form, but the complexity of the filter is increased because of the increase in the dimension of the observation vector. Furthermore, the interdependence of the pixel-specific signals must be quantified and incorporated into the estimation algorithm. This approach will be explored in more detail in the next section. Suffice it to say that this sensor-processor architecture preserves all of the sensor signals at the preprocessor level, and does the data aggregation in the subsequent estimator/predictor.

At an intermediate level of data aggregation, the measurement cohorts can be collected into data packages---frames of data---which can be interpreted by an image processor. This sequential data consolidation is particularly appealing when used to infer nonlocal properties of the target. When the image generating capability of the sensor is used to identify these holofeatures of the target, the tracker is said to be image-based.[10],[11] The implications for system design of using an imager-processor that explicates the pixel-specific signals will be explored in detail in what follows.

While not motivated by these sensor-particular issues, some recent studies have proposed to enhance the performance of tracking systems by appending to the dynamic state in Eq. (1) holofeatures such as orientation of the target with respect to the sensor. The observation retained the generic form given in Eq. (2), but it was assumed that the behavior of the appended states is directly reflected in the sensory data vector. For an encounter in which the target acceleration is continuous, a problem of this type was proposed and discussed by Kendrick, Maybeck and Reid in [12]. Their work suggests a fusing of conventional measurements (radar), and the image data in an "angle aspect Kalman filter." These ideas were extended by Andrisani et al. in [13], [14] and [15]. These latter papers consider tracking and prediction of the path of a helicopter or fixed wing aircraft using measurements which include angular orientation. Although the acceleration model does not contain a maneuver term, it is shown by an example that direct measurement of orientation provides a significant aid in a maneuvering environment.

Since the holostates manifest themselves in properties of target extent, they are properly measured by an imaging sensor-explicator. It is important, therefore, to properly represent the distinctive attributes of an imaging sensor in the dynamic encounter model. Any estimation algorithm converts a temporal sequence of measurements of target attributes into an estimate of the current position and motion of the target. In the references of the previous paragraph, target orientation was an element in the augmented state vector, and it was assumed that this feature could be measured in wide-band noise by the sensor-explicator. The output of the image processor is not well described in this fashion. In a FLIR, frames of data are created and interpreted at a discrete set of time points. A statement or mark from a vocabulary determined by an a priori partitioning of the image feature space is generated after the analysis of each data frame. Thus, if the image processor proffers a statement of orientation, the relevant sensor-processor errors are discrete misclassifications rather than additive orientation perturbations.

To illustrate how an estimation algorithm must be modified when an imager is used, consider the generic problem of estimating the orientation of an agile target from a sequence of noisy images. In summary the procedure for generating an analogue to Eq. (2) can be described as follows. Suppose the set of all possible target orientations are separated into L orientation bins, and denote the event that the current target orientation is in the i'th bin by $r_t=i$. Then the process $\{r_t\}$ displays a discrete orientation trajectory for the encounter. Suppose a FLIR sensor-explicator is used to provide an orientation measurement. After analyzing a frame of data, the image processor produces an orientation mark. If the sensor-explicator block were error free, the event $r_t=i$ would map to the mark $u_i \in U=\{u_1,...u_L\}$. Unfortunately, an actual sensor-image processor is subject to error, and its action is better described by a discernability matrix P:

$$P_{ij} = \Pr[\text{image processor reports mark } u_i | r_t=j] \quad (3)$$

Ideally, P would be the L×L identity matrix, but more realistically, P should represent the fallibility of both the sensor and the image processor.

In contrast to Eq. (2), the image-based observation process $\{y_t\}$ is given by the accumulation of the observation marks, or equivalently by a vector point process $\{\sigma_t\}$ with

$$\sigma_i(t) = \text{number of observations with mark i on the interval } [0,t];\ i=1,...,N \quad (4)$$

The orientation dynamics are not well modeled by Eq. (1) or a direct analogue. It is shown in [16] that if $\{r_t\}$ is a Markov process with transition rate matrix Q;

$$Pr(r_{t+dt}=j|r_t=i) = \begin{cases} q_{ij}dt+o(dt) & i=j \\ 1+q_{ii}dt+o(dt) & i\neq j \end{cases} \tag{5}$$

and the marks are generated at exponentially distributed random times with mean rate λ marks/second, then the filter which generates the conditional probability distribution (L-vector) for the current orientation state; $\hat{\phi}_t = [Pr(r_t=i)|\boldsymbol{Y}_t]$ takes the form[1]

$$d\hat{\phi}_t = Q^t\hat{\phi}_t dt + (\mathrm{diag}(\hat{\phi}_t)-\hat{\phi}_t\hat{\phi}_t^t)\Lambda^t \mathrm{diag}(\hat{\lambda}_t^{-1})d\sigma_t \tag{6}$$

where $\{\mathbf{Y}_t\}$ is the filtration generated by $\{y_t\}$ and

$$\Lambda = \lambda P = [\lambda_i(j)]; \quad \hat{\lambda}_i = \Sigma\, \Lambda_{ij}\hat{\phi}_j$$

It has been determined that volatile target motions require a high frame rate if the orientation path is to be followed with high fidelity.

An imaging sensor provides information that a nonresolving sensor cannot. Utilization of such a sensor increases potential responsiveness in guidance and tracking. However, this potential is not exploited when the imager is inserted into a classical architecture designed to suit point-target sensors Such systems do not have the inherent flexibility to enable them to respond expeditiously to sudden changes in the encounter. Novel analytical description which are compatible with maneuvering motions must be employed to develop the next generation of tracking and guidance algorithms. In the next sections, approaches which avail themselves of this enhanced sensor capability will be explored.

II. THE POINT TARGET TRACKING PROBLEM

Before discussing the design issues that arise when the target maneuvers at a relatively close range, consider the conceptually simpler problem of tracking a non-maneuvering target at a long range. At an extended distance from the sensor, target features that depend upon physical extent cannot be resolved by the imaging optics, and the generic model of the point-locations states given in Eq. (1) is a good representation of target dynamics. At such ranges, the target subtends a fraction of a pixel on the focal plane of the imager. The intervening atmosphere attenuates the signal from the target while the sensor both spreads and adds noise to it. The result is a low signal-to-noise ratio at the receiver. To track a target in this environment, a sequence of distorted and noisy images must be converted into an accurate estimate of the position and velocity of the target. This section presents an approach to determining just how well this can be done.

---

[1] If A is any matrix, A^t is its transpose.

A. AN OBSERVATION MODEL FOR POINT TARGET TRACKING

The front end of a tracking system consists of an EO sensor-preprocessor and a state estimation algorithm. Over its field-of-view, the sensor measures the distribution of the power within a specific spectral band. From these intensity measurements, the state estimator infers the location and motion of the target. As indicated earlier, a convenient algorithm for use in this application is the extended Kalman filter (EKF). The filter coefficients are derivable from the motion model and the observation equation. Since a point target is being considered, the conventional model given in Eq. (1) is adequate:

$$dx_t = Fx_t dt + Gdw_t \tag{7}$$

with $\{w_t\}$ a Brownian motion with intensity R_w; i.e., $dw_t(dw_t)^t = R_w dt$. The forcing function $\{w_t\}$ represents the external forces which create unpredictable motion in the target. At long ranges, maneuvers are not of great import, and consideration of maneuvers will be deferred to the next section in which resolved-target tracking is investigated.

The imager generates a discrete sequence of observations, and it is convenient to express target motion in a compatible manner. To simplify the notation in this section, denote the value of an arbitrary process $\{y_t\}$ at time $t=t_i$ by y_i. Then Eq. (7) can be approximated by a difference equation

$$x_{i+1} = \Phi(i+1,i)x_i + \omega_i \tag{8}$$

where the transition matrix Φ is a matrix exponential

$$\Phi(i+1,i) = \exp(F(t_{i+1}-t_i))$$

and the exogenous term $\{\omega_i\}$ has the statistical parameters,

$$E\{\omega_i\} = 0; \quad E\{\omega_i \omega_j^t\} = R_\omega(i)\delta_{ij}$$

$$R_\omega(i) = \int_{t_i}^{t_{i+1}} \Phi(i+1,\tau)GR_w(\tau)G^t\Phi(i+1,\tau)^t d\tau$$

Equation (8) is the motion model which delineates the behavior of the target at the sequential observation times.

The measurement model is a particularly important element in system design because it quantifies the peculiar attributes of a specific sensor. Suppose that the sensor is a FLIR containing an N×M set of detectors arrayed uniformly in a rectangular grid. The pixels can be labeled lexicographically to produce the nonlinear, vector observation sequence

$$z_i = I_o h(x_i,i) + n_i \tag{9}$$

where $(z_i)_j$ is the output of the j'th detector at time t_i, I_o is the signal intensity, and $\{h(x,i), i=1,...\}$ is the normalized transfer characteristic of the detector at time t_i when the target is in state x. The noise process $\{n_i\}$ aggregates both detector and background disturbances, and will be assumed to be a white sequence.

To deduce an explicit expression for h, a block diagram of a typical EO sensor can be constructed. To good approximation, each element can be represented by a spatial transfer function. The overall input-output relation is called the optical transfer function (OTF) of the sensor. The OTF of an incoherent imaging system with diffraction limited optics and a circular aperture is given by the expression

$$G(\rho) = \begin{cases} (2/\pi)[\cos^{-1}(\rho/2\rho_o)-(\rho/(2\rho_o))(1-(\rho/2\rho_o)^2)^{0.5}]; & \rho\leq 2\rho_o \\ 0 \text{ otherwise} \end{cases} \quad (10)$$

with $\rho = (\omega_1{}^2+\omega_2{}^2)^{0.5}$, $\rho_o = \pi D/(\lambda f)$, and with D = aperture diameter, f = focal length for distant target, λ = wavelength of the imaged light, $\{\omega_1, \omega_2\}$ = angular spatial frequencies in the horizontal and vertical directions respectively.

The signal from a detector must be preprocessed before it is in a form that can be utilized in an estimation algorithm. While there is a variety of detectors in current use, a generic model which closely approximates detector response in its linear region can be deduced. Let $I(\zeta_1,\zeta_2)$ be the image irradiance across the focal plane. The detected signal at position (ζ_i^*,ζ_j^*) is given by

$$d(\zeta_i^*,\zeta_j^*) = \iint_{-\infty}^{\infty} r_{det}(\zeta_i^*-\zeta_1,\zeta_j^*-\zeta_2)I(\zeta_i,\zeta_j)d\zeta_1 d\zeta_2$$
$$= r_{det}(\zeta_1,\zeta_2) * I(\zeta_1,\zeta_2)$$

where * is two dimensional convolution, and r_{det} is the responsivity of the detector. The responsivity of a uniform detector is constant across its active area. For such a detector

$$r_{det}(\zeta_1,\zeta_2) = \begin{cases} r_o; & |\zeta_i|\leq \Delta_i/2, \ i=1,2 \\ 0 \text{ otherwise} \end{cases}$$

where Δ_1 and Δ_2 are the dimensions of the detector in the ζ_1 and ζ_2 directions respectively. The transfer function of the detector is the Fourier transform of r_{det}:

$$R_{det} = r_o \mathrm{Sinc}(\Delta_1\omega_1/2)\mathrm{Sinc}(\Delta_2\omega_2/2))^{-1}$$

The detectors are sampled periodically, and the resulting signal is amplified and electronically processed to convert it into a form compatible with the subsequent stages of processing. Usually, the signal from the detector is AC coupled to block the bias in the background irradiance, and the ensuing stages of amplification are bandlimited to attenuate noise outside the signal band. The aggregate effect of these stages of preprocessing can be well represented by a second order relation given by

$$F_{elec}(\omega_1,\omega_2) = ([1+2\rho_1(j\omega_1/\omega_{1n})+(j\omega_1/\omega_{1n})^2][1+2\rho_2(j\omega_2/\omega_{2n})+(j\omega_2/\omega_{2n})^2])^{-1}$$

The values of ρ_i, ω_{in}; i=1,2 can be obtained from the dominant poles of the electronics. The indicated transfer function allows processing to be different in the horizontal and vertical directions, a useful degree of flexibility in EO scanning sensors.

The overall transfer function of the sensor-processor is given by the product of the individual transfer functions given above. A horizontal slice of the two-dimensional OTF is closely approximated by a Gaussian transfer function:

$$H(\omega_1,\omega_2) = K(\pi\sigma_1\sigma_2)^{-.5} \exp\{-.5(\sigma_1^2\omega_1^2+\sigma_2^2\omega_2^2)\} \tag{11}$$

This simple functional form is useful because it closely approximates the OTF, and is expressed in terms of two easily interpreted parameters. The inverse transform of the OTF is called the point spread function (PSF). To the extent that the Gaussian curve (11) is a satisfactory description of the OTF, the PSF is also Gaussian;

$$h(x_i;\zeta_i,\zeta_2) = (\pi\sigma_1\sigma_2)^{-0.5} \exp\{-0.5[(\zeta_1-x_1)^2\sigma_1^{-2} + (\zeta_2-x_2)^2\sigma_2^{-2}]\} \tag{12}$$

Based upon the foregoing analysis, the sensor-processor is characterized by the pair of parameters, σ_i and σ_2. They determine the effective width of the PSF. As they become larger, the image will become more blurred, and for this reason they are known as the sensor blur parameters. The observation at point (ζ_1,ζ_2) can be written from (9)

$$z(\zeta_1,\zeta_2) = I_o(\pi\sigma_1\sigma_2)^{-0.5} \exp\{-0.5[(\zeta_1-x_1)^2\sigma_1^{-2} + (\zeta_2-x_2)^2\sigma_2^{-2}]\} + n$$

B. TRACKING PERFORMANCE OF AN EO SENSOR

To see how the above sensor-preprocessor description may be used in system design, consider an estimation algorithm developed by Maybeck and Mercier in [17]. In this reference an EKF is proposed which translates a vector sequence of pixel intensities into an estimate of the target state. Equation (9) gives the general form of the observation. It will be supposed that the sensor-background noise at each pixel is independent and identically distributed after the bias component is removed. In this event, $R_n = \sigma_n^2 \mathbf{I}$ where $\mathbf{I}$ is the identity matrix. Define the sensor signal-to-noise ratio (SNR) as I_o/σ_n. The effect of pixel spacing is implicit in (9). A convenient way of specifying the interpixel spacing is in terms of the unitless quantities $\{\eta_i;i=1,2\}$; the number of pixels per blur radius (σ_i) in the i'th direction.

To relate the sensor parameters to tracker performance, a simply deduced index of comparison is helpful. A commonly used measure of estimation accuracy is the track error covariance. As part of the EKF algorithm a pseudo-covariance matrix $\{P_i;i=1,...\}$ is evaluated. It should be noted, however, that $\{P_i\}$ is a random process dependent upon the observation sequence $\{z_i\}$, and thus, cannot be precomputed. Further, the linearization of the observation equation implicit in the EKF implies that $\{P_i\}$ is not the actual covariance.

It would be advantageous to relate $\{P_i\}$ to a bound on the covariance, but this is difficult to do. It has been shown that in the absence of process noise, $\{P_i\}$ is the Cramer-Rao lower bound (CRLB) on the error covariance.[18] Unfortunately, the process noise is an essential part of the target motion model, and can not be neglected. An alternative bound, presented in [19], is more useful in this application. In the present context---linear motion dynamics and nonlinear observation---the performance of the EKF is related to a linear-Gaussian system given by (7) with x_o a $N(0,\Sigma_o)$ random vector, and observation equation

$$\varsigma_i = H_i x_i + \upsilon_i \tag{13}$$

where the gain H_i satisfies

$$H_i^t H_i = I_o^2 E\{[(\partial/\partial x_i)h(x_i)^t][(\partial/\partial x_i)h(x_i)]\} \tag{14}$$

The model (7),(13) has a covariance $\{\mathbf{P}_i\}$ that is simply computed as the solution to a matrix differential equation. This covariance is related to the true covariance of the state estimate generated by the EKF $\{\Pi_i\}$ by the equation

$$\mathbf{P}_i \leq \Pi_i \text{ for all i a.s.}$$

Thus, the performance of the EKF is bounded below by that of an associated linear Kalman filter. The bounding function $\{\mathbf{P}_i\}$ is called the Bobrovsky-Zakai lower bound (BZLB) for the estimator.

To illustrate how these notions can be used in system design, consider a baseline sensor with:

$\eta_1 = \sigma_1 = 1.0, \qquad \eta_2 = \sigma_2 = 2.0, \quad \text{SNR} = 10, \qquad R_v = 0.01\mathbf{I}$

Since the BZLB provides a deterministic constraint on performance, if the actual error covariance is nearly that given by $\{\mathbf{P}_i\}$, improvements in performance can only be achieved by improving the sensor acuity. Suppose for example, that the track error covariance in the x_2 direction must be less that 0.707. The nominal BZLB curve can be used to tentatively select a value of $\sigma_2 = 2.0$ as necessary to meet the specification. Since the BZLB curve is only a bound, the value of σ_2 so determined will be optimistic. If a subsequent simulation study indicates the true vale of the error variance is that shown in the figure, the BZLB curve can be used as an interpolation function to determine the next iteration on the sensor parameter.

This section considers the performance of a FLIR sensor-preprocessor in estimating the motion of a point target following a random path. The detector array provides a set of temporally concurrent measurements of the target centroid. Though the observation is perforce nonlinear, the covariance of the tracking error of a recursive filter can be bounded below with a precomputable matrix. This bounding function can be used in an interpolation algorithm which relates the natural sensor parameters to the performance of an EKF. This approach helps to isolate sensor-specific limitations from algorithmically induced constraints.

III. TRACKERS OF MANEUVERING TARGETS

A maneuvering target is difficult to track, and for this reason evasive maneuvers are an ever present factor in a hostile encounter. Early work on tracking utilized a Gauss-Markov model (1) to create a random target path. More contemporary acceleration models contain both continuous and discontinuous components; the former derived from Brownian motion by causal linear filtering, and the latter, a finite state process that jumps at unpredictable times. This second component is frequently thought to result from a maneuver, and will be so labeled.

Incorporation of discontinuous accelerations into the estimation algorithm leads to unorthodox estimation problems. The target trajectory is not well described as a Gauss-Markov process, and although the Kalman filter paradigm can be modified to accommodate this on an ad hoc basis, performance is degraded.[3] A more robust class of algorithms, Multiple Model Adaptive Filters (MMF), have been proposed to lessen some of the shortcomings of the maneuver detection based approaches.[20] The MMF estimator consists of a set of parallel Kalman filters, and a variable weighting for combining their outputs. Each filter is consonant with a specific dynamic hypothesis---a distinct maneuver acceleration in this case---and the filter residuals are combined to compute the a posteriori likelihoods of the various hypotheses. For example, if there are K possible maneuver accelerations, the MMF contains K Kalman filters, each with the dimension of the target state, and each processing the observation sequence. The MMF generates an approximation to the conditional probabilities of the alternative target models from an analysis of the filter residuals. Even when the operative hypothesis changes with time, the MMF functions surprisingly well considering that variation in acceleration is not incorporated into the calculation of the likelihoods. To be effective, the MMF should be tightly tuned, and is subject to behavior not "yet completely understood from a theoretical point of view."[21]

A fundamental difficulty with including acceleration variability in the hypotheses leading to the MMF is that the set of possible dynamic events increases quite rapidly. The MMF can be adapted to this situation in a case by case manner, but because the acceleration dynamics are not explicitly displayed in the MMF, it is not evident how sensitive the resulting algorithm is to variations in the tempo of the encounter. An approach which incorporates system variability into the initial analytical description of the encounter is the interactive multiple model (IMM) filter.[22] In the tracking application described above, the IMM filter would also utilize K concurrent Kalman filters, but a stage of data fusion takes place at the input to each of the filters which partially compensates for the variability in the dynamic model.

In both of the previous estimators, the inputs are the center-of-reflection states of an unresolved target, and the motion model is a differential equation describing the evolution of these states. Despite the obvious difficulties, maneuvers can and have been identified from such motion data. There is an attractive alternative when the sensor suite contains an imaging sensor-explicator. An architecture blending observations of holostates with classical ones becomes possible. Because the measurements of the center-of-reflection states have a different character than do those which are associated with a spatially distributed feature, the proper fusion of data has implications regarding the choice of a motion model. One possible modeling paradigm is described in this section, with tracker synthesis and performance analysis deferred to the next section.

A. THE ENCOUNTER MODEL AND ACCELERATION ESTIMATORS

To illustrate the generic issues which arise in the synthesis of image-based trackers, consider estimating the motion of a tank following an evasive path. Figure 1 shows a visual image of a tank at different angular orientations with respect to the sensor. Such a vehicle may maneuver by selecting a sequence of radial accelerations while maintaining a nearly constant speed. This maneuver pattern creates a path that approximates a set of nearly circular arcs joined to form a continuous curve. The effect of such paths on the design of fire control systems is explored by Burke in [23].

Equation (6) gives an image interpretation algorithm which can be used in this application. Note that a radial acceleration manifests itself in changes in orientation, and is not assessable from a single frame of data. A turn to the right with a given acceleration is revealed by a rotation through the orientation bins with a particular sense and rate. Even with an imaging sensor, inferences regarding acceleration lag the underlying event.

Fig. 1. A Visual image of a tank at several aspect angles

To derive an effective algorithm for maneuver estimation, the detailed structure of the encounter must be made explicit. It will be assumed that the target is moving with constant speed and performing jinking motions within the field-of-view of an imaging sensor. In this section only the rotational dynamics of the target will be discussed. Let $(\Omega, \mathbf{A}, \mathbf{P})$ be a probability space, and let $\{\mathbf{A}_t\}$ be a filtration on this space. The target maneuvers by changing the value of its radial acceleration at random points in time. Following [6], suppose that the maneuver acceleration, $\{a_t\}$, takes on a value in a set E_a with K members. Denote the indicator of a_t by α_t; i.e., if $a_t=i$, then $\alpha_t=e_i$.[2] It will be assumed that the maneuver dynamics are well described by an $\mathbf{A}_t$-Markov process with transition rate matrix Q^o (see (5)).

The orientation dynamics of the target are contingent upon radial acceleration. One such acceleration might rotate the vehicle to the right at a nearly fixed rate, while another might cause the vehicle to rotate hardly at all. Even if the maneuver acceleration were known, the orientation trajectory would not be known a priori because of wide band terms influencing rotational dynamics. Let the indicator of quantized orientation be $\{\rho_t\}$. It will be assumed that a specific acceleration produces an orientation process well described by a corresponding $\mathbf{A}_t$-Markov process. If for example, the acceleration is the i'th ($\alpha_t=e_i$),

---

[2] Let e_i be the i'th unit vector in R^K, or more generally in any appropriate finite dimensional space. In what follows $\rho \in R^L$ and $\alpha \in R^K$. The dimension of the associated unit vector will be clear from the context.

then the transition rate matrix for $\{\rho_t\}$ is Q^i. The set of possible transition rate matrices $\{Q^i;i=1,...K\}$ can be derived from the sojourn times in each orientation bin, and the bin sequence associated with each acceleration in E_a. These matrices are indexed by the acceleration, but the sample functions are responsive to the effect of the wide band accelerations as well. When the target initiates another phase of the maneuver, this is represented by a corresponding switch in the Q-matrix which describes the rotation process.

The combined sensor-image processor block generates and interprets frames of data from a scene containing the target. As was the case in Section 1, the sensor-explicator will be dealt with as an undifferentiated unit delineated by the discernability matrix P, the frame rate λ, and the mark set $U=\{u_1,...u_L\}$. The basic observation process $\{\sigma_t\}$ is composed of these marks (see (4)), and $\{\mathbf{Y}_t\}$ is the filtration generated by $\{\sigma_t\}$ with the "^" notation representing conditional expectation with respect to $\{\mathbf{Y}_t\}$. Thus, $\{\hat{\alpha}_t\}$ is the vector of conditional probabilities of the various possible maneuver accelerations.

Before presenting the equations of this filter a brief digression is worthwhile. Suppose, that the target has three possible acceleration states, say $a_t \in \{0,1,-1\}$. First consider the case in which the target is not maneuvering ($a_t=0$, $\alpha_t=e_1$). In this mode of operation, the orientation process would be nearly constant with the only deviations from constancy produced by the wide band perturbations. If there were three distinguishable orientations (L=3), then the Q-matrix associated with $a_t=0$ might be:

$$Q^1 = \begin{bmatrix} -0.2 & 0.1 & 0.1 \\ 0.1 & -0.2 & 0.1 \\ 0.1 & 0.1 & -0.2 \end{bmatrix} \tag{15}$$

Equation (15) should be interpreted as follows. When not maneuvering, the mean sojourn time in a given orientation is five seconds (1/(0.2)). If a transition in orientation takes place, it is equally likely to move to either of the other orientations. If the frame rate of the imager were 10 frames/second, there would be approximately fifty frames of data in each orientation state. In the absence of misclassification of the received images, the sensor-processor would create fairly long strings in which the marks were constant.

When the target maneuvers, the rotational changes become more rapid. For example, the maneuver associated with $a_t=1$ ($\alpha_t=e_2$) might be described by

$$Q^2 = \begin{bmatrix} -2.0 & 1.9 & 0.1 \\ 0.1 & -2.0 & 1.9 \\ 1.9 & 0.1 & -2.0 \end{bmatrix} \tag{16}$$

In (16) the mean sojourn time in each orientation state is reduced to only 0.5 seconds (1/(2.0)), and the rotation has become more regular. If $\rho_t=e_1$, the probability that it will make a $1\rightarrow2$ transition next is 0.95. The likelihood of moving in the other direction in only 0.05. Thus, $\alpha=e_2$ is distinguished by the transitions $1\rightarrow2$, $2\rightarrow3$, $3\rightarrow1$. When these transitions are observed, acceleration mode 2 is implied. The mean sojourn time in a given orientation state is now only five frames.

If, when $\alpha=e_3$ the transition rate matrix is the transpose of (16), a jinking maneuver is indicated by a reversal in the sense of the rotation. The data upon which a maneuver estimate is made is the sequence of observed marks. So, if the observation sequence is

$$u_1\rightarrow u_1\rightarrow u_1\rightarrow\ldots\ldots$$

this would imply that $\alpha=e_1$. On the other hand, the observation sequence

$$u_1\rightarrow u_1\rightarrow\ldots\rightarrow u_2\rightarrow u_2\ldots\rightarrow u_3\rightarrow u_3\rightarrow\ldots$$

would be a typical sample when $\alpha=e_2$.

Unfortunately, the sensor-image processor errors complicate the interpretation of the observations. In the image string

$$u_1\rightarrow u_1\rightarrow u_9\rightarrow u_1\rightarrow u_1\ldots$$

the anomalous frame of data would be properly recognized as a misreading of the underlying event. When the encounter is volatile, the detection of erroneous observation marks becomes a more onerous task. The possibility of errors in image interpretation creates an irresoluteness in data explication. This ambivalence is a product of the acceleration dynamics (exogenous), the rotation dynamics (exogenous and endogenous) and the sensor-image processor acuity (endogenous). The algorithm which maps the mark string into an estimate of the maneuver acceleration must rationally weight these influences, and arrive at $\{\hat{\alpha}_t\}$.

To present the equations of evolution for $\{\hat{\alpha}_t\}$, it is convenient to place both the acceleration and orientation states in a common space. This can be accomplished by defining the modal indicator KL-vector ϕ as follows;

$$\phi = (\alpha_1\rho_1,\alpha_1\rho_2,\alpha_1\rho_3,\ldots,\alpha_K\rho_{L-1},\alpha_K\rho_L)^t \tag{17}$$
$$= \alpha \otimes \rho$$

where " $\otimes$ " denotes Kronecker product(see [24]).[3] The process $\{\phi_t\}$ is the complete modal (acceleration $\otimes$ orientation) trajectory of the target.

---

[3] If $A(p\otimes q)$ and $B(m\otimes n)$ are matrices, $A\otimes B$ is a $pm\times qn$ matrix given by $A\otimes B=[A_{ij}B]$.

Since ρ and α are unit vectors, both orientation and acceleration can be deduced from ϕ in a natural way. For example

$$\alpha_1 = (\alpha_1\rho_1+\alpha_1\rho_2+...+\alpha_1\rho_L)$$

More generally, let $\mathbf{1}_L$ be an L-vector with every element equal to one

$$\mathbf{1}_L = (1,...,1)^t$$

and $\mathbf{I}_k$ the K×K Identity matrix. Then

$$\alpha = (\mathbf{I}_k \otimes \mathbf{1}_L^{\ t})\phi \tag{18}$$

and

$$\rho = (\mathbf{1}_k^{\ t} \otimes \mathbf{I}_L)\phi \tag{19}$$

The equations for $\{\hat{\phi}_t\}$ are developed in [25]. They take the form of (6) in KL-dimensional space. Define

$$Q^* = \begin{bmatrix} Q^1 & 0 & 0.. & 0 \\ 0 & Q^2 & 0.. & 0 \\ & & .... & \\ 0 & 0 & 0.. & Q^K \end{bmatrix}$$

and

$$Q = (Q^{\circ}\otimes\mathbf{I}_L) + Q^*$$

The algorithm for modal estimation can be written

$$d\hat{\phi}_t = Q^t\hat{\phi}_t dt + K(\hat{\phi}_t)d\sigma_t \tag{20}$$

where the nonlinear gain $K(\hat{\phi}_t)$ is given by

$$K(\hat{\phi}_t) = (\mathrm{diag}(\hat{\phi}_t)-\hat{\phi}_t\hat{\phi}_t^{\ t})(\mathbf{1}_K^{\ t}\otimes\mathbf{I}_L)\Lambda^t\mathrm{diag}(\Lambda\hat{\rho}_t)_{-1} \tag{21}$$

The vector of maneuver probabilities is a simple function of $\hat{\phi}_t$:

$$\hat{\alpha}_t = (\mathbf{I}_K\otimes\mathbf{1}_L^{\ t})\hat{\phi}_t \tag{22}$$

From (20), it is evident that the filter has an endogenous drift characterized by Q, and a weighting on the observation process given by $K(\hat{\phi}_t)$. The drift term depends upon the encounter dynamics. The observation gain is contingent on the sensor processor quality and the current state estimate. Equation (20) can be directly integrated to find the conditional probabilities of $(\alpha \otimes \rho)$. It has the form given in (6), but the factor $\Lambda^t\mathrm{diag}(\Lambda\hat{\phi}_t)^{-1}$ is only of dimension L×L rather than the higher dimension of ϕ.

For tracking and prediction, interest focuses on the K-vector of maneuver probabilities $\{\hat{\alpha}_t\}$.[26] To gain some insight into how $\{\hat{\alpha}_t\}$ evolves, consider the drift term in (20). Direct calculation shows that

$$E\{d\hat{\alpha}_t|\mathbf{Y}_t\} = (Q^{\circ})^t\,\hat{\alpha}_t dt$$

In the absence of new data, the estimate of acceleration moves in accord with the dynamics of $\{\alpha_t\}$.

The gain associated with $d\hat{\alpha}_t$ requires aggregation of several terms. For example, consider the probability that the target is in maneuver state 1 ($\alpha_t=e_1$). Direct calculation yields

$$d\hat{\alpha}_1-((Q^o)^t\hat{\alpha}_t)_1dt = \Sigma_{i\varepsilon L}\,[\hat{\phi}_i\Sigma_{j\varepsilon L}\,((\lambda_j(i)-\hat{\lambda}_j)/\hat{\lambda}_j)d\sigma_j] \qquad (23)$$

where the index set {1,...,L} is denoted by **L**. This can be written in a more evocative manner as

$$d\hat{\alpha}_1 = ((Q^o)^t\hat{\alpha}_t)_1dt - \hat{\alpha}_1\mathbf{1}_L{}^td\sigma + \Sigma_{j\varepsilon L}d\sigma_j\,\Sigma_{i\varepsilon L}((\lambda_j(i))/\hat{\lambda}_j)\hat{\phi}_i \qquad (24)$$

The first two terms on the right hand side of (24) are simple functions of the K-dimensional process $\{\hat{\alpha}_t\}$. The third term is the aberrant one. To evaluate it, the j'th mark is accorded a weight involving the correlation of orientation and acceleration. This term can not be rendered a function of $\{\hat{\alpha}_t\}$ alone, and consequently, the maneuver estimator does not have an exact K-dimensional form.

B. AN EXAMPLE

To illustrate the behavior of $\{\hat{\alpha}_t\}$, suppose that the target is a tank moving in the (x,y)-plane as indicated in Figure 2. A fixed sensor located at the origin of the coordinate plane is used to monitor the angular motion of the target. It will be assumed that the sensor-explicator works with an angular grid with bin size 10°; i.e., L=36. The corresponding values of the ρ's are shown as the indices in Figure 3. Thus, if the tank is moving directly toward the sensor with $\theta=0°$, then $\rho=e_1$, while if it is moving to the right at an angle perpendicular to the line-of-sight ($\theta=0°$, $\beta=90°$), $\rho=e_{10}$, and so on.

Ideally, if the unquantized orientation $r^*\in$ bin i ($\rho=e_i$), the observation would so indicate; i.e., the observation mark received would be the i'th one ($\Delta y=u_i$). Unfortunately, the acuity of the sensor is limited, and misclassification errors occur. For example, there are broad spectrum random errors (E#1) in the scene stimuli and in the sensor elements that will cause an essentially uniform distribution of responses from the image processor, if $r^*\in$ bin i, and an error of this type occurred, the resulting observation would be uniformly classified over the full range of bin locations. In the example this is a small effect, and amounts to a 1% total error; i.e,. $\Pr\{\Delta y=u_j|\rho=e_i$ and E#1 occurs$\}=1/36$ for all i,j.

Another error is associated with the act of quantizing the orientation. The true orientation of the target is a continuous variable. When the image classifier locates it by the statement that say $\rho=e_i$, it implies that the observed image features are such as to put it in the i'th bin. The nearer the actual orientation is to the edge of the bin, however, the more likely it is to be incorrectly classified as a member of the adjoining bin. To

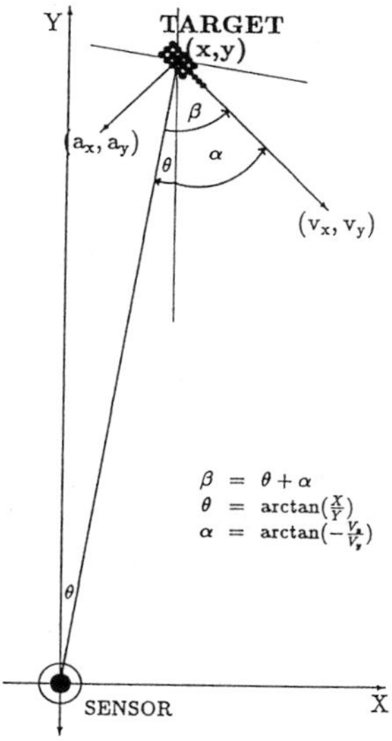

Fig. 2. Tracking diagram

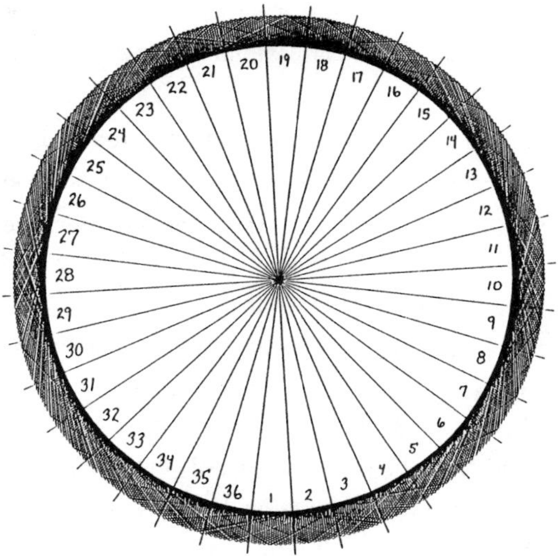

Fig. 3. Orientation bins when L=36i.e.,

capture this effect in a quantitative manner, suppose that r^* is the true orientation of the target when a frame of data is generated, and let I_i represent the angular interval of the i'th bin. Then the quantization effect is introduced by supposing that the image processor classifies a modified orientation r^c which is related to the true orientation by the equation $r^c = r^*+\nu$ where ν is a Gaussian random variable with mean 0 and standard deviation σ, and which is independently selected at each frame time. The image classifier then places the orientation in bin i if $r^c \in I_i$. If r^* is uniformly distributed over I_i and $\sigma=1°$, a direct calculation shows that this effect (E#2) will produce a symmetric misclassification with probability of 0.4/(bin size in degrees). Again, this is a small error source.

There is third source of classification error which is intimately related to an inherent limitation on the sensitivity of any image interpretation algorithm which emphasizes the features of the target silhouette. The target silhouette is symmetric about a line perpendicular to the line-of-sight. For example, if the line-of-sight angle is $\theta=0°$, then the outline of a target with heading $\beta=0°$ is the same as that at $\beta=180°$. In fact, the two headings $\beta=90°\pm\xi$ are indistinguishable from the edge data alone. In certain circumstances other properties of the image can be used to aid in making the angular inference. If visible, the internal features of the target may be useful in this regard, or perhaps a separate sensor may be used to reduce the ambiguity. In any event, it is reasonable to suppose that a significant residual projection ambiguity will remain. Define the projection error (E#3) as:

$$\text{Projection error} = \Pr(\Delta y=u_j|\rho=e_i \text{ and no E\#1 or E\#2})$$

when bins i and j are symmetrically placed about $\beta=90°$.[4]

The projection error is the major source of error considered in this example. It will be assumed that E#3 is 30%; i.e., the probability of erroneously classifying an image as its reflection about a line perpendicular to the line-of-sight is 0.3. This error engenders subtle difficulties in image interpretation. Consider the case in which $r^* \in$ bin 2 is mistaken for $r^* \in$ bin 18 on the basis of a single image; i.e., $\Delta y=u_{18}$ and $\rho=e_2$. If the sequence of images which preceded this observation had led the estimator to be confident that r^* was in a neighborhood of bin 1, the algorithm would hardly conclude that the target had suddenly rotated by nearly 180°. Rather, the measurement would add weight to the hypothesis that the target was near $\rho=e_2$. The estimator is "anchored" about $\rho=e_1$, and interprets the observed mark in this context.

If, however, the target had been located near in bin 10, then a move to bin 9 could easily be confused with a move to bin 11. Such mistaken inferences are quite natural because they are compatible the motion hypotheses which delineate the target trajectory. Furthermore, they become more common as the probability of E#3 increases. The influence of the projection ambiguity on target tracking when target rotation takes it to a neighborhood of $\beta=\pm 90°$ will become clearer in what follows. A composition of the three indicated sources of misclassification yields the total discernibility matrix P. This has been done here by assuming that the individual error sources are independent.

The assumed acceleration dynamics are elementary. The target has three distinct acceleration regimes (K=3). In regime one ($\alpha=e_1$) the radial acceleration is zero. This is the nonmaneuvering mode of the target. When the target maneuvers, it may make a turn to the left ($\alpha=e_2$) or to the right ($\alpha=e_3$). It will be assumed that the mean sojourn time in a specific acceleration regime is four seconds; i.e., the mean time between transitions in the target acceleration is 4. However, when the target is maneuvering, it will create a jinking path by switching between the accelerating modes. To quantify this suppose that the α:2→3 transition occurs with probability 0.9, and the α:2→1 transition occurs with probability 0.1. If the situation in $\alpha=e_3$ is symmetric, and the maneuver is equally likely to begin in either non-null acceleration state, the Q^o matrix can be generated directly.

The absence of radial acceleration is indicated by the constancy of the orientation marks. When the image-processor provides a mark that is not in accord with the current estimate of ρ, the possibility of an acceleration is suggested. The mode of acceleration

---

[4] For this example, it will be assumed that $\theta=0°$ throughout.

corresponds directly to the sense of the orientation change. Awareness of the error sources listed above induces a natural hesitancy in the estimation algorithm. It will not change its estimate of the target acceleration significantly until a change in the orientation rate is confirmed by several observations.

To infer acceleration from orientation, the dependence of the orientation dynamics upon $\{\alpha_t\}$ must be particularized. Suppose that if the target has a nonzero acceleration, the mean sojourn time in a given orientation bin is 0.5 seconds. The sense of change in orientation corresponds to the sign of the acceleration. If the acceleration is zero ($\alpha=e_1$), the orientation process is constant. For the parameters which delineate this encounter, there is time for about five observation frames in any given orientation when the target is accelerating. Further, the sojourn times in each non-null acceleration state are long enough that the orientation will change roughly eight times before the acceleration changes again. Thus, this example has parameters which encourage the belief that the acceleration can be reasonably clearly extracted from noisy measurements of the angular motion of the vehicle.

To exercise the algorithm, consider the following scenario. At the initial time (t=0) the target is moving directly toward the sensor ($\rho=e_1$,or $\theta=0°,\beta=0°$) without any rotational motion ($\alpha=e_1$). The target is thus in modal state $\phi=e_1$, and it remains there on the interval [0,30). At time t=30, it begins to move to the left ($\alpha=e_2$) until it reaches an orientation perpendicular to the line-of-sight ($\rho=e_{10}$). After this, it unwinds the motion with an overshoot to $\rho=e_{31}$. It then returns to the null angle and null acceleration.

The filter needed to track this motion has dimension 108 (K=3, L=36, KL=108). From this filter both $\{\hat{\rho}_t\}$ and $\{\hat{\alpha}_t\}$ can be deduced by suitably combining the elements of $\{\hat{\phi}_t\}$. This has been done, and sample functions displaying filter behavior are shown in the following figures. Figure 4 shows the indicator function of the first four orientation bins (shown piecewise constant) along with the associated element of $\{\hat{\phi}_t\}$ (the irregular function). Consider the initial interval $t\in[0,30)$ first. The filter identifies the true orientation classification of the target $\rho=e_1$ with high probability. There are some sharp excursions when false indications of angular motion are received, but these are corrected quickly in the main. A conspicuous occurrence is located near t=24 sec. The filter is convinced by the data that a transition to $\rho=e_2$ has occurred. Even in this case, the error is quickly corrected. When the maneuver actually begins at t=30, the filter begins to track the motion with some initial delay in moving to $\rho=e_2$. Once the rotation has been identified, the estimator follows the orientation sequence closely.

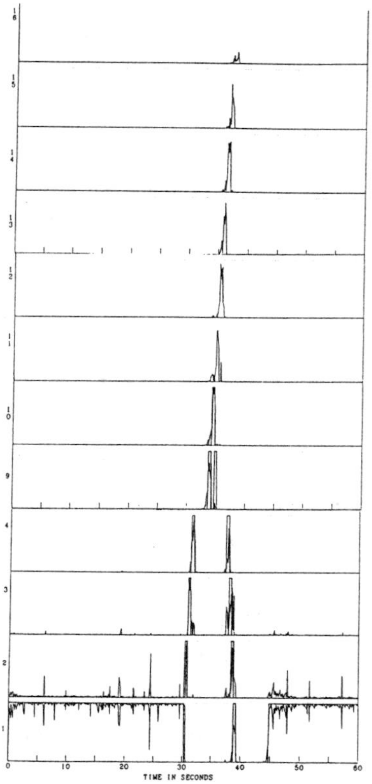

Fig.4. Orientation on a sample trajectory (bins 1-4 and 9-16)

Figure 4 also shows orientation states 9 through 12. There is little confusion as the orientation trajectory approaches β=90°. Indeed, $\rho=e_{10}$ is clearly identified. At this point, the target acceleration is reversed, and the orienta tion returns to $\rho=e_9$ at t=35$^-$ The estimator fails to identify this change in acceleration regime with the result that an interval of digression begins. The estimator output, $\{\hat{\rho}_t\}$, pursues the orientation path associated with the old maneuver rather than correctly switching the sense of rotation as would be appropriate for this new phase of the target motion. Thus, $\hat{\rho}_t \sim e_{11}$ after the change in $\{\alpha_t\}$, and markedly misses the true orientation bin. As the trajectory evolves, the digression decays. By the time $\rho_t \in$ bin 4, the filter has nearly eliminated the false trail.

For tracking purposes, the estimation of orientation is only an intermediary in the estimation of acceleration. Acceleration is the quantity upon which tracking and prediction is based, and the fidelity of $\{\hat{\alpha}_t\}$ is the most relevant performance measure. Figure 5 shows the sample behavior of $\{\hat{\alpha}_t\}$ along the scenario described above. In this figure, the sample function of α_i is shown piecewise constant, and $\hat{\alpha}_i$ is shown solid. On the interval [0,30) the target is not rotating. This lack of angular change is distinguished by a constancy of the mark sequence. The main source of error is the projection ambiguity, and this creates few instances of unwonted behavior when the target is in orientation e_1. This particular orientation is most commonly confused with e_{19}, but the corresponding mark ($\Delta y=u_{19}$) is correctly recognized as an aliased version of u_1. In this phase of the encounter, errors of type #2 are most troubling because they connote a permissible motion to the neighboring bins. Thus, at t=24, the orientation misclassification manifests itself as a strong, but brief, indication of an acceleration transition that could create a false alarm if a detection trigger were used.

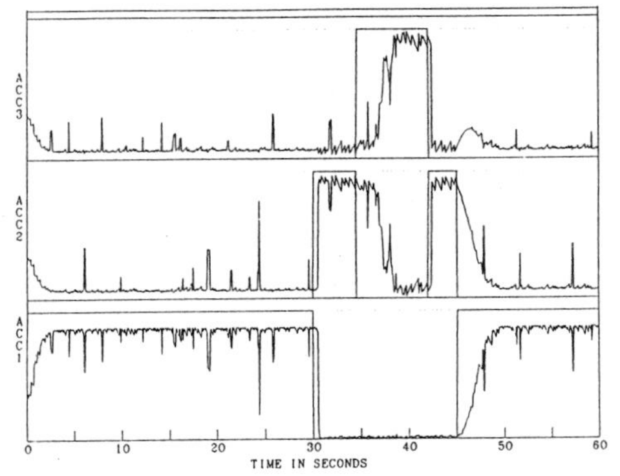

Fig. 5. Acceleration estimates

When the maneuver begins, it is quickly recognized. The α:1→2 transition is identified in about one second. The α:2→3 transition occurs at about t=34. As mentioned earlier, the time for this transition was selected intentionally to yield the maximum ambiguity in the data. The resulting misperception in the estimator is displayed quite clearly. The time at which the acceleration regime changed was not clearly indicated by a change in the sense of rotation of the $\{\hat{\rho}_t\}$ process. The delay in to identifying the correct orientation bin causes $\{\hat{\alpha}_t\}$ to be slow in reflecting the change in target maneuver.

C. INFLUENCE OF SENSOR-PROCESSOR ERRORS

In the previous section, a sample trajectory showing the response of the maneuver estimation algorithm is presented. In this example, the main source of misclassification error is the projection ambiguity (E#3), and the scenario is such as to accentuate the effect of this error. The algorithm is nonlinear, and its performance is scenario dependent; e.g., the jinking maneuver is more difficult to distinguish when β is near 90° than when it is near 0°. It is enlightening to consider how performance varies with the indicated sources of error.

Consider a scenario having the four transitions described earlier.

A: α:1→2 @ β=0°; This initiates the maneuver and takes place at a favorable orientation with respect to the sensor. The detection of this phase of the trajectory requires the processor to identify that a regular motion has begun.

B: α:2→3 @ β=90°; This is a change in the direction of radial acceleration and takes place at an unfavorable angle with respect to the sensor. The estimator must identify the fact that the sense of rotation has reversed in a region where the aliasing errors are hard to eliminate.

C: α:3→2 @ β=300°; This maneuver is similar to B, but takes place in a more favorable angular orientation.

D: α:2→1 @ β=0°; This terminates the maneuver, and takes place at a favorable angle with respect to the projection error. Because the signature of $\alpha=e_1$ is midway between its alternatives, this transition is fairly difficult to detect. Furthermore, based upon the Markov model, this is an unlikely modal transition.

The algorithm responds to change in acceleration in a hesitant, yet locally volatile, manner. A simple measure upon which to base a sensitivity study is the delay before a maneuver transition is detected. Define identification delay as the mean number of frames required for $\{\hat{\phi}_t\}$ to move from its operating state before a transition, to within 0.8 of the proper acceleration mode; i.e., if α:i→j at time 0, let δ be the first time that

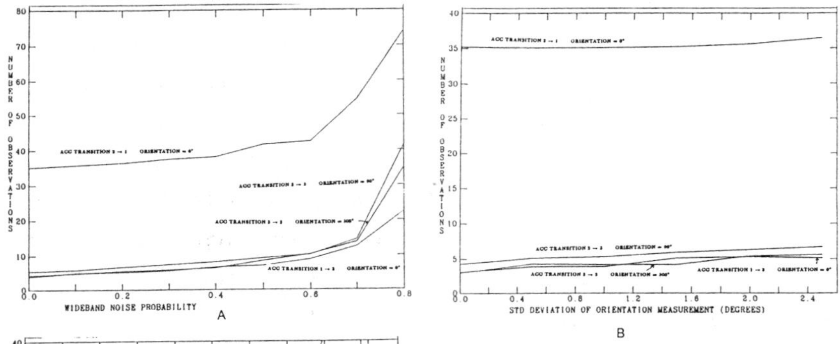

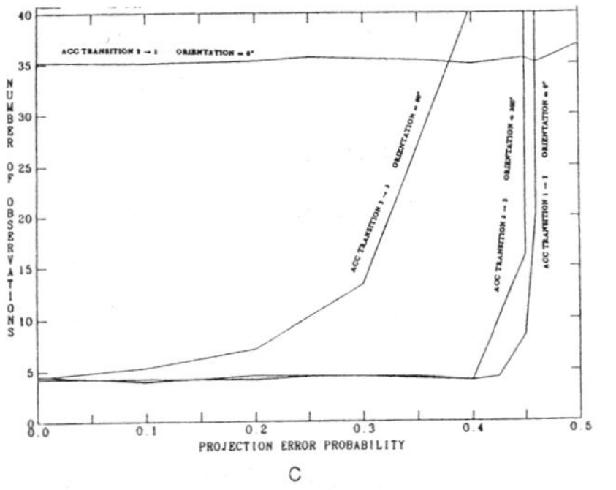

Fig. 6.
Mean observation count to detect a maneuver at different orientations

$(\hat{\phi}_{\delta})_j>0.8$, and then δ/λ is the number of frames for detection. Figures 6 shows the (sample) mean time delay for the indicated transitions as determined from a sample of size ten. The nominal conditions for all of the tests are; 1) wide-spectrum error = 0.01, 2) standard deviation of the quantization error = 1°, 3) projection error = 0.1. The nominal discernability matrix is near the identity in order that the influence of each individual error source be made as clear as possible. Note that various sources of error create a synergistic effect, and total performance degradation can not be determined by simply adding the results given in these figures.

Figure 6A shows the delays associated with wide spectrum errors. Not surprisingly, the detection of maneuvers are fairly insensitive to E#1 on the probability interval [0,0.7]. Nor are these delays particularly orientation dependent. The most distracting errors are those which place the target in a neighboring bin. Hence, most E#1 errors are correctly rejected by (20) as being outlyers, and they have negligible influence on $\{\hat{\alpha}_t\}$ unless the energy in this source of error is large. The α:3→1 transition is inherently hard to detect because it corresponds to an unlikely event. Still, its dependence upon the probability of E#1 is consistent with the other transitions.

The quantization error is more difficult to detect and excise. It acts to shift the apparent orientation of the target into an adjoining bin. This is compatible with the dynamic hypotheses upon which the estimation algorithm is based. Indeed, such errors are quite insidious because they connote precisely what the maneuver detection algorithms is looking for; a change in the sense of orientation. The sensor and image processing quality interrelate to produce the standard deviation of the classifier; e.g., a coarse sensor has no need of a sensitive explication algorithm, and L can be selected correspondingly small. Since L is set at 36, there is no reason to consider values if σ much beyond 2°. Larger values of σ could be accommodated more effectively with a smaller L. Figure 6B shows the detection delay as a function of σ on the interval [0.5,2.0]. This figure shows that timely detection of a maneuver is nearly independent of the size of E#2 in the indicated range. This may be surprising, but note that the quantization error has most effect on the creation of false alarms. A maneuver is displayed by a regular rotation. This type of error simply creates extraneous observations that are easily removed by (20).

The final source of misclassification is projection error, and this is much more dependent upon the orientation at the time of a change in the phase in the maneuver than are the other sources. Figure 6C shows the dependence of delay on E#3. The nominal power in E#1 and E#2 is so small, that E#3 creates no basic ambiguity in recognizing a transition in $\{\alpha_t\}$. Only when the value of ρ is unfavorable at the time of a change in the sense of rotation, does E#3 have an important impact. In this case, however, the aliasing of the sensory data causes a significant delay. To detect jinking maneuver in a timely manner at β=90°, the projection error must be less than 0.25. If the orientation is near 0°, the filter is much more forgiving, and permits this error source to grow to nearly 0.4 before performance degrades unacceptably. These results are for a fairly clean data stream and the indicated bound on projection error is optimistic in the sense that as the earlier error sources grow, E#3 must be made significantly smaller if expeditious maneuver detection is to be accomplished.

This section provides an indication of how performance degrades with increases in the various error sources. The nominal conditions were advantageous to making a timely detection of a maneuver. With this caveat, it appears that wide band errors are relatively unimportant, and projection error should be limited to 20%. The primary effect of σ is to introduce false alarms. A value of $\sigma=1°$ is compatible with an angular bin size of 10°. All of these restrictions introduce significant constraints on both the sensor, the processor, and the conditions which circumscribe the encounter; e.g., range, foreground clutter level, etc.

IV. TARGET TRACKING AND PREDICTION

The previous section gives an image-based interpretation algorithm which provides an estimate of the current acceleration of an agile target. A composite tracker must fuse this estimate with location measurements. The two categories of observation are processed in parallel, and the tracker architecture has a form like that shown in Figure 7. The details of an implementation for such a hybrid data tracker is provided next.

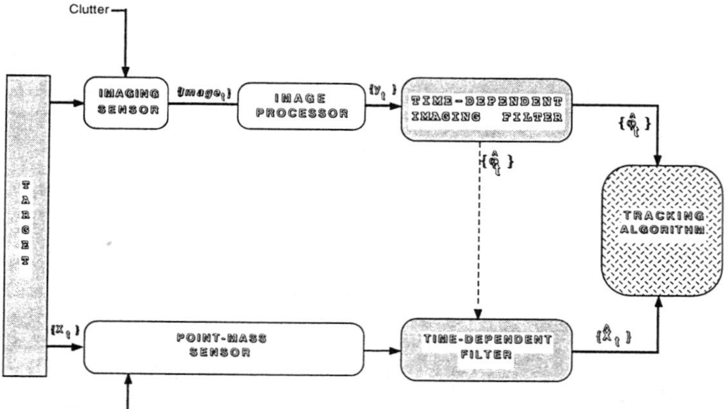

Fig. 7. Block diagram of a hybrid tracker.

A. THE AUGMENTED KALMAN FILTER

The image enhanced tracker shown in Figure 7 has two primary data paths. An imager provides an expeditious indication of maneuvers, and a conventional sensor locates the target. Suppose that the latter measures range and bearing. In the absence of image data, the target tracker would likely employ an EKF to estimate target motions. Suppose that the vehicle moves in the plane with motion described by

$$dx_t = Fx_t dt + Gdw_t + a_t(k \times v_t^*)dt \tag{25}$$

where x_t is a 4-vector containing the vehicle position (components 1 and 2) and velocity, v_t (components 3 and 4), v_t^* is a unit vector along v_t and k is a vertical unit vector. The first two terms retain their identity from (1). The final term delineates the maneuver acceleration. It has sense and magnitude given by a_t, and its direction is perpendicular to the velocity. The EKF neglects the last term in (25), and linearizes the range-bearing observation about the estimate of $\{x_t\}$. Prediction is accomplished by direct forward projection using the transition matrix. Maneuvers are difficult for the EKF to deal with because the model upon which it is based does not admit sharp changes in acceleration. Not only is the EKF slow to respond when a change in acceleration occurs, but it is slow to return to nominal operation after a maneuver ends because of the cumulative effect of large tracking residuals during the transient interval.

As in the preceding section suppose that a FLIR makes direct measurements of target orientation, and classifies them into L orientation bins. The augmented extended Kalman filter (AEKF) proposed here uses the basic EKF filter/predictor algorithm with the inclusion of the mean estimate of the third term in (25) in a modified drift. The AEKF has a rather prosaic implementation with its additive maneuver compensation. Despite its simplicity, the AEKF is quite responsive.[27], [28] The next section contains a comparison of the tracking and prediction accuracy of the EKF and the AEKF on a sample trajectory.

B. PERFORMANCE ON A SAMPLE SCENARIO

To illustrate the performance improvement attributable to an imaging sensor, consider a simple scenario. The initial location of the target is (1000,6400)[5] with velocity (5,-13.4) with respect to a stationary sensor at (0,0). After 30 sec. on the initial course, the target turns to the right at 0.5g. This acceleration persists for 8 sec at which time the target acceleration returns to zero. Range and bearing are measured ten times/sec with independent errors having standard deviations of 5 meters and 0.25 degrees respectively. For convenience, no wide-band acceleration was included in the simulation, but the EKF is such that these accelerations would cause a nominal 10 meter position error.

Figure 8 shows the estimation fidelity of the EKF. Primary concern centers on consistent errors in the filter, and this figure shows both the true target trajectory in the plane (the smooth curve) and the mean of 20 independent samples of the response of the EKF. While performance is good throughout, when the maneuver begins, the filter is slow to accommodate to it. It also overshoots slightly when the maneuver ends. Other

---

[5] Distances are in meters and angles are in degrees.

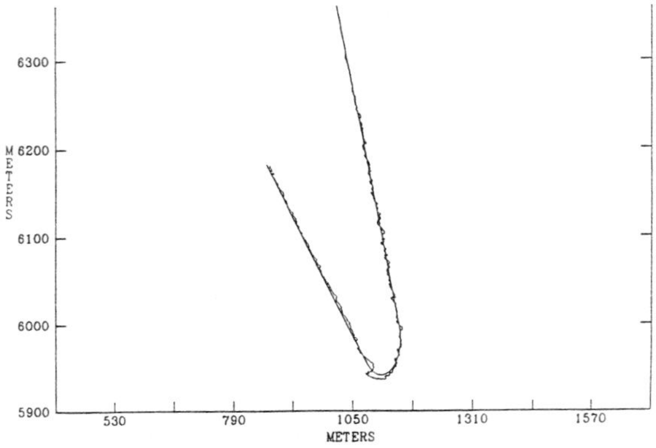

Fig. 8. Estimate of target location (EKF)

mean tracking error is about 45 meters, with nominal operation nearer 10 meters.

The behavior of the velocity estimate contrasts sharply with that of position. Location is measured directly, and the estimator performance is good even though the filter is using a target model which does not acknowledge the possibility of a maneuver. The velocity is, however, a slack variable. When the target maneuvers, the observed position residuals are assumed to emanate from variations in velocity, and there is no direct measurements which would disabuse the filter of this illusion. Because of the smoothing of the motion dynamics, the resulting velocity fluctuations can be relatively large. Figure 9 shows the sample-mean velocity trajectory in the velocity phase plane along with the true velocity path. The velocity estimate is poor, and even has the wrong sense of rotation about the origin. Data indicates that the mean velocity error peaks at 21 m/sec. and falls slowly after the maneuver ends.

Velocity errors have a major influence on the accuracy of predictions of future target position. A simple fixed-interval predictor propagates estimated position forward along the mean velocity vector for the prediction interval. Figure 10 shows the sample-mean trajectory of the 2-second predictor on the same scenario. The velocity errors mentioned earlier create large prediction errors. Failure to deduce the correct sense of velocity causes the predictor to overcorrect during the maneuver interval. The peak prediction error is approximately 70 meters, and the time required for the error to return to its quiescent level is over a minute.

When the EKF is augmented with an image-based link, estimator fidelity improves. The maneuver estimation algorithm (20) is parameterized by the rate of change in acceleration and orientation, the frame rate λ, and the discernability matrix P. Suppose that the mean sojourn time in each acceleration mode is 4 seconds. If $a_t=0$, it is equally

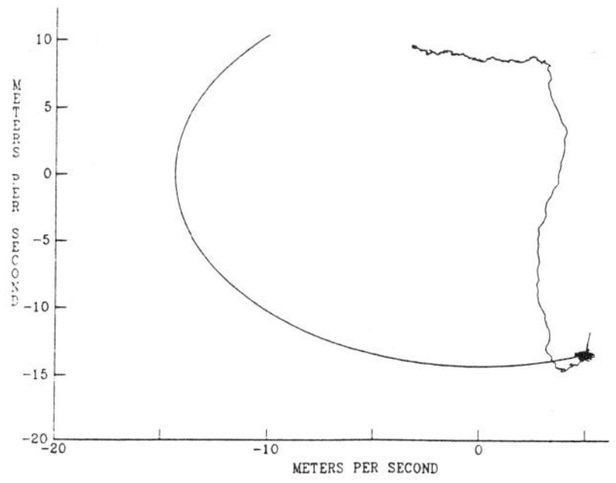

Fig. 9. Estimate of target velocity

(EKF)

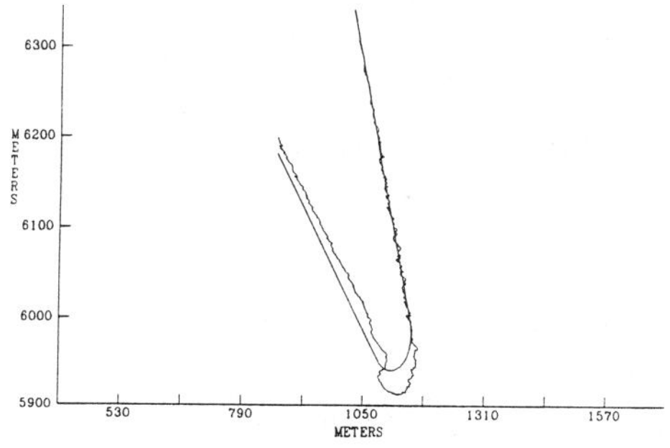

Fig. 10. Prediction of target location

(EKF)

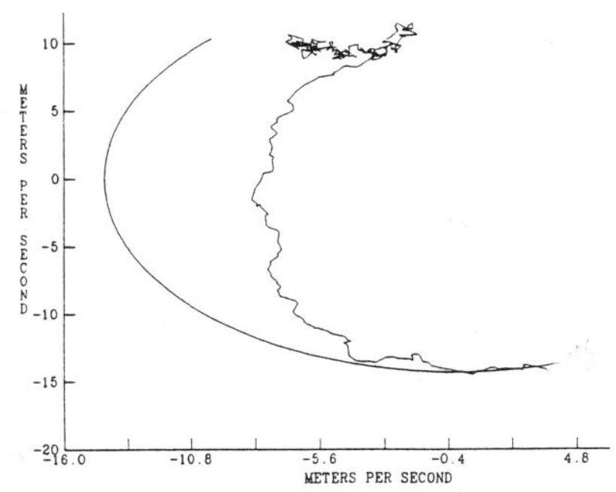

Fig. 11. Estimate of target velocity

(AEKF)

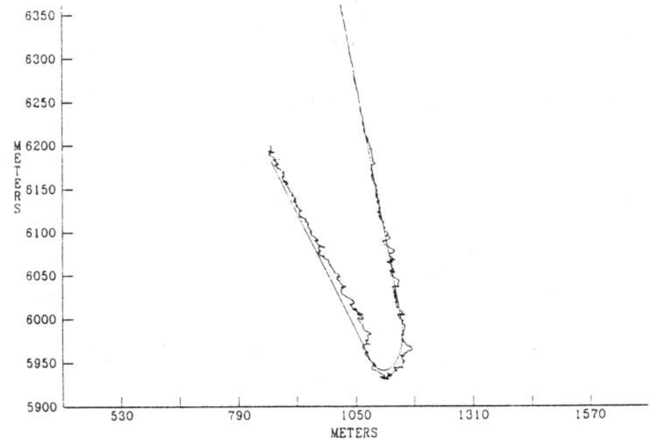

Fig. 12. Prediction of target location

(AEKF)

likely to next move to either of the two maneuver states; ±0.5g. If in one of the maneuver states, the next change would be to the other with probability 0.9; i.e., once a evasive maneuver begins, the target does not quickly return to quiescent operation. The frame rate is 10 frames/second.

To determine P, the previous taxonomy of imager errors is useful. In this example it will be supposed that E#1 occurs 1% of the time, the variance of the symmetric misclassification is $\sigma=1°$, and the projection error is 10%. If the listed error sources are assumed to be independent, P can be deduced and the AEKF implemented. For the indicated scenario, the tracking performance of the AEKF is indistinguishable from that of the EKF. The velocity phase plot shown in Fig. 11 is, however, far superior to that given in Fig. 9. The maximum velocity error of the AEKF is reduced to about 10 m/sec. This improvement is reflected directly in the 2-sec. prediction trajectory shown in Fig. 12. The maximum prediction error is reduced by the AEKF from 70 m. to less than 25 m. A similar performance enhancement has been observed on tests with other trajectories.

The size and hence computational load of (20) is determined by the number of acceleration states K, and the number of orientation states L. Reducing the number of orientation bins leads to a lower dimensional AEKF, but at the cost of a coarser orientation grid. This coarser grid tends to make the identification of a maneuver slower, and the AEKF more hesitant in its response. It is through the reduction in velocity error that the AEKF achieves most of it performance improvement. For the indicated trajectory, Figure 13 shows the magnitude of the error in the velocity estimate for different orientation grids (The EKF can be thought of as an AEKF with L=0). With 36 orientation bins, the AEKF is far superior to the EKF as shown earlier. A coarser classification of 22.5° is not as good, but still reduces the peak error of the EKF by a factor of two. Even a 45° angular grid is an improvement over the EKF. The coarse grids do lead to a significant delay in the elimination of velocity error when the maneuver ceases. This is not unexpected since if the grid is too loose, it is difficult to discern when angular motion ceases.

V. CONCLUSIONS

The ability to detect maneuvers expeditiously permits the design of tracking systems with enhanced following qualities. Because of the time delay that occurs between the initiation of a maneuver and the time that this maneuver is unambiguously reflected in the motion of the center-of-reflection states of the target, a jinking target is difficult to track and predict. This paper presents a study of post-processors which convert a sequence of sensor outputs into an estimate of the location of a maneuvering target. Although notable performance improvements have been made, these results are preliminary. They do illustrate the positive effect of using imaging arrays in tracking and guidance. The newly evolving system architectures will have a profound influence on performance in rapidly changing encounters.

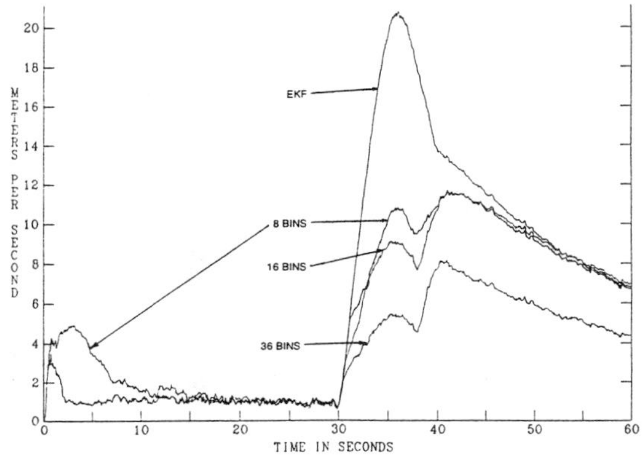

Fig. 13. Velocity errors vs. bin size

ACKNOWLEDGEMENTS

This research was supported by a grant from the Hughes Aircraft Co., and by the MICRO Program of the State of California under Project No. 89-021.

REFERENCES

1. R.J. Elliott, *Stochastic Calculus and Applications*, Springer-Verlag, New York,1982.

2. P.S. Maybeck, *Stochastic Models, Estimation, and Control*, Vol. 1, Academic Press, 1979.

3. J.R Cloutier, J.H. Evers and J.J. Feeler, "Assessment of Air-to-Air Missile Guidance and Control Technology," *IEEE Control Systems Magazine*, Oct. 1989, 27-34.

4. R.A. Singer, "Estimating Optimal Tracking Filter Performance for Manned Maneuvering Targets," *IEEE Trans. on Aerospace and Electronic Systems*, Vol. AES-10, July 1970, 473-482.

5. R.D. Russ, "No Sitting Ducks," *Air Force Magazine*, July 1988, 92-97.

6. R.L. Moose, V.F. Vanlandingham, and D.H. McCabe, "Modeling and Estimation for Tracking Maneuvering Targets, *IEEE Trans. on Aerospace and Electronic Systems*, Vol AES-15, May 1979, 448-455.

7. P.L. Bolger, "Tracking a Maneuvering Target Using Input Estimation, *IEEE Trans on Aerospace and Electronic Systems*, Vol AES-23, May 1987, 298-310.

8. T.L. Song, J.Y. Ahn and C. Park, "Suboptimal Filter Design with Pseudomeasurements for Target Tracking," *IEEE Trans. on Aerospace and Electronic Systems*, Vol. AES-24, Jan. 1988, 28-39.

9. D.E. Williams, and B.Friedland, "Target Maneuver Detection and Estimation," *Proc of the 27th IEEE Conf. on Decision and Control*, Austin, TX, Dec. 1988, 851-855.

10.D.D. Sworder, "On the Relative Advantage of Imaging versus Continuous Trackers," *Proceedings of the 20th Asilomar Conference on Circuits, Systems and Computers*, (Pacific Grove, California, November 1986). IEEE Computer Society Press, (1987), 585-589.

11. D.D. Sworder, and R. G. Hutchins, "Algorithms for Image Based Trackers of Maneuvering Targets," in *Control and Dynamic Systems, Advances in Theory and Applications*, Vol. 28, Part 1, Academic Press, Inc., New York, (1988), 261-302.

12. J.D. Kendrick, P.S. Maybeck, and J.G. Reid, "Estimation of Aircraft Target Motion Using Orientation Measurements," *IEEE Trans. on Aerospace and Electronic Systems*, Vol. AES-17, March 1981, 254-259.

13. D. Andrisani, F.P. Kuhl, and D. Gleason, "A Nonlinear Tracker Using Attitude Measurements," *IEEE Trans. on Aerospace and Electronic Systems*, Vol AES-22,Sept. 1986, 533-539.

14. F. Kuhl, T. Hutchins, and D. Andrisani, "Aircraft Lead Angel Prediction," *Proc. of the Eighth Meeting of the Coordinating Group on Modern Control*, Oakland Univ. Feb, 1987, 35-39.

15. D. Andrisani and J.D. Schierman, "Tracking Maneuvering Helicopters Using Attitude and Rotor Angle Measurements," *Proc. of the 21'st Asilomar Conf. on Circuits, Systems, and Computers*, Pacific Grove, CA, November, 1987.

16. D.D. Sworder and R. G. Hutchins, "Image-Enhanced Tracking," *IEEE Trans.on Aerospace and Electronic Systems*, Vol. AES-25, 5, Sept. 1989, 701-710.

17. P.S. Maybeck and D.E. Mercer, "A Target Tracker Using Spatially Distributed Infrared Measurements," *IEEE Trans. on Automatic Control*, Vol. AC-25, Feb. 1980, 222-225.

18. S.H. Taylor, "The Cramer-Rao Estimation Error Lower Bound Computation for Deterministic Nonlinear Systems," *IEEE Trans. on Automatic Control*, Vol. AC-24, April 1979, 343-344.

19. B.Z. Bobrovsky and M. Zakai, "A Lower Bound on the Estimation Error for a Markov Process," *IEEE Trans. on Automatic Control*, Vol AC-20, Dec. 1975, 786-788.

20. P.S. Maybeck and R.I. Suizu, "Adaptive Tracker Field-of-View variation Via Multiple Model Filtering," *IEEE Trans. on Aerospace and Electronic Systems*, Vol AES-21, July 1985, 529-539.

21. C.B. Chang, and J.A. Tabaczynski, "Application of State Estimation to Target Tracking, *IEEE Trans. on Automatic Control*, Vol. AC-29, Feb. 1984, 98-109.

22. H.A.K. Blom and Y. Bar-Shalom, "The Interacting Multiple Model Algorithm for Systems with Markovian Switching Coefficients," *IEEE Trans. on Automatic Control*, Vol.33, Aug. 1988, 780-783.

23. H. Burke, "CAAM, Circular Arc Aimed Munitions," *Proc. of the 21'st Asilomar Conf. on Circuits Systems and Computers*, Pacific Grove, CA, November, 1987.

24. J.W. Brewer, "Kronecker Products and Matrix Calculus in Systems Theory," *IEEE Trans. on Circuits and Systems*, Vol. CAS-25, Sept. 1978, 772-781.

25. D.D. Sworder and R.G. Hutchins, "Maneuver Estimation Using Measurements of Orientation," *IEEE Trans. on Aerospace and Electronic Systems*, Vol. AES-26, 5 (1990),

26. D.D. Sworder, "Improved Target Prediction Using an IR Imager," *Proceedings of SPIE: Infrared Systems and Components*, 750, (1987) 105-108.

27. D.D. Sworder, and R.G. Hutchins, "Fusion Algorithms for Prediction," *Proc. of the 23rd Asilomar Conf. on Signals, Systems, and Computers*, (Monterey CA), Oct. 1989.

28. R.G. Hutchins, and D.D. Sworder, "Image Fusion Algorithms for Tracking Maneuvering Targets," *AIAA Journal of Guidance, Control and Dynamics*, 1990 (to appear).

CONTINUOUS TIME PARAMETER ESTIMATION: ANALYSIS VIA A LIMITING ORDINARY DIFFERENTIAL EQUATION

DOUGLAS G. DEWOLF

Hughes Aircraft Company
Los Angeles, California 90009

I. INTRODUCTION AND BACKGROUND

The techniques introduced here extend Ljung's Ordinary Differential Equation (ODE) technique for analysis of discrete time parameter estimators to continuous time stochastic parameter estimators [1]. With the ODE method, the asymptotic behavior of a stochastic parameter estimator is determined by the asymptotic behavior of a related deterministic ordinary differential equation. The ODE method differs from other analysis techniques in that it is well suited for general convergence analysis, comparison of different algorithms (by comparing their ODE's), and numerical analysis of specific instances of an estimator. To illustrate the use of the ODE method, the convergence properties of the continuous time Recursive Prediction Error Method (RPEM) will be analyzed, both for the general state space model and specific instances of that model. The convergence of RPEM is of independent interest, since other more complex parameter estimators (in particular the Wiberg Estimator [2]) can be shown asymptotically equivalent to RPEM. In fact the equivalence is shown by proving RPEM and the Wiberg estimator have the same associated ODE. The theoretical results were validated with numerical simulation of the stochastic differential equations. The simulation highlights the advantages and disadvantages of the ODE technique.

The parameter estimation problem will be formulated in terms of state space. Consider the system of stochastic differential equations (SDE's) :

$$dx = A_o x\,dt + B_o u\,dt + \sigma_o dw_1 \qquad \Sigma_o = \sigma_o \sigma_o^T$$
$$dy = Cx\,dt + \rho\,dw_2 \qquad R = \rho\rho^T \tag{1}$$

The Kalman Filter problem is to estimate the states, $x(t)$, based upon the observations $\{y(s) : 0 \le s \le t\}$. For the Kalman Filter, A_o, B_o, and σ_o are assumed known. The parameter estimation problem is to estimate A_o, B_o, and σ_o, as well as $x(t)$ given the measured data $\{y(s) : 0 \le s \le t\}$. It is assumed there is a model for the system (1) given by

$$dx = A(\theta)x\,dt + B(\theta)u\,dt + \sigma(\theta)dw_1 \qquad \Sigma(\theta) = \sigma(\theta)\sigma(\theta)^T$$
$$dy = Cx\,dt + \rho\,dw_2 \qquad R = \rho\rho^T \tag{2}$$

where the functions $A(\cdot)$, $B(\cdot)$ and $\sigma(\cdot)$ are known, and the constants C and ρ are known. The problem reduces to determining a parameter value $\theta(t)$ conditioned on $\{y(s) : 0 \le s \le t\}$ which optimizes some performance index. If there is a θ_o such that $A(\theta_o)=A_o$, $B(\theta_o)=B_o$, and $\sigma(\theta_o)=\sigma_o$, the performance index could be $\theta(t) \to \theta_o$ wp1 as $t \to \infty$. Another possible performance index is $\theta(t)$ converging to a local minimum of a cost function involving the output estimation error, $y(t) - \hat{y}(t;\theta)$, or the state estimation error, $x(t) - \hat{x}(t;\theta)$. Many parameter estimators (e.g. RPEM , Extended Kalman Filter, Extended Least Squares or the Wiberg Estimator) can be put in the form

$$d\theta = \frac{1}{t} f_1(\theta, z)dt + \frac{1}{t} f_2(\theta, z)dw_1(t)$$
$$dz = h_1(\theta, z)dt + h_2(\theta, z)dw_2(t) \tag{3}$$

where θ is the parameter estimate and z is a vector whose components contain the states (x), the measurements (y) and other auxiliary quantities (e.g. solutions of Riccati and sensitivity equations). Assume the $z(\cdot)$-process *with θ held constant* has a unique stationary probability measure $P^{\theta}(\cdot)$, and define $f_o(\cdot)$ by $f_o(\theta) = \int f_1(\theta, z)P^{\theta}(dz)$. Then the primary theoretical result presented here is that the asymptotic behavior of $\theta(t)$ is determined by the asymptotic behavior of $\theta^o(t)$, where $\theta^o(t)$ satisfies the ODE

$$\dot{\theta}^o = f_o(\theta^o) \tag{4}$$

In particular, it will be shown:

(a) If, for any initial condition, $\theta^o(t)$ converges to a point θ_1 , then $\theta(t)$ converges to θ_1 wp1

(b) If , for any initial condition, $\theta^o(0)$, we have $\theta^o(\ln(t+T)) - \theta^o(\ln(t)) \to 0$, then $\theta(t+T) - \theta(t) \to 0$ wp1

(c) If $\theta(t) \to \theta_1$ with probability greater than 0, then θ_1 is a stable stationary point of $f_o(\cdot)$

(a) and (b) follow from a more general result which is stated in Section II and proved in Appendix B. The proofs of (a)-(c) are sketched in Appendix B and presented in detail in [3]. (a) is used to show convergence. The condition in (b) is used to show stability of adaptive controllers based on the parameter estimator. (c) determines the possible convergence points. The two major advantages of the ODE method are that the ODE, Eq. (4), is of much lower dimension than the system Eq. (3), and the ODE is deterministic while Eq. (3) is stochastic.

While L. Ljung's work on the discrete time ODE method ([1],[4]) inspired the development of the continuous time ODE method, the mathematics required to carry out the development is based upon the work of Kushner [5] and his colleagues Clark [6], Shwartz [7], and Huang [8]. In particular, the invariant measure approach introduced by Shwartz [7]. The appropriate way to relate $\theta(\cdot)$ and $\theta^o(\cdot)$ is by the weak convergence of $\theta(e^{\cdot+t_\varepsilon})$ to $\theta^o(\cdot)$ as $\varepsilon \to 0$ (t_ε is a sequence such that $t_\varepsilon \to \infty$ as $\varepsilon \to 0$). The theorems needed to justify the weak convergence and characterize the asymptotics of $\theta(\cdot)$ are similar to theorems found in [5]. There are several difference between the present work and the theorems in [5]. Kushner uses a t/ε time scaling rather than the exponential scaling and there are no diffusion term in the $d\theta$ equation of Eq. (3). Papanicolaou, Stroock and Varadhan develop a weak convergence theory with the diffusion term but still use the t/ε time scaling. Finally in [5] and [9], only the $z(\cdot)$ equations are accelerated, while in the present work, both $\theta(\cdot)$ and $z(\cdot)$ are accelerated. These differences increase the complexity of the proofs of the weak convergence theorems. For the asymptotic theorems, Kushner considers diffusion limits (of which ODE limits are a special case). In the present work only ODE limits are considered and more specialized asymptotic results are proved. The differences between the present work and the previous results arise from fundamental differences in the problems to be solved. Kushner (and colleagues) are mainly concerned with approximation of "wideband noise" systems by diffusions. The correctness of these approximations is central to modelling physical processes by diffusions. The t/ε time scaling is the appropriate and expedient method of generating a diffusion from the physical model. Other time scaling could be used but the theory would be more difficult. This thesis is concerned with the asymptotics of a diffusion, and the diffusion equations are considered to be an exact problem formulation. The exponential time scaling is implicit in the problem formulation and not an approximation. It should be mentioned that the present theory can be applied to time scales to other than the exponential. The exponential time scaling arises from the $1/t$ gain sequence in Eq. (3). Minor changes to the proofs allow other gain sequences, $\gamma(t)$, as large as $o(t^{-1/2})$ provided $\int_1^\infty \gamma(t)dt = \infty$. Each gain sequence leads to its own time scaling. Requirements on the gain sequence are discussed in greater detail in Section II.

Convergence of specific continuous time parameter estimators has been shown by van Schuppen [10], Chen [11], Moore [12], and Goodwin, Gevers and Wertz [13,14]. These estimators are related to Extended Least Squares algorithms, applied to various signal models. The relation between these results and the present ones are briefly discussed in Section II.

Discrete time Recursive Prediction Error Methods (RPEM) have been extensively studied by ODE and other methods. A continuous time version is derived in [3] and the ODE analysis of this estimator shows that it converges to a local minimum of the cost function

$$J(\theta) = \text{trace}\left(C^T R^{-1} C\, P_{\tilde{x}}(\theta)\right) \tag{5}$$

where $P_{\tilde{x}}(\theta)$ is the covariance matrix of the steady state state estimation error of a related filter. The related filter is the Kalman Filter for Eq. (1) using $A(\theta)$, $B(\theta)$, and $\sigma(\theta)$ (θ held constant) in the filter equations instead of the true values. The asymptotic optimality of the "true" Kalman Filter implies $J(\theta) \geq J(\theta_o)$ for $\theta \neq \theta_o$, but the existence of other local minima depends upon the exact signal model ("ARX" vs "ARMAX" etc). In addition to general convergence proofs, the ODE limit can be calculated for specific plant and measurement equations. This was done for a one dimensional example and the true parameter value is shown to be the unique local minima for Eq. (5). Thus RPEM will converge with probability one to the true parameter value in this case. In addition to the numerical analysis of the ODE's, the stochastic differential equations for RPEM were simulated. The simulation results supported the theoretical ones and highlighted the efficiency of ODE analysis.

The weak convergence and asymptotic theorems are stated in Section II and proved in Appendix B. These theorems apply to a general class of stochastic processes and are not tied to parameter estimation. These theorems are used in Section III to derive the general convergence properties of RPEM. An example in Section IV demonstrates the ODE limit calculation in a specific case (i.e. a particular choice of state parameters) and compares the results with Monte Carlo simulation. Appendix A lists some theorems from probability theory useful for understanding the weak convergence results. Appendix C is a collection of previously known results on Runge-Kutta approximation of SDE's highlighted by some simulation results.

II. STATEMENT OF THE WEAK CONVERGENCE THEOREMS

This section is devoted to stating the theorems that characterize the asymptotic behavior of certain time scaled stochastic systems. Both the notation and assumptions appear daunting, but the results are fairly simple to state and easy to use. The assumptions for Theorem 1 are set up to facilitate the proofs rather than physical plausibility. Theorem 2 gives conditions that are easier to verify. Theorem 3 characterizes the behavior of the accelerated process, $x^\varepsilon(t)$, for small ε and large t. Theorem 4 recasts Theorem 3 in terms asymptotic behavior of the original $x(\cdot)$ process and is the theorem used in practice. The notation, statements and proofs of the theorems draw heavily on the work of Kushner and Shwartz ([7], [5]). The statement and proof of Theorem 1 is similar to Theorem 5.6 in [5], but in the present case, the dynamics of $x^\varepsilon(\cdot)$ are different, complicating the theorem and its proof.

The parameter estimators of interest can be put in the form

$$dx = \frac{1}{t}[F_1(x,y)dt + F_2(x,y)dw] \qquad x(t_s) = x_o$$
$$dy = H_1(x,y)dt + H_2(x,y)dw \qquad y(t_s) = y_o \tag{1}$$

x and y are vector valued stochastic processes, w is a vector valued standard Wiener process. Let $\varepsilon \in (0,1]$. Let $t_\varepsilon \to \infty$ as $\varepsilon \to 0$ (note that $t_{(\cdot)}$ is not required to be continuous). The accelerated system, $(x^\varepsilon(\cdot), y^\varepsilon(\cdot))$, is defined by:

$$x^\varepsilon(t+s) - x^\varepsilon(t) =$$
$$\int_t^{t+s} F_1(x^\varepsilon(\tau), y^\varepsilon(\tau))d\tau + \int_{e^{t+t_\varepsilon}}^{e^{t+s+t_\varepsilon}} F_2(x^\varepsilon(\log(\tau)-t_\varepsilon), y^\varepsilon(\log(\tau)-t_\varepsilon))\frac{dw(\tau)}{\tau}$$

$$d\tilde{y}^\varepsilon(t) = H_1(x^\varepsilon(\log(t)-t_\varepsilon), \tilde{y}^\varepsilon(t))dt + H_2(x^\varepsilon(\log(t)-t_\varepsilon), \tilde{y}^\varepsilon(t))dw(t)$$
$$y^\varepsilon(t) = \tilde{y}^\varepsilon(e^{t+t_\varepsilon})$$
$$x^\varepsilon(0) = x_\varepsilon \qquad y^\varepsilon(0) = y_\varepsilon \qquad (\therefore \tilde{y}^\varepsilon(e^{t_\varepsilon}) = y_\varepsilon) \tag{2}$$

x_ε and y_ε will be random variables independent of $w(t)$ for $t > e^{t_\varepsilon}$ and whose values may or may not be tied to the $(x(\cdot), y(\cdot))$ process. Loosely, $(x^\varepsilon(\cdot), y^\varepsilon(\cdot))$ is equal to

$$x^\varepsilon(t) = x(e^{t+t_\varepsilon}) \qquad x^\varepsilon(0) = x_\varepsilon$$
$$y^\varepsilon(t) = y(e^{t+t_\varepsilon}) \qquad y^\varepsilon(0) = y_\varepsilon \tag{3}$$

with initial conditions specified at e^{t_ε} *rather than* t_s, so the existence of solutions to Eq. (1) implies the existence solutions to Eq (2). Two more processes needed for the theorem are the fixed x process:

$$d\tilde{y}(t;x) = H_1(x, \tilde{y}(t;x))dt + H_2(x, \tilde{y}(t;x))dw(t) \tag{4}$$

and the process defined by

$$d\tilde{y}_z(t) = H_1(z(t), \tilde{y}_z(t))dt + H_2(z(t), \tilde{y}_z(t))dw(t) \tag{5}$$

where $z(\cdot)$ is a bounded nonanticipative process. Usually $z(\cdot)$ will be a time scaled segment of $x^\varepsilon(\cdot)$.

The transition probability measures for these processes are:

$$\tilde{P}^{\varepsilon}(x,y,t,t_o,A) = P\{\tilde{y}^{\varepsilon}(t) \in A | x^{\varepsilon}(\log(t_o) - t_{\varepsilon}) = x, \tilde{y}^{\varepsilon}(t_o) = y\} \tag{6}$$

$$\tilde{P}_z(y,t,t_o,A) = P\{\tilde{y}_z(t) \in A | \tilde{y}_z(t_o) = y, z(t_o)\} \tag{7}$$

$$\tilde{P}(y,t,t_o,A|x) = P\{\tilde{y}(t;x) \in A | \tilde{y}(t_o;x) = y\} \tag{8}$$

$$P^{\varepsilon}(x,y,t,t_o,C) = P\{(x^{\varepsilon}(t),y^{\varepsilon}(t)) \in C \mid (x^{\varepsilon}(t_o),y^{\varepsilon}(t_o)) = (x,y)\} \tag{9}$$

The last probability measure to be defined is the stationary measure, $P^x(\cdot)$, of the fixed x system, Eq. (4). This measure plays a central role in both the theory and application and its existence is assured by hypothesis. Define the function $\underline{F}(\cdot)$ by

$$\underline{F}(x) = \int F_1(x,y)P^x(dy) \tag{10}$$

Define $x^o(\cdot)$ by $\dot{x}^o = \underline{F}(x^o)$. Theorem 1 is concerned with the relation between $x^{\varepsilon}(\cdot)$ and $x^o(\cdot)$. There is no reference to an underlying $x(\cdot)$ process. Theorem 2 gives conditions on the $x(\cdot)$ process which imply that the related $x^{\varepsilon}(\cdot)$ process satisfies the hypothesis of Theorem 1. Assume $x(\cdot)$ and $y(\cdot)$ take values in $\mathbf{R}^{n_x}$ and $\mathbf{R}^{n_y}$. $E_t\{\cdot\}$ denotes expectation with respect to $\{w(s) : s \le t\}$. The following assumptions are needed to prove the weak convergence theorems.

(A1) *For each initial condition* $(x_1,y_1) \in \mathbf{R}^{n_x} x \mathbf{R}^{n_y}$ *, the system Eq. (2) has a unique (in the sense of distributions) solution defined for all time. For each C and* t_o *its transition probability* P^{ε} *is Borel measurable in* (x,y,t)*.*

(A2) *For each compact* $K \subseteq \mathbf{R}^{n_x} x \mathbf{R}^{n_y}$ *and* $\eta > 0$ *there exists a* $\delta > 0$ *such that if* (x,y), $(x',y) \in K$ *and* $\|x - x'\| < \delta$ *then* $\|F_1(x,y) - F_1(x',y)\| < \eta$ *(i.e. uniform continuity in the x variable)*

(A3) *Assume that either*

(a) $F_1(\cdot,\cdot)$ *is bounded*

or

(b) *for any T>0 and* $\delta > 0$ *the set of random variables*

$$\left\{ \sup_{\substack{0 \le s \le \delta \\ x \in \mathbf{R}^{n_x}}} \left\| F_1(x,y^{\varepsilon}(t+s)) \right\| : t \in [0,T], \varepsilon \in (0,1] \right\}$$

is uniformly integrable

(A4) *Assume that either*

(a) $F_2(\cdot,\cdot)$ *is bounded*

or

(b) *for any* $T>0$ *and* $\delta>0$ *the set of random variables*

$$\left\{ \sup_{\substack{0\le s\le\delta \\ x\in\mathbf{R}^{n_x}}} E_{e^{t+t_\varepsilon}} \left\| F_2(x,y^\varepsilon(t+s)) \right\|^2 : t\in[0,T],\ \varepsilon\in(0,1] \right\}$$

is bounded

(A5) (a) $\{y^\varepsilon(0):\varepsilon>0\}$ *is tight*

(b) *for any* $T>0$ *if* $\{y^\varepsilon(0):\varepsilon>0\}$ *is tight then* $\{y^\varepsilon(t):t\le T,\ \varepsilon>0\}$ *is tight*

(A6) *For each* $x\in\mathbf{R}^{n_x}$, *there is a unique invariant measure* $P^x(\cdot)$ *for the fixed x system, Eq. (4). For each compact* $K\subseteq\mathbf{R}^{n_x}$ *the collection* $\{P^x(\cdot):x\in K\}$ *is tight.*

(A7) *either*

(a) $\{y^\varepsilon(t):t\ge0\ \varepsilon>0\}\subseteq B(0,N)$ *for some N*

or

(b) *For any* $\varepsilon>0$ *there is a set function* $\beta_\varepsilon(\cdot)$ *such that for any compact subset* $B\subseteq\mathbf{R}^{n_y}$ *we have* $\varepsilon\beta_\varepsilon(B)\to0$ *as* $\varepsilon\to0$, *and for fixed* $T,K>0$ *the collection of measures*

$$\left\{ \tilde{P}_{z^\varepsilon}(y,s,e^{t_\varepsilon},\cdot)I_B(y) : \left\{ \begin{array}{l} \text{compact}\ B\subseteq\mathbf{R}^{n_y} \\ y\in\mathbf{R}^{n_y},\ \varepsilon\in(0,1] \\ e^{\varepsilon\beta_\varepsilon(B)+t_\varepsilon}\le s\le e^{T+t_\varepsilon} \\ z^\varepsilon(.)\ \text{such that}\ \sup_s\left|z^\varepsilon(s)\right|\le K \end{array} \right\} \right\}$$

is tight.

(A8) (a) $P(y,t,t_o,B\,|\,x)$ *is Borel measurable in* (x,y,t) *for each Borel B and* $t_o>t_s$

(b) *the fixed x system Eq. (4) has a unique solution for each x and* $y(0;x)$.

(A9) (a) *for* $f\in C_0(\mathbf{R}^{n_x})$ *and* $u>0$, $\int f(\tilde{y})\tilde{P}(y,u,t_o,d\tilde{y}|x)$

is bounded and continuous in $(x,y) \in \mathbf{R}^{n_x} x \mathbf{R}^{n_y}$

(b) *if* $z^\varepsilon(\cdot)$ *is a bounded continuous nonanticipative process and*

$$\lim_{\varepsilon \to 0} \sup_{t_o \le s \le u} \left\| z^\varepsilon(s) - x \right\| = 0 \ \ wp1$$

then

$$\lim_{\varepsilon \to 0} \int f(\tilde{y}) \tilde{P}_{z^\varepsilon}(y,u,t_o,d\tilde{y}) = \int f(\tilde{y}) \tilde{P}(y,u,t_o,d\tilde{y} \mid x)$$

with the convergence uniform on compact x sets

(A10) (a) $\int \tilde{P}(y,u,t_o,d\tilde{y} \mid x) F_1(x,\tilde{y})$ *is continuous on* $\mathbf{R}^{n_x} x \mathbf{R}^{n_y}$

(b) *if*

$$\lim_{\varepsilon \to 0} \sup_{t_o \le s \le u} \left\| z^\varepsilon(s) - \tilde{x}_\varepsilon \right\| = 0 \ \ wp1 \ \ and \ \left\| x - \tilde{x}_\varepsilon \right\| \to 0$$

then

$$\lim_{\varepsilon \to 0} \int F_1(\tilde{x},\tilde{y}) \tilde{P}_{z^\varepsilon}(y,u,t_o,d\tilde{y}) = \int F_1(x,\tilde{y}) \tilde{P}(y,u,t_o,d\tilde{y} \mid x))$$

and the convergence is uniform on compact (x,y) *sets*

The assumptions most deserving of comment are (A2), (A3), (A4), and (A10). The others are either technical in nature or not restrictive in practice. (A2) is important in that continuity is only required in the first variable, x. In many any adaptive algorithms the measurements (part of $y(\cdot)$ process) pass through limiters or other discontinuous devices. Continuity after averaging with respect to a transition probability measure is all that is needed, i.e. (A10). With regards to (A3) and (A4), it has been recognized in the parameter estimation literature that $F_1(x(\cdot),y(\cdot))$ and $F_2(x(\cdot),y(\cdot))$ should be bounded. H.F. Chen, in his analysis of of an Extended Least Squares algorithm [11 pp.135 & 174] made explicit assumptions on the boundedness of his equivalents of $F_1(x(\cdot),y(\cdot))$ and $F_2(x(\cdot),y(\cdot))$. Goodwin, Gevers and Wertz [14] felt this was unnatural since it places a priori restrictions on the data $(x(\cdot),y(\cdot))$, rather than the functions $F_1(\cdot,\cdot)$ and $F_2(\cdot,\cdot)$ alone. Hypothesis on $(x(\cdot),y(\cdot))$ are often difficult to verify since $(x(\cdot),y(\cdot))$ are solutions to complex systems of stochastic differential equations. In [14] the authors went on to describe the convergence of an estimator of their own devising in which $F_1(\cdot,\cdot)$ and $F_2(\cdot,\cdot)$ were bounded. (A3) and (A4) encompass both of these approaches. These assumptions either posit the boundedness of $F_1(\cdot,\cdot)$ and $F_2(\cdot,\cdot)$ or a less restrictive, but harder to verify data dependent uniform integrability or boundedness assumption. Many parameter estimators have matrix inverses as part of $F_j(\cdot,\cdot)$ and singularities can occur at finite values of x. The most expedient ways to handle this is to alter $F_j(\cdot,\cdot)$ to remove the singularities. Kushner uses truncations to prove a theorem for $F_j(\cdot,\cdot)$ that are unbounded at ∞ [5]. For RPEM, we will use projections to keep $x(\cdot)$ bounded, which in turn will assure that (A3) and (A4) are satisfied. (A5) and (A7) require that $y^{\varepsilon}(\cdot)$ is either bounded or sufficiently stable so large initial conditions die out fairly quickly.

Let $\Rightarrow$ denote weak convergence for either probability measures on $C(\mathbf{R},\mathbf{R}^{n_x})$ or for random vectors in $\mathbf{R}^{n_x}$. The first convergence theorem is

THEOREM 1: *Let* $\{t_\varepsilon\}_{\varepsilon\in(0,1]}$ *such that* $t_\varepsilon \to \infty$ *as* $\varepsilon \to 0$. *Define the accelerated system* $\{x^{\varepsilon}(\cdot)\}$ *via Eq. (2). If:*

(a) A1 - A10 *hold*

(b) *For each initial condition* $x^o(0)$, $\dot{x}^o = \underline{F}(x^o)$ *has a unique solution defined on all* $\mathbf{R}^+$

(c) $x^{\varepsilon}(0) \Rightarrow x^o(0)$

then

(a) $\underline{F}(\cdot)$ *is continuous*

(b) $x^{\varepsilon}(\cdot) \Rightarrow x^o(\cdot)$

The proof of Theorem 1 is lengthy and will be presented in Appendix B. As a simple example of how Theorem 1 is used consider the equations:

$$\dot{x}(t) = \frac{1}{t}y(t)^2 x(t)$$
$$dy(t) = -\frac{1}{2}y(t)dt + \sigma^2 dt \tag{11}$$

The stationary distribution for $y(\cdot)$is $N(0,\sigma^2)$. $x^{\varepsilon}(\cdot)$ is defined by

$$\dot{x}^{\varepsilon}(t) = -y^{\varepsilon}(t)^2 x^{\varepsilon}(t) \qquad y^{\varepsilon}(t) = y(e^{t+t_\varepsilon})$$
$$dy(t) = -\frac{1}{2}y(t)dt + \sigma^2 dw \tag{12}$$

Theorem 1 says if $x^{\varepsilon}(0) \Rightarrow x^o(0)$ then $x^{\varepsilon}(\cdot) \Rightarrow x^o(\cdot)$ where $x^o(\cdot)$ satisfies

$$\dot{x}^o(t) = -E_{x^o(t)}\left\{y^2\right\}x^o(t) = -\sigma^2 x^o(t) \tag{13}$$

In this case, we can explicitly solve Eqs. (12) and (13) to get

$$x^{\varepsilon}(0)\exp(-\int_o^{(\cdot)} y^{\varepsilon}(s)^2 ds) \Rightarrow x^0(0)\exp(-\sigma^2 \times (\cdot)) \tag{14}$$

The averaging in Equation (14), taken in isolation, is not completely obvious since $y^{\varepsilon}(\cdot)$ is neither stationary nor ergodic.

Theorem 2 is gives conditions on Eq. (1) which imply (A1)-(A10).

THEOREM 2: *if* $H_1(\cdot,\cdot)$ *and* $H_2(\cdot,\cdot)$ *of Eq. (2) are such that*

$$y(t) = \begin{bmatrix} y_1(t) \\ y_2(t) \end{bmatrix}$$

$$dy_1 = A_1(x,y_2)y_1\,dt + B_1(x)u\,dt + S_1(x,y_2)\,dw$$
$$\dot{y}_2 = A_2(x,y_2)$$

where

(a) $F_1(\cdot,\cdot)$, $F_2(\cdot,\cdot)$, $A_1(\cdot,\cdot)$ $A_2(\cdot,\cdot)$, $B_1(\cdot)$ *and* $S_1(\cdot,\cdot)$ *are locally bounded, and have locally bounded* 2nd *derivatives.* $S_1(\cdot,\cdot)$ *is bounded.*

(b) $F_1(\cdot,\cdot)$ and $F_2(\cdot,\cdot)$ *have compact x-support.*

(c) $A_1(\cdot,\cdot)$, $A_2(\cdot,\cdot)$, $B_1(\cdot)$, $F_1(\cdot,\cdot)$ and $F_2(\cdot,\cdot)$, *satisfy*

$$\max\left(x^T F_1(x,y_1,y_2),\ \ y_1^T(A_1(x,y_2)y_1 + B_1(x)u(t)),\ \ y_2^T A_2(x,y_2)\right) \le C\left(1+\|x\|^2+\|y_1\|^2+\|y_2\|^2\right)$$

$$\sup_{\|z\|=1}\left\|z^T F_2(x,y_1,y_2)\right\|^2 \le C\left(1+\|x\|^2+\|y_1\|^2+\|y_2\|^2\right)$$

(d) $\exists\ \lambda_o<0$ *such that*

$$\sup_{(x,y)\in\mathbf{R}^{n_x}\times\mathbf{R}^{n_y}} \mathrm{Real}\{\lambda_{\max}(A_1(x,y))\} \le 2\lambda_o < 0$$

and either $u(t) \to 0$ *as* $t\to\infty$ *or* $u(\cdot)$ *is a bounded asymptotically stationary stochastic process.*

(e) *for any fixed* $x\in\mathbf{R}^{n_x}$, *the* y_2 *equation has a unique steady state solution* $y_{2\infty}(x)$, $y_2(t;x)\to y_{2\infty}(x)$ *uniformly on compact x-sets, and* $y_{2\infty}(\cdot)$ *is continuous*

(f) *for each initial condition* (x_o,y_o) *and* $C>0$, *if* $\|z(\cdot)\|\le C$ *then equation* $\dot{y}_2 = A_2(z(t),y_2(t))$ *has a unique solution which is bounded for all time. For each C, the solution bound is uniform on compact* y_o *-sets.*

Let $t_\varepsilon\to\infty$ *and define the accelerated system via Eq. 2 with* $y_\varepsilon=y(e^{t_\varepsilon})$ *and* $\{x_\varepsilon\}$ *a tight collection of random variables such that* x_ε *is independent of* $\{w(s) : s\ge e^{t_\varepsilon}\}$ *Then* A1-A10, *are satisfied and* $\{x^\varepsilon(t) : t\ge 0, \varepsilon>0\}$ *is tight.*

The proof of Theorem 2 is in Appendix B. The boundedness of $u(t)$ in (d) is not essential. If $u(\cdot)$ were, for instance, the solution to a stable stochastic differential equation, it could be included as a state in the "y_1" process.

Weak convergence alone is not enough to capture the asymptotics of $x^\varepsilon(\cdot)$ for small ε. Kushner proves a theorem that does characterize the asymptotics of $x^\varepsilon(\cdot)$ [5, p.155]. His results apply when $x^o(\cdot)$ is a stochastic process and has a unique invariant measure. Kushner's proofs can be specialized to the ODE case to get stronger conclusions. The hypotheses on $x^o(\cdot)$ are

(B1) $x^o(\cdot)$ *is solution to an ODE,* $\dot{x}^o = \underline{F}(x^o)$, *which has a unique solution for each initial condition.*

(B2) (a) *There exists* R_c *has the property:* $x^o(t) \to R_c$ *for any* $x^o(0) \in \mathbf{R}^{n_x}$ *and the convergence is uniform on compact* $(x^o(0))$ *sets.*

(b) Let $\lambda(\cdot)$ *be monotone increasing function, differentiable for* $t>0$. *with* $\lambda(0)=0$. $\forall\ x^o(0) \in \mathbf{R}^{n_x}$ *and fixed* $T>0$,

$\left\| x^{\varepsilon}(\lambda(t+T)) - x^{\varepsilon}(\lambda(t)) \right\| \to 0$ *as* $t \to \infty$ *with the convergence uniform on compact* $x^o(0)$*-sets*

(B3) (a) *If* $\underline{F}(\cdot)$ *is twice continuously differentiable and* $\exists\ \varepsilon_o > 0$ *such that*

$$\forall \rho > 0 \qquad \lim_{T \to \infty} P\left\{ \sup_{\varepsilon < \varepsilon_o} \sup_{t \geq T} \left\| x^{\varepsilon}(t) - x_o \right\| < \rho \right\} \geq \pi_o > 0$$

(b) $x^{\varepsilon}(t)$ *has positive definite covariance for each* $t \geq 0$ *and* $\varepsilon > 0$,

The hypotheses on $\{x^{\varepsilon}(\cdot)\}$ are

(C1) *if* $x^{\varepsilon}(0) \Rightarrow x^o(0)$ *then* $x^{\varepsilon}(\cdot) \Rightarrow x^o(\cdot)$

(C2) *if* $s_{\varepsilon} \to s_o \leq \infty$ *and* $x^{\varepsilon}(s_{\varepsilon}) \Rightarrow x^o(0)$ *then* $x^{\varepsilon}(s_{\varepsilon} + \cdot) \Rightarrow x^o(\cdot)$

(C3) $\exists\ \varepsilon_o > 0$ *such that* $\{\ x^{\varepsilon}(t) : \varepsilon \in (0, \varepsilon_o]$ *and* $t \geq 0\ \}$ *is tight*

Theorem 1 implies both (C1) and (C2). The set in (C3) is a set of random variables, not a set of random processes. (C3) is a strong hypothesis and can be difficult to prove directly, especially if $H_1(\cdot,\cdot)$ and $H_2(\cdot,\cdot)$ depend explicitly upon x variable. For the RPEM analysis we shall assume $F_1(\cdot,\cdot)$ and $F_2(\cdot,\cdot)$ are projected to insure $x^{\varepsilon}(\cdot)$ bounded. It should be noted that Theorem 3 only assumes weak convergence properties of $\{x^{\varepsilon}(\cdot)\}$. No assumption is made on the dynamics of $\{x^{\varepsilon}(\cdot)\}$ and $x^{\varepsilon}(\cdot)$ does not have to be an "accelerated" process.

THEOREM 3: *If* $\{x^{\varepsilon}(\cdot)\}$ *satisfies* C1,C2,C3 *and* $x^o(\cdot)$ *satisfies* B1, *then*

(a) *if* $x^o(\cdot)$ *satisfies* B2(a), *then* $\exists\ \varepsilon_o > 0$ *such that* $\forall\ \varepsilon < \varepsilon_o$ $x^{\varepsilon}(t) \to R_c$ *wp1 as* $t \to \infty$

if $x^o(\cdot)$ *satisfies* B2(b), *then* $\exists\ \varepsilon_o > 0$ *such that* $\forall\ \varepsilon < \varepsilon_o$ *and fixed* $T>0$ $\left\| x^{\varepsilon}(\lambda(t+T)) - x^{\varepsilon}(\lambda(t)) \right\| \to 0$ *wp1 as* $t \to \infty$

(b) *if* B3(a) *holds then* $\underline{F}(x_o) = 0$. *If in addition* B3(b) *holds then* $\underline{F}_x(x_o)$ *has all its eigenvalues in the (closed) left half plane.*

THEOREM 4: *Assume $x(\cdot)$ satisfies the hypothesis of Theorem 2 and for each initial condition the ODE $\dot{x}^o = \underline{F}(x^o)$ has a unique solution defined for all time .*

(a) *if for any initial conditions $x^o(t) \rightarrow R_c$, then $x(t) \rightarrow R_c$ wp1 as $t \rightarrow \infty$*

(b) *if for any initial conditions $\left\| x^o(\log(t+T)) - x^o(\log(t)) \right\| \rightarrow 0$, then $\|x(t+T) - x(t)\| \rightarrow 0$ wp1 as $t \rightarrow \infty$*

(c) *if $\underline{F}(\cdot)$ is twice continuously differentiable, $x(t) \rightarrow x_\infty$ with probability strictly greater than 0 and $x(t)$ has positive definite covariance for each t, then x_∞ is a stable stationary point for the $x^o(\cdot)$-ODE equation (i.e. $\underline{F}(x_\infty)=0$ and $\underline{F}_x(x_\infty)$ has all its eigenvalues in the closed left half plane).*

Theorem 4 characterizes the asymptotics of $x(\cdot)$ in terms of the ODE limit $x^o(\cdot)$. 4a concerns convergence and 4c can be used to show a point is not a convergence point. 4b is useful for showing the stability of adaptive controllers. Of course, proofs similar to that of Theorem 3 can be used to derive other relationships between $x^\varepsilon(\cdot)$ and $x^o(\cdot)$.

Theorem 3 depends only upon the weak convergence $x^\varepsilon(\cdot) \Rightarrow x^o(\cdot)$ and not upon the gain sequence $1/t$. The natural question is whether the theory still works for other gain sequences $\gamma(\cdot)$. If $\Gamma(\cdot)$ satisfies

$$t = \int_{\Gamma(0)}^{\Gamma(t)} \gamma(\tau)\, d\tau$$

and $x(\Gamma(\cdot + t_\varepsilon)) \Rightarrow x^o(\cdot)$, where $x^o(\cdot)$ is the solution of an ODE, then Theorems 3 and 4 may still apply with $\Gamma^{-1}(t)$ instead of *log*(t) in Theorem 4(b). Clearly $\gamma(t) \rightarrow 0$ is needed for an ODE limit and $\Gamma(t) \rightarrow \infty$ is needed to characterize asymptotics. Work by Baldi [15] indicates $\gamma(t) = o((\log t)^{-1/2})$ is the slowest gain sequence for an ODE limit. It is not clear how to modify the proof of Theorem 1 to get weak convergence for $\gamma(t) = o((\ln t)^{-1/2})$, but the modification is straightforward for $\gamma(t) = o(t^{-1/2})$. All that is involved is sharpening some bounds in Lemmas B.2 and B.5, and modifying the choice of δ_ε preceding Lemma B.2.

Theorem 4 comprises the continuous time ODE method, relating $x(.)$ to its ODE limit $x^o(\cdot)$, and is the ultimate goal of this section. Theorems 1 and 2 give conditions for the existence of an ODE limit and specify how to calculate $x^o(\cdot)$. It was indicated that convergence theorems can be proved for gain sequences $\gamma(t)$ other than $1/t$, provided $\gamma(t) = o(t^{-1/2})$ and that $\gamma(t) = o(t^{-1/2})$ probably is not the slowest possible sequence.

III. CONVERGENCE OF THE CONTINUOUS TIME RPEM

Recursive Prediction Error Methods (RPEM) are popular discrete time estimators whose continuous time analogues have not been analyzed in detail. Using his ODE technique, Ljung has shown that discrete time RPEM parameter estimators converge a local minima of certain cost functionals [16]. In this section, the continuous time RPEM applied to the state space model will be shown to converge to a local minimum of $J(\theta) = Trace(C^T R^{-1} C\, P_x)$, where P_x is the asymptotic state error covariance of a related Kalman Filter. Thus, in some sense, RPEM chooses the parameters to minimize the state estimation error. The true parameter values (if they exist) are shown to be a global minimum of $J(\theta)$. RPEM is useful for comparison to other more complex parameter estimators, for instance the Wiberg Filter [2]. A direct analysis of parameter estimators such a the Extended Kalman Filter and the Wiberg Filter is difficult since these filters are approximations to the optimal filter and themselves have no inherent optimality properties. Since RPEM is derived via a cost criterion minimization, the convergence is easier to prove.

The equations for RPEM are ([3])

$$dx = A_o x\, dt + B_o u\, dt + \sigma_o dw_1 \qquad \Sigma_o = \sigma_o \sigma_o^T$$

$$dy = Cx\, dt + \rho\, dw_2 \qquad R = \rho\rho^T$$

$$d\varepsilon = dy - C\hat{x}\, dt$$

$$d\hat{x} = A(\theta)\hat{x}\, dt + B(\theta)u\, dt + K\, d\varepsilon \quad K = PC^T R^{-1}$$

$$d\theta = \frac{1}{t} V^{-1} (C\underline{W})^T \Lambda^{-1} d\varepsilon$$

$$\dot{V} = \frac{1}{t}\left((C\underline{W})^T \Lambda^{-1} (C\underline{W}) \;+\; V_o \;-\; V\right)$$

$$\dot{P} = A(\theta)P + PA(\theta)^T + \sigma(\theta)\sigma(\theta)^T - PC^T R^{-1} CP$$

$$\underline{W} = \begin{bmatrix} W_1 & \cdots & W_{n_p} \end{bmatrix}$$

$$dW_i = (A - KC)W_i\, dt + \frac{\partial A}{\partial \theta_i}\hat{x}\, dt + \frac{\partial B}{\partial \theta_i} u\, dt + \frac{\partial K}{\partial \theta_i} d\varepsilon \qquad \left(W_i = \frac{\partial \hat{x}}{\partial \theta_i}\right)$$

$$\frac{\partial K}{\partial \theta_i} = \Pi_i C R^{-1} \qquad \Pi_i = \frac{\partial P}{\partial \theta_i}$$

$$\dot{\Pi}_i = \frac{\partial A}{\partial \theta_i} P + P \frac{\partial A^T}{\partial \theta_i} + A\Pi_i + \Pi_i A^T - \Pi_i C^T R^{-1} CP - PC^T R^{-1} \Pi_i$$
$$+ \sigma \frac{\partial \sigma^T}{\partial \theta_i} + \frac{\partial \sigma}{\partial \theta_i} \sigma^T \tag{1}$$

A_o, B_o , σ_o, ρ, and R are the state space model parameters. ρ and R are known while A_o , B_o and σ_o are unknown. It is also assumed that all possible A's, B's and σ's are parameterized by known twice differentiable functions, $A(\cdot)$, $B(\cdot)$, $\sigma(\cdot)$. If there exists a θ_o such that $(A_o, B_o, \sigma_o)=(A(\theta_o),B(\theta_o),\sigma(\theta_o))$, then θ_o will be referred to as the true parameter values. Equation (1) is a Kalman Filter with additional equations describing the evolution of the parameter estimates $\theta(t)$. Note that A and σ in the Riccati equations are functions of θ and therefore time varying. Unlike the discrete time case, there is no a priori assurance that Eq. (1) has a unique solution or any solutions at all. Fortunately, Theorem II.2 will assure existence and uniqueness of solutions as well as allowing the use of Theorems II.1 and II.4. To use Theorem II.1, we need to express Eq. (1) in terms of the underlying Wiener processes, $w_1(\cdot)$ and $w_2(\cdot)$, and make the following associations:

$$\begin{aligned} \underline{x}^T &= \begin{bmatrix} \theta^T & V^T \end{bmatrix} \\ \underline{y_1}^T &= \begin{bmatrix} W_1^T & \cdots & W_{n_p}^T & \tilde{x}^T & x^T \end{bmatrix} \\ \underline{y_2}^T &= \begin{bmatrix} P & \Pi_1 & \cdots & \Pi_{n_p} \end{bmatrix} \\ \underline{y}^T &= \begin{bmatrix} \underline{y_1}^T & \underline{y_2}^T \end{bmatrix} \end{aligned} \tag{2}$$

Two things should be noted here. $\underline{x}$ refers to the "x" in Chapter II and Appendix B, while x refers to the state vector of Eq. (1) and similarly for $\underline{y}$ and y. The $\underline{x}$ and $\underline{y_2}$, as written above have matrices as entries. This is for notational convenience and should not cause any confusion. Using this notation, the functions of Theorems II.1 and II.2 are

$$F_1(\underline{x},\underline{y}) = \begin{bmatrix} V^{-1}\underline{W}^T C^T R^{-1} C\tilde{x} \\ \underline{W}^T C^T R^{-1} C\underline{W} + V_o - V \end{bmatrix}$$

$$F_2(\underline{x},\underline{y}) = \begin{bmatrix} V^{-1}\underline{W}^T C^T R^{-1}\rho \\ 0 \end{bmatrix}$$

$$A_1(\underline{x},\underline{y_2}) = \begin{bmatrix} A-KC & \cdots & \underline{0} & -\left(\frac{\partial A}{\partial \theta_1} - \frac{\partial K}{\partial \theta_1} C\right) & \frac{\partial A}{\partial \theta_1} \\ \vdots & \ddots & \vdots & \vdots & \vdots \\ \underline{0} & \cdots & A-KC & -\left(\frac{\partial A}{\partial \theta_{n_p}} - \frac{\partial K}{\partial \theta_{n_p}} C\right) & \frac{\partial A}{\partial \theta_{n_p}} \\ \underline{0} & \cdots & \underline{0} & A-KC & A_o - A \\ \underline{0} & \cdots & \underline{0} & \underline{0} & A_o \end{bmatrix}$$

$$B_1(\underline{x}) = \begin{bmatrix} \frac{\partial B}{\partial \theta_1} \\ \vdots \\ \frac{\partial B}{\partial \theta_{n_p}} \\ B - B_o \\ B_o \end{bmatrix}$$

$$S_1(\underline{x}, \underline{y_2}) = \begin{bmatrix} \underline{0} & -\Pi_1 C^T R^{-1} \rho \\ \vdots & \vdots \\ \underline{0} & -\Pi_{n_p} C^T R^{-1} \rho \\ \sigma_o & -PC^T R^{-1} \rho \\ \sigma_o & \underline{0} \end{bmatrix}$$

$$A_2(\underline{x}, \underline{y_2}) = \begin{bmatrix} (A-KC)P + P(A-KC)^T + \sigma\sigma^T + PC^T R^{-1} CP \\ (A-KC)\Pi_i + \Pi_i(A-KC)^T + \\ \left(\frac{\partial A}{\partial \theta_i} P + P \frac{\partial A^T}{\partial \theta_i} + \frac{\partial \sigma}{\partial \theta_i} \sigma^T + \sigma \frac{\partial \sigma^T}{\partial \theta_i} \right) \\ 1 \le i \le n_p \end{bmatrix} \quad (3)$$

We need to verify that RPEM satisfies the hypotheses of Theorem II.2. Hypothesis (a) requires that all the functions be locally bounded and have continuous locally bounded 2nd derivatives. This is true of all the functions except $F_1(\cdot,\cdot)$ and $F_2(\cdot,\cdot)$ at V=0. The differential equation for V forces $V \ge V_o$, so the solution never approaches the singularity. We can force $F_1(\cdot,\cdot)$ and $F_2(\cdot,\cdot)$ to satisfy (a) by multiplying each by a C^∞ scalar function which is 0 on $\{V : V \le V_o/2\}$ and 1 on $\{V : V \ge V_o\}$ ($\le$ is taken in the sense of symmetric quadratic forms). This projection is purely formal and does not affect the realizability of the filter. Similar formal projections can be used to enforce hypothesis (b) which requires $F_1(\cdot,\cdot)$ and $F_2(\cdot,\cdot)$ have compact $\underline{x}$-support. The boundedness of $S_1(\cdot,\cdot)$ is a greater problem since it is not accessible to the filter designer and a simple projection would not be realizable. Projections on $A_2(\cdot,\cdot)$ can be implemented by the filter designer, hence we can enforce the boundedness of P and Π_i. Then, as before, use a "formal" projection on $S_1(\cdot,\cdot)$. This also enforces hypothesis (f).

Hypothesis (c) is

$$\max\left(\underline{x}^T F_1(\underline{x}, \underline{y_1}, \underline{y_2}),\ y_1^T\left(A_1(\underline{x}, \underline{y_2}) y_1 + B_1(\underline{x}) u(t)\right), y_2^T A_2(\underline{x}, \underline{y_2})\right) \le C\left(1 + \|\underline{x}\|^2 + \|\underline{y_1}\|^2 + \|\underline{y_2}\|^2\right)$$

$$\sup_{\|z\|=1} \left\| z^T F_2(\underline{x}, \underline{y_1}, \underline{y_2}) \right\|^2 \le C\left(1 + \|\underline{x}\|^2 + \|\underline{y_1}\|^2 + \|\underline{y_2}\|^2\right) \quad (4)$$

With the projections of the previous paragraph $F_1(\cdot,\cdot)$ and $F_2(\cdot,\cdot)$ are bounded in $\underline{x}$, are at worst quadratic in $\underline{y}_1$, linear in $\underline{y}_2$ and therefore satisfy Eq. (4). If Hypothesis (d) holds, then the $A_1(\cdot,\cdot)$ term in Eq. (4) is strictly negative and satisfies Eq. (4). Since $F_1(\cdot,\cdot)$ and $F_2(\cdot,\cdot)$ have compact x-support $A(\theta)$, $B(\theta)$, and $\sigma(\theta)$ can be made bounded by formal projections. Thus $\underline{x}$ does not contribute any cubic cross terms in Eq. (4). The $A_2(\cdot,\cdot)$ term has cubic factors in the variables P and Π_i and could potentially violate Eq. (4). Examination of Eq. (3) shows the cubic terms in $\underline{y_2}^T A_2(\underline{x},\underline{y_2})$ will have "-" signs, hence if P and Π_i are projected to remain non-negative definite, the $A_2(\cdot,\cdot)$ term will also satisfy Eq. (4). To satisfy Hypothesis (d), $A_1(\cdot,\cdot)$ must be a strictly stable matrix for any value of $\underline{x}$ or $\underline{y}_2$. The eigenvalues of $A_1(\cdot,\cdot)$ are the eigenvalues of $A(\theta)$-$K(\theta)C$ and A_o. Thus only stable plants can be analyzed by the ODE method and conditions to insure the stability of $A(\theta)$-$K(\theta)C$ are needed. To satisfy hypothesis (e), θ can be projected into a closed subset of $\{\theta : (A(\theta),\sigma(\theta))$ stabilizable and $(C,A(\theta))$ detectable$\}$. Techniques for doing this are derived in [17]. The continuity requirements in hypothesis (e) follow from [18]. To summarize, the required projections are:

(1) P must be kept positive semidefinite and bounded and Π_i must be kept bounded

(2) $F_1(\cdot,\cdot)$ and $F_2(\cdot,\cdot)$ must be truncated as functions of θ and V.

(3) $A(\theta)$-$K(\theta)C$ must be kept stable

(4) $(A(\theta),\sigma(\theta))$ stabilizable and $(C,A(\theta))$ detectable.

Given these projections the RPEM satisfy the hypotheses of Theorem II.2, which in turn implies RPEM can be analyzed using Theorems II.1 and II.4.

The ODE for RPEM is

$$\dot{\theta}^o = (V^o)^{-1} E_\theta\left\{\underline{W}^T C^T R^{-1} \tilde{x}\right\}$$
$$V^o = E_\theta\left\{\underline{W}^T C^T R^{-1} C \underline{W} + V_o\right\} \tag{5}$$

E_θ is expectation with respect to the asymptotic probability measure of the system:

$$d\begin{bmatrix} W_1 \\ \vdots \\ W_{n_p} \\ \tilde{x} \\ x \end{bmatrix} = A_1(\theta,P,\Pi_i)\begin{bmatrix} W_1 \\ \vdots \\ W_{n_p} \\ \tilde{x} \\ x \end{bmatrix} dt + B_1(\theta)u(t)dt + S_1(\theta,P,\Pi_i)\begin{bmatrix} dw_1 \\ dw_2 \end{bmatrix} \tag{6}$$

where θ *is held constant* and (P,Π_i) satisfy the steady state Riccati and Lyapunov equations (with the same θ):

$$0 = (A(\theta) - K(\theta)C)P + P(A(\theta) - K(\theta)C)^T + \sigma(\theta)\sigma(\theta)^T + PC^T R^{-1} CP$$

$$0=(A(\theta)-K(\theta)C)\Pi_i+\Pi_i(A(\theta)-K(\theta)C)^T+\left(\frac{\partial A(\theta)}{\partial\theta_i}P+P\frac{\partial A(\theta)^T}{\partial\theta_i}+\frac{\partial\sigma(\theta)}{\partial\theta_i}\sigma(\theta)^T+\sigma(\theta)\frac{\partial\sigma(\theta)^T}{\partial\theta_i}\right) \tag{7}$$

Since $W_i=-\frac{\partial\tilde{x}}{\partial\theta_i}$ the first part of Eq. (5) is equivalent to

$$\dot{\theta}^o=-\tfrac{1}{2}(V^o)^{-1}\frac{\partial}{\partial\theta}E_\theta\left\{\tilde{x}^TC^TR^{-1}C\tilde{x}\right\}=-(V^o)^{-1}\frac{\partial}{\partial\theta}J(\theta)$$
$$J(\theta)=\tfrac{1}{2}Trace\left(C^TR^{-1}C\,P_x(\theta)\right) \tag{8}$$

where $P_x(\theta)=E_\theta\left\{\tilde{x}\,\tilde{x}^T\right\}$, the asymptotic state error covariance for the fixed θ system. Since $(V^o)^{-1}$ is positive definite, the ODE in (8) converges to a local minima of the function $J(\cdot)$. Thus Theorem II.4 says RPEM can only converge to a local minima of $J(\cdot)$ (with positive probability).

It is easy to show that the true parameter values, θ_o, are a global minimum of $J(\cdot)$. For each fixed θ, Eq. (6) defines a linear estimator for $x(t)$ based upon $\{y(s) : s\le t\}$ and this estimator is the Kalman Filter (initialized with the steady state covariance) if $\theta=\theta_o$. The asymptotic optimality of the Kalman Filter implies $P_x(\theta_o)\le P_x(\theta)$ for any θ. This implies $J(\theta_o)\le J(\theta)$.

Now consider the case where the true parameters, θ_o, are an isolated stationary point for $J(\cdot)$ and R_c is a compact neighborhood of θ_o. If the gradient of $J(\cdot)$ points inward on the boundary of R_c and $\theta(t)$ is projected to remain in R_c, then $\theta(t)$ will converge to the true parameters with probability one.

These results are summarized in

THEOREM 1: *Assume that A_o is stable and the RPEM parameter estimator given in Eq. (1) is projected such that:*

(1) *P is positive semidefinite and bounded and Π_i is bounded*

(2) *$A(\theta)$-$K(\theta)C$ is kept stable*

(3) *$(A(\theta),\sigma(\theta))$ stabilizable and $(C,A(\theta))$ detectable*

(4) *$F_1(\cdot,\cdot)$ and $F_2(\cdot,\cdot)$ must be truncated as functions of θ and V. (i.e. forced to have compact support)*

Then:

(a) *$\theta(t)$ can only converge (with positive probability) to local minima of the cost function $J(\theta)=\tfrac{1}{2}Trace\left(C^TR^{-1}C\,P_x(\theta)\right)$*

(b) *If there are true parameter values, θ_o, then they are global minima of $J(\cdot)$.*

(c) *if*

(1) *θ_o, is an isolated local minima of $J(\cdot)$*

(2) *R_c is a compact neighborhood of θ_o with smooth boundary*

(3) $$\left(\frac{\partial J}{\partial \theta}\right)^T (\theta - \theta_o) < 0 \quad \forall \in \partial R_c$$

(4) *RPEM is projected so that*

$$\theta(t) \in R_c \qquad \forall t \geq 0 \quad wp1$$

then $\theta(t) \to \theta_o \qquad wp1$

Theorem 1(a) characterizes the potential convergence points of RPEM algorithms. Theorem 1(b,c) says that RPEM is asymptotically unbiased, at least in a local sense. Unfortunately this does not tell us if $\theta(t)$ is an unbiased estimate of θ_o for finite t.

A detailed examination of the local minima of $J(\theta)$ is beyond the scope of this dissertation. Whether there are local minima other than θ_o depends upon many details of the state space model. For instance B_o could not be estimated if $u(t)$ were identically zero. Furthermore, in discrete time, it is known that the input/output model structure (e.g. ARX, ARARX, ARMAX or ARMA) which leads to the state space model influences the nature of the local minima. In some cases there are no local minima other than the global minimum (ARX), in other cases this is only true for sufficiently high SNR (ARARX), or there are no definitive results (ARMAX) [19]. In the next section,the ODE will be calculated for a specific instance of RPEM with more satisfying results.

In this section, the potential convergence points of the continuous time RPEM (applied to the state space model) have been shown to be the local minima of $J(\theta) = Trace(C^T R^{-1} C\, P_x)$,where P_x is the asymptotic state error covariance of a related Kalman Filter. The true parameter values (if they exist) were shown to be global minima of this cost function. Finally, a condition for wp1 convergence was given.

IV. NUMERICAL ODE EXAMPLES

This section is a brief introduction to numerical calculation of the ODE limit and its utility in analyzing the asymptotics of a specific implementation of a parameter estimator. One of the advantages of the ODE approach is that it gives incisive results where Monte Carlo simulation results are obscure. A general discussion on calculating the ODE for RPEM is followed by a specific case.. The RPEM ODE is evaluated for a scalar plant case and the results contrasted with Monte Carlo simulation results.

As shown in Section III, the ODE for RPEM equations are

$$\dot{\theta}^o = (V^o)^{-1} E_\theta\left\{\underline{W}^T C^T R^{-1} C\tilde{x}\right\} = \underline{F}(\theta^o)$$
$$V^o = E_\theta\left\{\underline{W}^T C^T R^{-1} C\underline{W} + V_o\right\} \tag{1}$$

E_θ is expectation with respect to the asymptotic probability measure of the system:

$$d\begin{bmatrix} W_1 \\ \vdots \\ W_{n_p} \\ \tilde{x} \\ x \end{bmatrix} = A_1(\theta, P, \Pi_i)\begin{bmatrix} W_1 \\ \vdots \\ W_{n_p} \\ \tilde{x} \\ x \end{bmatrix} dt + B_1(\theta)u(t)dt + S_1(\theta, P, \Pi_i)\begin{bmatrix} dw_1 \\ dw_2 \end{bmatrix} \tag{2}$$

where θ *is held constant* and (P, Π_i) satisfy the steady state Riccati and Lyapunov equations (with the same θ):

$$0 = (A(\theta) - K(\theta)C)P + P(A(\theta) - K(\theta)C)^T + \sigma(\theta)\sigma(\theta)^T + PC^T R^{-1} CP$$
$$0 = (A(\theta) - K(\theta)C)\Pi_i + \Pi_i(A(\theta) - K(\theta)C)^T +$$
$$\left(\frac{\partial A(\theta)}{\partial \theta_i}P + P\frac{\partial A(\theta)^T}{\partial \theta_i} + \frac{\partial \sigma(\theta)}{\partial \theta_i}\sigma(\theta)^T + \sigma(\theta)\frac{\partial \sigma(\theta)^T}{\partial \theta_i}\right) \tag{3}$$

The problem is to calculate the function $\underline{F}(\cdot)$, defined in Eq. (1), in the vicinity of the true parameter values, then use Theorem II.4 or III.1 to analyze the convergence. To calculate $\underline{F}(\theta)$, first find the solutions to Eq. (3), the steady state Riccati and Lyapunov equations, for this fixed θ. The next step is to find the steady state probability distribution for Eq. (2). For simplicity assume that $u(\cdot)$ is deterministic and $u(t)$ goes to zero as t goes to ∞. The process described by Eq. (2) is Gaussian and asymptotically $\sim N(0,Z)$, where Z solves

$$A_1(\theta, P, \Pi_j, j = 1, \ldots, n_p)Z + ZA_1(\theta, \cdots)^T = -S_1(\theta, \cdots)S_1(\theta, \cdots)^T \tag{4}$$

Once Z is found, the expectations in Eq. (1) can be calculated. Hand calculation of these quantities is daunting even in low dimensional cases, but numerical calculation with MATLAB or similar tools is straightforward.

As an example, consider the case

$$dx = a_o x\,dt + b_o u(t)dt + \sigma_o dw_1$$
$$dy = x\,dt + dw_2$$
$$a_o = -1 \qquad b_o u(t) = 2\cos(2\pi(.2)t + \varphi) \qquad \sigma_o = 1 \tag{5}$$

where a_o is to be estimated and (b_o, σ_o) are assumed known. Equations (2) and (3) become

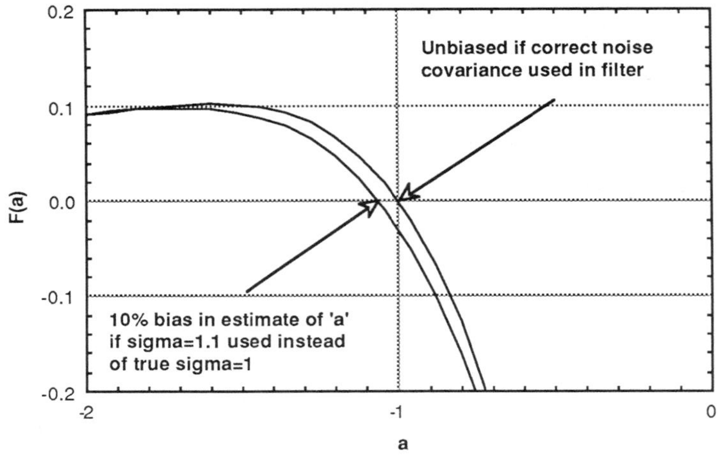

Fig. 1: The ODE method shows that RPEM will converge to the true parameter value (a=-1) provided there is no mismodeling. Slight mismodeling of the state noise covariance causes biases in the 'a' estimate. The ODE is $\dot{a} = \underline{F}(a)$.

$$d\begin{bmatrix} W_a \\ \tilde{x} \\ x \end{bmatrix} = \begin{bmatrix} a-p & 1-\pi_a & 1 \\ 0 & a-p & 1 \\ 0 & 0 & -1 \end{bmatrix}\begin{bmatrix} W_a \\ \tilde{x} \\ x \end{bmatrix} dt + \begin{bmatrix} 0 \\ 0 \\ 2 \end{bmatrix} u(t)dt + \begin{bmatrix} 0 & -\pi_a \\ \sigma_1 & -p \\ \sigma_1 & 0 \end{bmatrix}\begin{bmatrix} dw_1 \\ dw_2 \end{bmatrix} \tag{6}$$

$$\begin{aligned} 0 &= 2(a-p)p - p^2 + \sigma_1^2 \\ 0 &= 2(a-p)\pi_a + 2ap \end{aligned} \tag{7}$$

MATLAB was used to find the steady state covariance for Eq. (6) and to calculate $\underline{F}$(a). The results are shown in Fig. 1. The upper curve corresponds to σ_1=σ_o=1, the true state noise standard deviation. $\underline{F}$(a) crosses the x axis at a=-1=a_o and the slope is negative, hence the true parameter value is a stable stationary point. If RPEM converges it must converge to -1. Provided RPEM is projected to remain in a neighborhood of -1, this will be the only stationary point (in fact, a closed form analysis of the ODE shows -1 is the only zero). Theorem II.4 tells us that, in this case, RPEM will converge wp1 to the true parameter value.

To illustrate another use for the ODE method, σ_1 in Eqs.(6) and (7) was set to 1.1 rather than the correct value σ_o=1. This is a case of mismodeling in the filter design. The lower curve in Fig. 1 shows that mismodeling causes RPEM to be asymptotically biased. If the bias over the range of anticipated σ's is unacceptable, then a more complex estimator for both a and σ could be

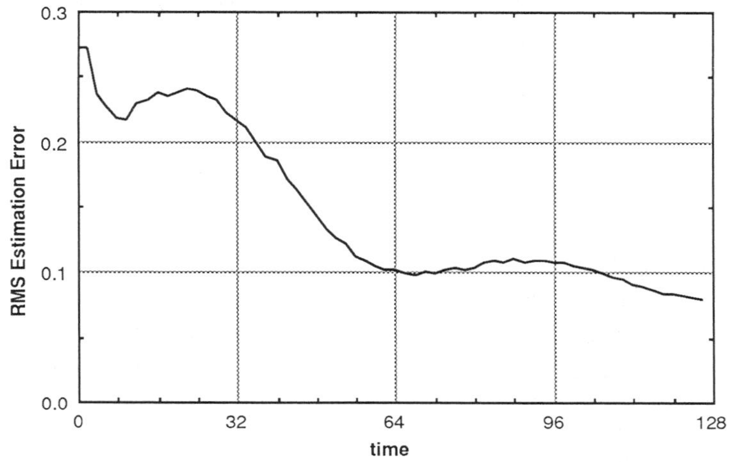

Fig. 2: RMS error estimating 'a' using RPEM. The true value of a is -1. The results are averages of 100 Monte Carlo trials.

designed. That mismodeling causes biases is expected, but the ODE method gives a simple way to evaluate the severity of the problem.

As a check on the theoretical results, the stochastic differential equations for RPEM were simulated. 100 Monte Carlo trials of 128 seconds each were run and the RMS estimation error for 'a' was calculated. The system parameters in Eq. (5) were used. The initial state (x) estimation error was ~$N(0,1)$, while the initial parameter estimation error was ~$N(0,.3)$. 512 Wiener process increments per second are used in the stochastic differential equation approximation and the estimated error in the approximation is .1%. The methods used and theory of approximating stochastic differential equations are discussed in Appendix C.

Fig. 2 shows the RMS error going to zero as the theory requires. Since the error is has only reached ~10% at 128 seconds, this level of Monte Carlo simulation is not sensitive enough for quantifying small biases that might arise through mismodeling. This could be overcome by simulating longer time segments, but as it was 100 Monte Carlo runs were taking 12 hours on a Macintosh II computer (68020 CPU with 68881 coprocessor at 16 mhz). In contrast, the ODE analysis took ~30 seconds using MATLAB on the same computer.

This section indicates how the ODE method could be used to analyze specific implementations of a parameter estimator. The ODE limit is relatively easy to calculate and an ODE analysis consumes much less computer time than Monte Carlo simulation. In the simple scalar example, RPEM will converge to the true parameter value with probability one. This is heartening since general convergence results of this strength escaped us in Section III.

V. SUMMARY AND CONCLUSIONS

The primary result presented here is the formulation and proof of the ODE method for analyzing continuous time parameter estimators. To illustrate the use of the ODE method the Recursive Prediction Error Method was analyzed, both in the general case and numerically in a scalar example.

The majority of the work is proving an abstract weak convergence result relating the process defined by:

$$dx = \gamma(t)(F_1(x,y)dt + F_2(x,y)dw) \quad (1)$$

$$dy = H_1(x,y)dt + H_2(x,y)dw \quad (2)$$

to the solutions of the ODE:

$$\dot{x}^o = \underline{F}(x^o) \qquad \underline{F}(x) = E_x\{F_1(x,y)\} \quad (3)$$

$E_x\{\ \}$ is expectation with respect to the stationary measure of Eq. (2) when x is held constant. The weak convergence was explicitly proved for the case $\gamma(t) = 1/t$, but it was pointed out that the proof goes through with $\gamma(t) \sim o(t^{-1/2})$ provided $\int_1^t \gamma(s)ds = \infty$. Extending the range of possible gain sequences is one avenue for further research.

Once we have weak convergence, the tightness of $\{\ x(t) : t \geq 0\}$ will give the asymptotic results:

(a) If, for any initial $x^o(0)$, $x^o(t) \to x_1$, then $x(t) \to x_1$ wp1

(b) If $x(t) \to x_1$ with probability greater than 0, then x_1 is a stable stationary point of $\underline{F}(\cdot)$

(c) If $\gamma(t) = 1/t$ and if for any initial $x^o(0)$, we have that if $x^o(\ln(t+T)) - x^o(\ln(t)) \to 0$, then $x(t+T) - x(t) \to 0$ wp1

(a) and (b) say that the asymptotics of a convergent parameter estimator are completely characterized by its limiting ODE. In itself, (c) is useful for proving stability of adaptive controllers. The proof of (c) can be adapted to other convergence relationships. Kushner does this for stochastic limits ([5], p. 155) and there is no reason this could not be done for deterministic ODE limits.

The abstract theory was applied to Recursive Prediction Error parameter estimators. RPEM can only converge local minima of a cost function derived in Section III and the true parameter values (if there are any) are a global minimum of this cost function. One avenue of research is to analyze the cost function for various signal models and attempt to derive stronger wp1 convergence results. Another avenue is to analyze other parameter estimators using the ODE approach and compare the results to existing martingale convergence results ([11], [12}, [13],[14]).

In Section IV, the theoretical convergence results were compared to Monte Carlo simulation results. A numerical ODE analysis was shown to be much more economical than Monte Carlo simulation for analyzing the (asymptotic) bias due to mismodelling (30 CPU seconds vs 12 CPU hours). In this particular example, the ODE analysis showed RPEM converges wp1 to the true parameter value. This, and other examples not reported here, indicate RPEM has stronger convergence properties than derived in Section III and further theoretical analysis is warranted.

APPENDIX A: MATHEMATICAL BACKGROUND

The mathematical background for the weak convergence theorems includes the notions of weak convergence, tightness, relative compactness and theorems for characterizing them in different probability spaces. The definitions will be made for probability measures on metric spaces. Theorems for characterizing tightness will be stated for random vectors in $\mathbf{R}^n$ and random processes in $C(\mathbf{R},\mathbf{R}^n)$. All the definitions and theorems in this appendix are standard and are reproduced for convenient reference. Most of the theorems are taken from [20].

Let $(B,d(\cdot,\cdot))$ be a metric space with the σ algebra of Borel sets $\mathfrak{B}$. Let $\{P_\alpha\}_{\alpha\in A}$ (or just $\{P_\alpha\}$) be a set of measures on $(B,\mathfrak{B})$. If A is countable the collection will be denoted by $\{P_k\}$.

DEFINITION 1: $\{P_\alpha\}$ *is* tight *if* $\forall\ \eta>0\ \exists$ *a compact set* B_η *such that* $\inf\{P_\alpha(B_\eta) : \alpha\in A\}>1-\eta$.

DEFINITION 2: $\{P_k\}$ converges weakly *to* P ($P_k\Rightarrow P$) *if* $\forall\ f\in C(B)$

$$\lim_{k\to\infty}\int f(x)P_k(dx)=\int f(x)P(dx) \tag{1}$$

DEFINITION 3: $\{P_\alpha\}$ *is* relatively compact *if every subsequence has a weakly convergent (sub)subsequence.*

A set of random variables on B, $\{X_\alpha\}$ or $\{X_k\}$, is said to be tight, weakly convergent or relatively compact if their corresponding measures are tight, weakly convergent or relatively compact respectively. The definition of tightness makes sense (and is important) even if $\{P_\alpha\}$ consists of a single measure.

THEOREM 1:([20], p.10) *If* B *is separable and complete then each probability measure is tight.*

THEOREM 2: *If* $B=\mathbf{R}^n$ *and* (1) *holds for* $f\in C^k_o(\mathbf{R}^n)$, *then* $P_k\Rightarrow P$.

THEOREM 3:([20], p.20) *Let* $B=B_1\times B_2$ *and assume* B *is separable. If* $\{P_{1n}\}$ *and* $\{P_{2n}\}$ *are measures on* B_1 *and* B_2 *converging to* P_1 *and* P_2. *Then* $P_{1n}\times P_{2n}\Rightarrow P_1\times P_2$ *if and only if* $P_{1n}\Rightarrow P_1$ *and* $P_{2n}\Rightarrow P_2$.

THEOREM 4:([20], p.30) *Let* $X_k\Rightarrow X$ *(with associated measures* $P_k\Rightarrow P$*) and* $f(\cdot)$ *be a measurable function whose set of discontinuities has P-measure* 0. *Then*

(a) $f(X_k)\Rightarrow f(X)$

(b) *if $f(\cdot)$ is bounded then* $Ef(X_k) \rightarrow Ef(X)$

(c) *if* A *is Borel and* $P(\partial A)=0$ *then* $P_k\{X_k \in A\} \rightarrow P\{X \in A\}$

THEOREM 5:([20], p.25)

(a) *If* $X_k \Rightarrow X$ *and* $d(X_k, Y_k) \rightarrow 0$ *in probability then* $Y_k \Rightarrow X$.

(b) *Assume* $X_{mn} \Rightarrow X_n$ *as* $m \rightarrow \infty$ *and* $X_n \Rightarrow X$ *as* $n \rightarrow \infty$. *If*

$$\forall \eta > 0 \qquad \lim_{m\to\infty} \limsup_{n\to\infty} P\{d(X_{mn}, Y_n) > \eta\} = 0$$

then $Y_n \Rightarrow X$ *as* $n \rightarrow \infty$.

THEOREM 6:([20], p.37)

(a) *If* $\{P_\alpha\}$ *is tight then* $\{P_\alpha\}$ *is relatively compact.*

(b) *If* B *is separable and complete then relative compactness implies tightness*

THEOREM 7:([5], p.29) *Let* $(B, d(\cdot,\cdot))$ *be a complete separable metric space. Let* $X_k \Rightarrow X$ *be* B *valued random variables (with associated measures* $P_k \Rightarrow P$*). Then there is a probability space* (Ω, W) *with* B *valued random variables* X'_k *and* X' *with corresponding measures* P'_k *and* P' *such that*

(a) *for any Borel set* $A \subseteq B$, $P'_k\{X'_k \in A\} = P_k\{X_k \in A\}$

$$P'\{X' \in A\} = P\{X \in A\}$$

(b) $d(X'_k, X') \rightarrow 0$ *wp1* (P')

Theorem 6 is Prohorov's theorem and Theorem 7 is the Skorohod embedding.

The metric space B in most cases will be either $\mathbf{R}^n$ (random vectors) or $C(\mathbf{R},\mathbf{R}^n)$ (sample continuous random processes). The topology on $C(\mathbf{R},\mathbf{R}^n)$ is induced by uniform convergence on bounded intervals. This can be metrized by:

$$\rho_t(x(.), y(.)) = \sup_{s\in[o,t]} \|x(s) - y(s)\|$$

$$\rho(x(.), y(.)) = \int_0^\infty e^{-t} \max(1, \rho_t(x(.), y(.)))\, dt$$

With this metric $C(\mathbf{R},\mathbf{R}^n)$ is both complete and separable. Thus the Skorohod embedding says that weak convergence is equivalent to uniform convergence on compact intervals.

DEFINITION 4: $\{X_\alpha\}$ *is* uniformly integrable *if*

$$\lim_{r\to\infty} \sup_{\alpha\in A} \int_{|x|\geq r} |x| P_\alpha(dx) = \lim_{r\to\infty} \sup_{\alpha\in A} E\{|x_\alpha| \mid |x_\alpha| \geq r\} = 0$$

One condition that implies uniform integrability is

$$\sup_{\alpha \in A} \mathrm{E}\left\{ |x_\alpha|^{1+\varepsilon} \right\} < \infty \quad \text{for some } \varepsilon > 0 .$$

THEOREM 8:([5], p.51) *If* $\dot{x}^{\varepsilon}(t) = F^{\varepsilon}(t)$ *and, for every* $T>0$, $\{ F_{\varepsilon}(t) : \varepsilon \in (0,1],\ t \in [0,T] \}$ *is uniformly integrable, then* $\{x^{\varepsilon}(\cdot)\}$ *is tight.*

THEOREM 9: *Let* $\{P_k\}$ *and* P *be probability measures on* $C(\mathbf{R},\mathbf{R}^n)$. *If the finite dimensional distributions of* P_k *converge to those of* P *and* $\{P_k\}$ *is tight then* $P_k \Rightarrow P$

Define the modulus of continuity as

$$w_x(\delta, T) = \sup_{\substack{|t-s|<\delta \\ t \in [0,T]}} \|x(t) - x(s)\|$$

Then the following theorem characterizes tightness in $C(\mathbf{R},\mathbf{R}^n)$.

THEOREM 10:([5, p.30]) $\{x^{\varepsilon}(\cdot)\}$ *is tight if*

(a) $\{x^{\varepsilon}(0)\}$ *is tight*

(b) $\forall\ \eta, v, T>0\ \exists\ \delta' \in (0,1]$ *and* $\varepsilon' \in (0,1]$ *such that for* $\delta<\delta'$ *and* $\varepsilon<\varepsilon'$, $P\{ x^{\varepsilon}(\cdot) : w_{x^{\varepsilon}}(\delta,T) \geq v \} \leq \eta$

The condition in 10(b) is a form of equicontinuity and the theorem is proven using the Arzela-Ascoli theorem.

THEOREM 11:([20], p.56) *The condition* 10(b) *holds if for each* t

$$P\left\{ \sup_{0 \leq s \leq \delta} \left\| x^{\varepsilon}(t+s) - x^{\varepsilon}(t) \right\| \geq v \right\} \leq \delta\eta$$

A useful inequality is

$$P\left\{ \sup_{0 \leq \sigma \leq s} \left\| \int_t^{t+\sigma} g(\tau)\, dw(\tau) \right\| \geq v \,|w(s) \quad s \leq t \right\}$$

$$\leq \frac{1}{v^2} \mathrm{E}_t \left\{ \left\| \int_t^{t+s} g(\tau)\, dw(\tau) \right\|^2 \right\} \leq \frac{1}{v^2} \int_t^{t+s} \mathrm{E}_t\left[\| g(\tau) \|^2 \right] d\tau \tag{2}$$

A useful theorem for showing existence of solutions is:

THEOREM 12: *If* $a(\cdot)$ *and* $\sigma(\cdot)$ *are locally bounded, with continuous locally bounded second derivatives and satisfy*

$$\sup_{\|z\|=1} \left\| z^T \sigma(x) \right\|^2 \leq C\left(1 + \|x\|^2\right) \qquad \text{and} \qquad x^T a(x) \leq C\left(1 + \|x\|^2\right) \tag{3}$$

Then the SDE $dx = a(x)dt + \sigma(x)dw$ *has a unique solution (in the sense of distributions) defined for all time (i.e. no finite explosions). The transition probability measures are measurable, Feller continuous functions of* (t,x).

This theorem is adapted from [21, Theorems 6.34 and 10.22]

APPENDIX B: PROOF OF THE WEAK CONVERGENCE THEOREMS

This appendix holds the proof of Theorems II.1-4. The proof of Theorem II.1 is similar to that of Theorem 5.6 of [5], but is more complicated since the time scaling is nonlinear and the there is a Wiener process term in the $x^{\varepsilon}(\cdot)$ equation. The ends of proofs of lemmas will be marked by a double diamond (♦♦) and the end of proofs of theorems by a triple diamond (♦♦♦). Theorems and equations from other sections and appendices will be identified by section number, e.g. Theorem II.2 or Eq. A.3. The assumptions from Section II will be enclosed in parentheses and will not have the section identified, e.g. (A4)

Proof of Theorem II.1:

The proof is presented as a series of 6 lemmas.

LEMMA 1: *$\underline{F}(\cdot)$ is continuous*

Let $x_k \to x$ be a sequence in converging to a point in $\mathbf{R}^{n_x}$. We will first show that the invariant measures $P^{x_k}(\cdot)$ converge weakly to $P^x(\cdot)$. By (A6) the collection $\{P^{x_k}(\cdot)\}$ is tight and every subsequence has a weakly convergent (sub)subsequence. Let $\{P^{x_{k_j}}(\cdot)\}$ be an arbitrary weakly convergent subsequence and denote its limit by $P^I(\cdot)$. Let $f(\cdot)$ be a continuous function on $\mathbf{R}^{n_y}$. Since each $P^{x_{k_j}}(\cdot)$ is invariant for the fixed x process we have

$$\int f(y)P^{x_{k_j}}(dy) = \iint f(\tilde{y})\tilde{P}(y,u,t_o,d\tilde{y}|x_{k_j})P^{x_{k_j}}(dy) \tag{1}$$

By (A9a) the integrand in the second integral is a continuous function of (x_{k_j},y). The weak convergence of $\{P^{x_{k_j}}(\cdot)\}$, (A7b) and Eq. (1) imply the the following equality:

$$\int f(y)P^I(dy) = \iint f(\tilde{y})\tilde{P}(y,u,t_o,d\tilde{y}|x)P^I(dy) \tag{2}$$

This implies $P^I(\cdot)$ is an invariant measure for the fixed x process. The uniqueness condition in (A6) implies $P^I(\cdot)=P^x(\cdot)$. Thus every convergent subsequence of $\{P^{x_k}(\cdot)\}$ has a weakly convergent (sub)subsequence which converges to $P^x(\cdot)$. This implies $P^{x_k}(\cdot)$ converges weakly to $P^x(\cdot)$.

Now we are ready to show $\underline{F}(x_n)\to \underline{F}(x)$.

$$\begin{aligned}\|\underline{F}(x_n)-\underline{F}(x)\| &\le \left\|\int F_1(x_n,y)P^{x_n}(dy) - \int F_1(x,y)P^{x_n}(dy)\right\| + \\ &\qquad \left\|\int F_1(x,y)P^{x_n}(dy) - \int F_1(x,y)P^{x}(dy)\right\| \\ &= I_1 + I_2\end{aligned} \tag{3}$$

Since $P^{x_n}(\cdot)$ is an invariant measure we can expand the integrals in I_1 to get:

$$I_1 = \left\|\iint F_1(x_n,\tilde{y})\tilde{P}(y,t_o,u,d\tilde{y}|x_n)P^{x_n}(dy) - \iint F_1(x,\tilde{y})\tilde{P}(y,t_o,u,d\tilde{y}|x)P^{x_n}(dy)\right\|$$

(4)

The inner integral of each term in Eq. (4) is a continuous function of (x_n,y) and (x,y) respectively and the convergence of the first inner integral to the second inner integral is uniform on compact sets (A10). The uniform convergence on compact sets and the tightness of the single measure $P^{x_n}(\cdot)$ imply that I_1 goes to 0. Similarly we can expand the integrals in I_2 as

$$I_2 = \left\| \iint F_1(x,\tilde{y})\tilde{P}(y,t_o,u,d\tilde{y}|x_n)P^{x_n}(dy) - \iint F_1(x,\tilde{y})\tilde{P}(y,t_o,u,d\tilde{y}|x)P^{x}(dy) \right\| \tag{5}$$

Again the inner integrals are bounded and continuous in (x,y) by (A10a). The weak convergence then implies that I_2 goes to zero. I_1 and I_2 going to zero and Eq. (4) gives the continuity of $\underline{F}(\cdot)$.

♦♦(LEMMA 1).

For Lemma 2 we need some additional notation. Let $\delta_\varepsilon \to 0$ be a sequence with $\delta_o = 1$. $\{\delta_\varepsilon\}$ should also have the property: $\delta_\varepsilon e^{t_\varepsilon} \to \infty$. This insures δ_ε goes to zero slowly. Let $f(\cdot)$ be a real valued compactly supported differentiable function on $\mathbf{R}^{n_x}$. Define the following difference and differential operators:

$$A_d^\varepsilon f(x^\varepsilon(t)) = \mathrm{E}_{e^{t+t_\varepsilon}}\left\{ \frac{1}{\delta_\varepsilon} \int_t^{t+\delta_\varepsilon} f_x^T(x^\varepsilon(\tau))F_1(x^\varepsilon(\tau),y^\varepsilon(\tau))d\tau \right\} \tag{6}$$

$$A_s^\varepsilon f(x^\varepsilon(t)) = \mathrm{E}_{e^{t+t_\varepsilon}}\left\{ \frac{1}{\delta_\varepsilon} \int_{e^{t+t_\varepsilon}}^{e^{t+\delta_\varepsilon+t_\varepsilon}} f_x^T(x^\varepsilon(\log(\tau)-t_\varepsilon))\cdot F_2(x^\varepsilon(\cdots),y^\varepsilon(\cdots)\frac{dw(\tau)}{\tau} \right\} \tag{7}$$

$$A^\varepsilon f(x^\varepsilon(t)) = A_d^\varepsilon f(x^\varepsilon(t)) + A_s^\varepsilon f(x^\varepsilon(t)) \tag{8}$$

$$Af(x^o(t)) = f_x^T(x^o(t))\underline{F}(x^o(t)) \tag{9}$$

$f_x(\cdot)$ denotes the gradient of $f(\cdot)$ which in this section is a column vector. The expectation E_t is taken wrt the σ-algebra generated by $\{w(\tau) : 1 \le \tau \le t\}$. A^ε is a difference quotient operator and

$$\mathrm{E}_{e^{t+t_\varepsilon}}\left\{ \frac{f(x^\varepsilon(t+\delta_\varepsilon)) - f(x^\varepsilon(t))}{\delta_\varepsilon} \right\} \approx A^\varepsilon f(x^\varepsilon(t)) \tag{10}$$

when δ_ε is small. The crux of the theory is that A^ε converges to A (in some sense) as ε goes to zero. Since $x^\varepsilon(\cdot)$ is not differentiable, the rate at which δ_ε goes to zero is critical. We also need the integrals of Eqs. (6),(7),(8):

$$I_d^\varepsilon(t) = \int_0^t A_d^\varepsilon f(x^\varepsilon(\tau))\,d\tau \tag{11}$$

$$I_s^\varepsilon(t) = \int_0^t A_s^\varepsilon f(x^\varepsilon(\tau))\,d\tau \tag{12}$$

$$I^\varepsilon(t) = I_d^\varepsilon(t) + I_s^\varepsilon(t) \tag{13}$$

For convenience, $x^\varepsilon(\cdot)$ is split into two processes:

$$x_d^\varepsilon(t+s) - x_d^\varepsilon(t) = \int_t^{t+s} F_1(x^\varepsilon(\tau), y^\varepsilon(\tau))\,d\tau \qquad x_d^\varepsilon(0) = x^\varepsilon(0) \tag{14}$$

$$x_s^\varepsilon(t+s) - x_s^\varepsilon(t) = \int_{e^{t+t_\varepsilon}}^{e^{t+s+t_\varepsilon}} F_2(x^\varepsilon(\log(\tau) - t_\varepsilon), y^\varepsilon \log(\tau) - t_\varepsilon))\frac{dw(\tau)}{\tau}$$

$$x_s^\varepsilon(0) = 0 \tag{15}$$

We are now prepared to state and prove Lemma 2.

LEMMA 2: *$\{x^\varepsilon(\cdot), I^\varepsilon(\cdot)\}$ is tight in* $C(\mathbf{R},\mathbf{R}^n) \times C(\mathbf{R},\mathbf{R})$

Since $(x^\varepsilon(\cdot), I^\varepsilon(\cdot)) = (x^\varepsilon{}_d(\cdot) + x^\varepsilon{}_s(\cdot), I^\varepsilon{}_d(\cdot) + I^\varepsilon{}_s(\cdot))$, it suffices to show that each of families, $\{x^\varepsilon{}_d(\cdot)\}$, $\{x^\varepsilon{}_s(\cdot)\}$, $\{I^\varepsilon{}_d(\cdot)\}$, $\{I^\varepsilon{}_s(\cdot)\}$, are tight. Recall (from Appendix A) that a family of random processes, $\{z^\varepsilon(\cdot)\}$, in $C(\mathbf{R},\mathbf{R}^n)$ is tight if

(a) $\{z^\varepsilon(0)\}$ is tight in $\mathbf{R}^n$

(b) $\forall\ \eta, v, T > 0\ \exists\ \varepsilon', \delta'$ such that for $\varepsilon \le \varepsilon'$, $\delta \le \delta'$ we have
$P\{\, w_{z^\varepsilon}(\delta, T) \ge v \,\} \le \eta$ where

$$w_{z^\varepsilon}(\delta, T) = \sup_{\substack{t \le T \\ |s| \le \delta}} \left\| z^\varepsilon(t+s) - z^\varepsilon(t) \right\|$$

or

(b') $\forall\ \eta, v, T > 0\ \exists\ \varepsilon', \delta'$ such that for $\varepsilon \le \varepsilon'$, $\delta \le \delta'$, $t \le T$ we have

$$P\left\{ \sup_{0 \le s \le \delta} \left\| z^\varepsilon(t+s) - z^\varepsilon(t) \right\| \ge v \right\} \le \delta\eta$$

(a) is tightness of the initial conditions and (b) or (b') are equicontinuity conditions. (b') implies (b) and is sometimes easier to prove.

$x^\varepsilon{}_d(0) = x^\varepsilon(0)$ and $\{x^\varepsilon(0)\}$ is tight since, by assumption, it is weakly convergent. $x^\varepsilon{}_s(0) = I^\varepsilon{}_d(0) = I^\varepsilon{}_s(0) = 0$ wp1 which trivially implies tightness. Thus (a) is satisfied by each of the families of processes.

Consider the following inequalities:

$$\sup_{0\le s\le\delta}\left\| x_d^{\varepsilon}(t+s)-x_d^{\varepsilon}(t)\right\| = \sup_{0\le s\le\delta}\left\| \int_t^{t+s} F_1(x^{\varepsilon}(\tau),y^{\varepsilon}(\tau))d\tau \right\|$$
$$\le \delta \sup_{\substack{0\le s\le\delta \\ x\in\mathbf{R}^{n_x}}}\left\| F_1(x,y^{\varepsilon}(t+s))\right\| \tag{17}$$

Now we have

$$P\left\{\sup_{0\le s\le\delta}\left\| x_d^{\varepsilon}(t+s)-x_d^{\varepsilon}(t)\right\|\ge v\right\}$$
$$\le P\left\{\sup_{\substack{0\le s\le\delta \\ x\in\mathbf{R}^{n_x}}}\left\| F_1(x,y^{\varepsilon}(t+s))\right\|\ge \frac{v}{\delta}\right\}$$
$$\le \frac{\delta}{v}\, E\left\{\sup_{\substack{x\in\mathbf{R}^{n_x} \\ 0\le s\le\delta}}\left\| F_1(x,y^{\varepsilon}(t+s))\right\| \,\middle|\, \sup_{\substack{x\in\mathbf{R}^{n_x} \\ 0\le s\le\delta}}\left\| F_1(x,y^{\varepsilon}(t+s))\right\|\ge\frac{v}{\delta}\right\}$$
$$=\frac{\delta}{v}\; O(\delta) \tag{18}$$

The first inequality in Eq. (18) follows from Eq. (17), while the second is the Markov inequality. The last follows from the uniform integrability in Assumption (A3). Thus $\{x^{\varepsilon}{}_d(\cdot)\}$ satisfies (b') and is tight. When $F_1(\cdot,\cdot)$ is bounded the proof is simpler and will not be shown.

For $x^{\varepsilon}{}_s(\cdot)$ we have

$$w_{x_s^{\varepsilon}}(\delta,T)=\sup_{\substack{t\le T \\ |s|\le\delta}}\left\| \int_{e^{t+t_\varepsilon}}^{e^{t+s+t_\varepsilon}} F_2(x^{\varepsilon}(\log(\tau)-t_\varepsilon),y^{\varepsilon}(\log(\tau)-t_\varepsilon))\frac{dw(\tau)}{\tau}\right\| \tag{19}$$

which implies that

$$P\left\{w_{x_s^{\varepsilon}}(\delta,T)\ge v\right\}\le$$
$$\frac{1}{v^2}\sup_{t\le T}\int_{e^{t+t_\varepsilon}}^{e^{t+\delta+t_\varepsilon}} \mathrm{E}\left\{\left\| F_2(x^{\varepsilon}(\log(\tau)-t_\varepsilon),y^{\varepsilon}(\log(\tau)-t_\varepsilon))\right\|^2\right\}\frac{d\tau}{\tau^2} \tag{20}$$

By (A4), the integrand in Eq. (20) is bounded and we have

$$P\left\{w_{x_s^{\varepsilon}}(\delta,T)\ge v\right\}\le\frac{K}{v^2}\sup_{t\le T}\frac{e^{t+\delta+t_\varepsilon}-e^{t+t_\varepsilon}}{\left(e^{t+t_\varepsilon}\right)^2}$$
$$=\frac{K}{v^2}\sup_{t\le T} e^{-t_\varepsilon}\, e^{-t}\left(e^{\delta}-1\right)$$

$$= \frac{K}{\nu^2} e^{-t_\varepsilon}\left(e^{\delta} - 1 \right) \tag{21}$$

Thus $\{x^\varepsilon{}_s(\cdot)\}$ satisfies (b) and is tight.

For $\{I^\varepsilon{}_d(\cdot)\}$, with $f \in C^k{}_0(\mathbf{R}^{n_x})$, we have

$$\sup_{0\le s\le\delta} \left\| I_d^\varepsilon(t+s) - I_d^\varepsilon(t) \right\|$$

$$= \sup_{0\le s\le\delta} \left\| \frac{1}{\delta_\varepsilon} \int_t^{t+s} \int_\tau^{\tau+\delta_\varepsilon} f_x^T(x^\varepsilon(\sigma)) F_1(x^\varepsilon(\sigma), y^\varepsilon(\sigma))\, d\sigma\, d\tau \right\|$$

$$= \delta \sup_{0\le s\le\delta} \left\| f_x^T(x^\varepsilon(t+s)) F_1(x^\varepsilon(t+s), y^\varepsilon(t+s)) \right\|$$

$$\le \delta \sup_{\substack{0\le s\le\delta \\ x\in\mathbf{R}^{n_x}}} \left\| f_x^T(x) F_1(x, y^\varepsilon(t+s)) \right\| \tag{22}$$

Arguments similar to those for $\{x^\varepsilon{}_d(\cdot)\}$ show $\{I^\varepsilon{}_d(\cdot)\}$ is tight.

For $\{I^\varepsilon{}_s(\cdot)\}$, with $f \in C^k{}_0(\mathbf{R}^{n_x})$, we have

$$w_{I_s^\varepsilon}(\delta, T)$$

$$= \sup_{\substack{t\le T-s-\delta_\varepsilon \\ |s|\le\delta}} \left\| \frac{1}{\delta_\varepsilon} \int_t^{t+s} \int_{e^{\tau+t_\varepsilon}}^{e^{\tau+\delta_\varepsilon+t_\varepsilon}} f_x^T(x^\varepsilon(\log(\sigma)-t_\varepsilon)) F_2(x^\varepsilon(\cdots), y^\varepsilon(\cdots)) \frac{dw(\sigma)}{\sigma} d\tau \right\|$$

$$\le \sup_{\substack{t\le T-s-\delta_\varepsilon \\ |s|\le\delta}} \sup_{t\le\tau\le t+s} \left\| \frac{s}{\delta_\varepsilon} \int_{e^{\tau+t_\varepsilon}}^{e^{\tau+\delta_\varepsilon+t_\varepsilon}} f_x^T(x^\varepsilon(\log(\sigma)-t_\varepsilon)) \cdot F_2(x^\varepsilon(\cdots), y^\varepsilon(\cdots)) \frac{dw(\sigma)}{\sigma} \right\|$$

$$\le \sup_{t\le T-\delta_\varepsilon} \frac{\delta}{\delta_\varepsilon} \left\| \int_{e^{t+t_\varepsilon}}^{e^{t+\delta_\varepsilon+t_\varepsilon}} f_x^T(x^\varepsilon(\log(\sigma)-t_\varepsilon)) \cdot F_2(x^\varepsilon(\cdots), y^\varepsilon(\cdots)) \frac{dw(\sigma)}{\sigma} \right\| \tag{23}$$

Using the Inequality A.2, we have

$$P\left\{ w_{I_s^\varepsilon}(\delta, T) \ge \nu \right\} \le$$

$$\left(\frac{\delta}{\nu\delta_\varepsilon} \right)^2 \sup_{t\le T} \int_{e^{t+t_\varepsilon}}^{e^{t+\delta+t_\varepsilon}} E\left\{ \left\| f_x^T(x^\varepsilon(\log(\tau)-t_\varepsilon) \cdot F_2(x^\varepsilon(\cdots), y^\varepsilon(\cdots)) \right\|^2 \right\} \frac{d\tau}{\tau^2} \tag{24}$$

Using (A4) and the boundedness of $f_x(\cdot)$ we get

$$P\left\{w_{I_s^\varepsilon}(\delta,T) \geq v\right\} \leq \left(\frac{\delta}{v\,\delta_\varepsilon}\right)^2 K\,\frac{e^{t+\delta_\varepsilon+t_\varepsilon} - e^{t+t_\varepsilon}}{\left(e^{t+\delta_\varepsilon+t_\varepsilon}\right)^2}$$

$$= \frac{K\delta^2}{v^2}\,e^{-2\delta_\varepsilon}\,e^{-t}\,\frac{1}{\delta_\varepsilon e^{t_\varepsilon}}\,\frac{e^{\delta_\varepsilon}-1}{\delta_\varepsilon} \tag{25}$$

Since $\delta_\varepsilon e^{t_\varepsilon} \to \infty$, the rhs of Eq. (25) goes to zero as ε or δ go to zero. Thus (b) is satisfied and $\{I^\varepsilon{}_s(\cdot)\}$ is tight.

♦♦(LEMMA 2)

Since $\{x^\varepsilon(\cdot), I^\varepsilon(\cdot)\}_{\varepsilon\in(0,1]}$ is tight, we know that every ε-subsequence taken from $\{x^\varepsilon(\cdot), I^\varepsilon(\cdot)\}_{\varepsilon\in(0,1]}$ has a convergent (sub)subsequence as $\varepsilon \to 0$. If we can show all convergent subsequences have the same limit, then we can conclude that $\{x^\varepsilon(\cdot), I^\varepsilon(\cdot)\}_{\varepsilon\in(0,1]}$ converges to a unique limit process as $\varepsilon \to 0$. Let $J_o \subseteq (0,1]$, with $\inf J_o = 0$, denote the ε-indices of a convergent subsequence and denote the subsequence as $\{x^\varepsilon(\cdot), I^\varepsilon(\cdot)\}_{\varepsilon\in J_o}$. Denote the limit process by $(x^o(\cdot), I^o(\cdot))$. The Skorohod embedding lets us assume that $\{x^\varepsilon(\cdot), I^\varepsilon(\cdot)\}_{\varepsilon\in J_o}$ converges to $(x^o(\cdot), I^o(\cdot))$ uniformly wp1 on $[0,T]$ for any finite T. Also, the weak convergence of $\{x^\varepsilon(\cdot), I^\varepsilon(\cdot)\}_{\varepsilon\in J_o}$ to $(x^o(\cdot), I^o(\cdot))$ implies the weak convergence of the two components to $x^o(\cdot)$ and $I^o(\cdot)$ individually.

For the next lemma we need some additional notation and definitions. Let $T>0$. The tightness of assumption (A5) says there exist compact sets $B_{\varepsilon,T}$ such that

$$\lim_{\substack{\varepsilon\to 0\\ \varepsilon\in J_o}}\ \inf_{t\leq T} P\left\{y^\varepsilon(t) \in B_{\varepsilon,T}\right\} = 1 \tag{26}$$

By assumption (A7b), there is a set function $\beta_\varepsilon(\cdot)$ such that

$$\lim_{\substack{\varepsilon\to 0\\ \varepsilon\in J_o}} \varepsilon\beta_\varepsilon(B_{\varepsilon,T}) = 0 \tag{27}$$

With this $\beta_\varepsilon(\cdot)$ we may also assume that δ_ε satisfies

$$\lim_{\substack{\varepsilon\to 0\\ \varepsilon\in J_o}} \frac{\varepsilon\beta_\varepsilon(B_{\varepsilon,T})}{\delta_\varepsilon} = 0 \tag{28}$$

and $\varepsilon\beta_\varepsilon(B_{\varepsilon,T}) \leq \delta_\varepsilon$. This is consistent with the assumptions on δ_ε in Lemma 2 as both assumptions upper bound the speed at which δ_ε goes to zero. For brevity of notation denote $\beta_\varepsilon(B_{\varepsilon,T})$ by β_ε. Denote the indicator function of a set B by $\mathrm{I}_B(\cdot)$. Define the averaged probability measures

$$Q(t,\varepsilon,A) = \frac{1}{\delta_\varepsilon}\int_t^{t+\delta_\varepsilon} P^\varepsilon(x^\varepsilon(t), y^\varepsilon(t), \tau, t, \mathbf{R}^{n_x} \times A)\,d\tau \tag{29}$$

$$Q'(t,\varepsilon,A) = \frac{1}{\delta_\varepsilon} \int_{t+\varepsilon\beta_\varepsilon(B)}^{t+\delta_\varepsilon} P^\varepsilon(x^\varepsilon(t), y^\varepsilon(t), \tau, t, \mathbf{R}^{n_x} \times A)\, d\tau \tag{30}$$

$$Q_B(t,\varepsilon,A) = I_B(y^\varepsilon(t))Q(t,\varepsilon,A) \tag{31}$$

$$Q'_B(t,\varepsilon,A) = I_B(y^\varepsilon(t))Q'(t,\varepsilon,A) \tag{32}$$

Note that if one of these sequences of measures has a weak limit as $\varepsilon \to 0$, then each of the others will have the same limit. For each t, define the set of measures

$$M(t) = \left\{ Q'_{B_{\varepsilon,T}}(t,\varepsilon,\cdot) : \varepsilon \in J_o \right\} \tag{33}$$

LEMMA 3: *for each t, M(t) is tight*

We need to show that $\forall\ \eta\ \exists$ compact K_η and ε_o such that

$$\inf_{\substack{\varepsilon \le \varepsilon_o \\ \varepsilon \in J_o}} Q'_{B_{\varepsilon,T}}(t,\varepsilon,K_\eta) \ge 1 - \eta \tag{34}$$

Let $\eta > 0$ and consider the two alternative hypothesis of assumption (A7). If the range of $y^\varepsilon(\cdot)$ has compact closure then set K_η equal to its closure to get

$$Q'_{B_{\varepsilon,T}}(t,\varepsilon,K_\eta) = I_{B_{\varepsilon,T}}(y^\varepsilon(t)) \cdot \frac{1}{\delta_\varepsilon} \int_{t+\varepsilon\beta_\varepsilon}^{t+\delta_\varepsilon} P^\varepsilon(x^\varepsilon(t), y^\varepsilon(t), \tau, t, \mathbf{R}^{n_x} \times K_\eta)\, d\tau$$

$$= I_{B_{\varepsilon,T}}(y^\varepsilon(t)) \cdot \left(1 - \frac{\varepsilon\beta_\varepsilon}{\delta_\varepsilon}\right) \tag{35}$$

Since for each fixed $T > 0$, the range of $y^\varepsilon(\cdot) \subseteq \cup B_{\varepsilon,T}$, we have the rhs of Eq. (35) $\to 1$ as $\varepsilon \to 0$, which proves the tightness.

If (A7b) holds we know that for any t_o, T and K that

$$\left\{ \tilde{P}_{z^\varepsilon}(y, t, e^{t_o+t_\varepsilon}, \cdot) I_B(y) : \left\{ \begin{array}{l} \text{compact}\, B \subseteq \mathbf{R}^{n_y} \\ y \in \mathbf{R}^{n_y},\ \varepsilon \in (0,1] \\ e^{t_o+\varepsilon\beta_\varepsilon(B)+t_\varepsilon} \le t \le e^{t_o+T+t_\varepsilon} \\ z^\varepsilon(.)\ \text{such that} \sup\limits_{t \le T} \left| z^\varepsilon(t) \right| \le K \end{array} \right\} \right\} \tag{36}$$

is tight. Letting $z^\varepsilon(t) = x^\varepsilon(\log(t) - t_\varepsilon)$ we have

$$P^\varepsilon\left(x^\varepsilon(t), y^\varepsilon(t), \tau, t, \mathbf{R}^{n_x} \times \cdot \right) = \tilde{P}_{z^\varepsilon}\left(y^\varepsilon(\log(t) - t_\varepsilon), e^{\tau+t_\varepsilon}, e^{t+t_\varepsilon}, \cdot \right) \tag{37}$$

Thus, for fixed $T > \delta_\varepsilon > 0$, the tightness in Eq. (36) implies the tightness of

$$\left\{ P_\varepsilon(x^\varepsilon(t), y^\varepsilon(t), \tau, t, \mathbf{R}^{n_x} \times \cdot) I_{B_{\varepsilon,T}}(y^\varepsilon(t)) : \left\{ \begin{array}{l} \varepsilon \in J_o \\ t + \varepsilon\beta_\varepsilon \le \tau \le t + \delta_\varepsilon \end{array} \right\} \right\} \tag{38}$$

since the measures in $M(t)$ are averages of measures in Eq. (38), $M(t)$ is tight also.

♦♦(LEMMA 3)

LEMMA 4: *$M(t)$, when considered as an ε-sequence of measures, converges weakly to the invariant measure $P^{x^{o}(t)}(\cdot)$ as $\varepsilon \to 0$ in J_o.*

By the tightness, it suffices to show that any weakly converging subsequence of $M(t)$ converges to $P^{x^{o}(t)}(\cdot)$. Let $J_1 \subseteq J_o$ be a subsequence with $\inf J_1 = 0$ and the weak limit

$$\lim_{\substack{\varepsilon \to o \\ \varepsilon \in J_1}} Q'_{B_{\varepsilon,T}}(t,\varepsilon,\cdot) = Q_o(\cdot) \tag{39}$$

Let $g(\cdot)$ be a bounded continuous function on $\mathbf{R}^{ny}$. Then

$$\lim_{\substack{\varepsilon \to 0 \\ \varepsilon \in J_1}} \int g(y) Q'_{B_{\varepsilon,T}}(\varepsilon,t,dy)$$

$$= \lim_{\substack{\varepsilon \to 0 \\ \varepsilon \in J_1}} \frac{1}{\delta_\varepsilon} \int_y g(y) \left(\int_{t+\varepsilon\beta_\varepsilon}^{t+\delta_\varepsilon} I_{B_{\varepsilon,T}}(y^\varepsilon(t)) P^\varepsilon(x^\varepsilon(t), y^\varepsilon(t), \tau, t, \mathbf{R}^{nx} \times dy)\, d\tau \right)$$

$$= \lim_{\substack{\varepsilon \to 0 \\ \varepsilon \in J_1}} \frac{1}{\delta_\varepsilon} \int_{(\tilde{x},\tilde{y})} \int_{t+\varepsilon u+\varepsilon\beta_\varepsilon}^{t+\delta_\varepsilon} \int_y g(y) P^\varepsilon(\tilde{x},\tilde{y},\tau,\tau-\tau_\varepsilon, \mathbf{R}^{nx} \times dy) I_{B_{\varepsilon,T}}(y^\varepsilon(t)) \cdot P^\varepsilon(x^\varepsilon(t), y^\varepsilon(t), \tau-\tau_\varepsilon, t, d\tilde{x}\, d\tilde{y})\, d\tau \tag{40}$$

where $\tau_\varepsilon = -\log(1 - e^{-(\tau+t_\varepsilon)}u)$. Notice $\tau_\varepsilon \to 0$ as $\varepsilon \to 0$, but u is a nonzero constant. The second equality in Eq. (40) is a consequence of the Chapman-Kolmogorov equation. With $z^\varepsilon(t) = x^\varepsilon(\log(t+\exp(\tau-\tau_\varepsilon+t_\varepsilon)) - t_\varepsilon)$, heuristically, we have $z^\varepsilon(0)=x^\varepsilon(\tau-\tau_\varepsilon)$ and $z^\varepsilon(u)=x^\varepsilon(\tau)$. Given these definitions, we have

$$\int g(y) P^\varepsilon(\tilde{x},\tilde{y},\tau,\tau-\tau_\varepsilon,\mathbf{R}^{nx}\times dy)$$
$$= \int g(y)\tilde{P}^\varepsilon(\tilde{x},\tilde{y},e^{\tau+t_\varepsilon},e^{t-\tau_\varepsilon+t_\varepsilon},dy)$$
$$= \int g(y)\tilde{P}^\varepsilon(\tilde{x},\tilde{y},e^{\tau+t_\varepsilon},e^{\tau+t_\varepsilon}-u,dy)$$
$$= \int g(y)\tilde{P}_{z^\varepsilon}(\tilde{y},u,0,dy) \tag{41}$$

Assumption (A9b) applied to Eq. (41) gives us

$$\lim_{\substack{\varepsilon \to o \\ \varepsilon \in J_1}} \int g(y) P^\varepsilon(\tilde{x},\tilde{y},\tau,\tau-\tau_\varepsilon,\mathbf{R}^{nx}\times dy) = \int g(y)\tilde{P}(\tilde{y},u,0,dy \mid \tilde{x}) \tag{42}$$

Combining Eqs. (40), (41) and (42) we get

$$\lim_{\substack{\varepsilon \to 0 \\ \varepsilon \in J_1}} \int g(y) Q'_{B_{\varepsilon,T}}(\varepsilon,t,dy) = \lim_{\substack{\varepsilon \to 0 \\ \varepsilon \in J_1}} \int \left[\int g(y)\tilde{P}(\tilde{y},u,0,dy \mid x^\varepsilon(t))\right] Q'_{B_{\varepsilon,T}}(\varepsilon,t,d\tilde{y}) \tag{43}$$

The weak convergence of the "Q"s to Q_o implies the lhs of Eq. (43) is equal to the lhs of Eq. (44). Assumption (A9a) says that the bracketed term in Eq. (43) is bounded and continuous in $(x,\tilde{y})$. Since t is fixed, the weak convergence of $x^{\varepsilon}(\cdot)$ and the Skorohod embedding imply the wp1 convergence of $x^{\varepsilon}(t)$ to $x^{o}(t)$ for as $\varepsilon \to 0$ in J_o. The convergence of $x^{\varepsilon}(t)$ to $x^{o}(t)$, the continuity and the weak convergence of the "Q"s to Q_o imply the rhs of Eq. (43) is equal to the rhs of Eq. (44).

$$\int g(y) Q_o(.,dy) = \int \left[\int g(y) \tilde{P}(\tilde{y},u,0,dy \mid x^{o}(t))\right] Q_o(d\tilde{y}) \tag{44}$$

This means $Q_o(\cdot)$ is an invariant measure for the fixed x process. Assumption (A6) says the invariant measures are unique, hence $Q_o(\cdot) = P^{x^{o}(t)}(\cdot)$.

♦♦(LEMMA 4)

LEMMA 5: *Given the subsequence* $\{x^{\varepsilon}(\cdot), I^{\varepsilon}(\cdot)\}_{\varepsilon \in J_o}$ *and its weak limit* $(x^{o}(\cdot), I^{o}(\cdot))$ *we have*

$$I^{o}(t) = \int_0^t Af(x^{o}(\tau))\,d\tau = \int_0^t f_x^T(x^{o}(t))\underline{F}(x^{o}(t))\,d\tau \tag{45}$$

Recall that $I^{\varepsilon}(\cdot) = I^{\varepsilon}{}_d(\cdot) + I^{\varepsilon}{}_s(\cdot)$. Theorem A.5 implies that if $I^{\varepsilon}{}_s(\cdot) \to 0$ in probability, uniformly on compact sets, we can conclude that $I^{\varepsilon}{}_d(\cdot) \Rightarrow I^{o}(\cdot)$. Thus to show Eq. (45) it suffices to show that $I^{\varepsilon}{}_s(\cdot) \to 0$ in probability uniformly on compact intervals, and $A^{\varepsilon}{}_d f(\cdot)$ converges to $Af(x^{o}(\cdot))$ uniformly wp1 on compact intervals.

Let $f(\cdot) \in C^k{}_o(\mathbf{R}^{n_x},\mathbf{R})$ and define $G_2(t) = f_x{}^T(x^{\varepsilon}(t))F_2(x^{\varepsilon}(t),y^{\varepsilon}(t))$. Equations (7) and (12) can be combined as

$$I_s^{\varepsilon}f(t) = \int_0^t \mathrm{E}_{e^{\tau+t_\varepsilon}}\left\{\frac{1}{\delta_\varepsilon}\int_{e^{\tau+t_\varepsilon}}^{e^{\tau+\delta_\varepsilon+t_\varepsilon}} G_2(\sigma)\frac{dw(\sigma)}{\sigma}\right\}d\tau \tag{46}$$

The obvious approximation gives

$$\left|I_s^{\varepsilon}f(t)\right| \le t \sup_{\tau \le t}\left\{\mathrm{E}_{e^{\tau+t_\varepsilon}}\left\{\frac{1}{\delta_\varepsilon}\int_{e^{\tau+t_\varepsilon}}^{e^{\tau+\delta_\varepsilon+t_\varepsilon}} G_2(\sigma)\frac{dw(\sigma)}{\sigma}\right\}\right\} \tag{47}$$

Applying the Inequality A.2 and (A4), we get

$$P\left\{\left|I_s^{\varepsilon}f(t)\right| \ge v\right\} \le P\left\{t\sup_{\tau \le t}\left\{\mathrm{E}_{e^{\tau+t_\varepsilon}}\left\{\frac{1}{\delta_\varepsilon}\int_{e^{\tau+t_\varepsilon}}^{e^{\tau+\delta_\varepsilon+t_\varepsilon}} G_2(\sigma)\frac{dw(\sigma)}{\sigma}\right\}\right\} \ge v\right\}$$

$$\leq \left(\frac{t}{\nu\delta_\varepsilon}\right)^2 \int_{e^{\tau+t_\varepsilon}}^{e^{\tau+\delta_\varepsilon+t_\varepsilon}} E\left|G_2(\sigma)\right|^2 \frac{d\sigma}{\sigma^2} \leq K\left(\frac{t}{\nu\delta_\varepsilon}\right)^2 \frac{e^{\tau+\delta_\varepsilon+t_\varepsilon}-e^{\tau+t_\varepsilon}}{\left(e^{\tau+\delta_\varepsilon+t_\varepsilon}\right)^2}$$

$$\leq K\left(\frac{1}{\nu}\right)^2 \left(\frac{e^{\delta_\varepsilon}-1}{\delta_\varepsilon}\right)\left(\frac{1}{\delta_\varepsilon e^{t_\varepsilon}}\right) t^2 e^{-t} \tag{48}$$

Each term in the last line of Eq. (48) is uniformly bounded in ε and t and the last bracketed term goes to zero as $\varepsilon \to 0$, hence $I^\varepsilon{}_s(\cdot) \to 0$ in probability uniformly on compact sets.

Define

$$G_1(x,y) = f_x^T(x) F_1(x,y) \tag{49}$$

and $\tau_\varepsilon = -\log(1 - e^{-(t+t_\varepsilon)}u)$. Notice $\tau_\varepsilon \to 0$ as $\varepsilon \to 0$, but u is a nonzero constant. As before, with $z^\varepsilon(t)=x^\varepsilon(\log(t+\exp(\tau-\tau_\varepsilon+t_\varepsilon)) - t_\varepsilon)$, we have $z^\varepsilon(0)=x^\varepsilon(\tau-\tau_\varepsilon)$ and $z^\varepsilon(u)=x^\varepsilon(\tau)$. Now we can show $A^\varepsilon{}_d f(\cdot)$ converges to $Af(x^o(\cdot))$ uniformly wp1 on compact intervals. We have

$$\lim_{\substack{\varepsilon\to 0\\ \varepsilon\in J_o}} A_d^\varepsilon f(x^\varepsilon(t)) = \lim_{\substack{\varepsilon\to 0\\ \varepsilon\in J_o}} \frac{1}{\delta_\varepsilon} \mathrm{E}_{e^{t+t_\varepsilon}} \left\{ \int_t^{t+\delta_\varepsilon} G_1(x^\varepsilon(\tau),y^\varepsilon(\tau))\,d\tau \right\}$$

$$= \lim_{\substack{\varepsilon\to 0\\ \varepsilon\in J_o}} \frac{1}{\delta_\varepsilon} \int_t^{t+\delta_\varepsilon} \int_{(x,y)} G_1(x,y) P^\varepsilon(x^\varepsilon(t),y^\varepsilon(t),\tau,t,dxdy)\,d\tau$$

$$= \lim_{\substack{\varepsilon\to 0\\ \varepsilon\in J_o}} \frac{1}{\delta_\varepsilon} \int_t^{t+\delta_\varepsilon} \int_{(\tilde{x},\tilde{y})} \int_{(x,y)} G_1(x,y) P^\varepsilon(\tilde{x},\tilde{y},\tau,\tau-\tau_\varepsilon,dxdy)\cdot P^\varepsilon(x^\varepsilon(t),y^\varepsilon(t),\tau-\tau_\varepsilon,t,d\tilde{x}d\tilde{y})\,d\tau$$

$$= \lim_{\substack{\varepsilon\to 0\\ \varepsilon\in J_o}} \frac{1}{\delta_\varepsilon} \int_t^{t+\delta_\varepsilon} \int_{(\tilde{x},\tilde{y})} E\left\{ G_1(x^\varepsilon(\tau),y^\varepsilon(\tau)) \,\middle|\, \begin{matrix} x^\varepsilon(\tau-\tau_\varepsilon)=\tilde{x} \\ y^\varepsilon(\tau-\tau_\varepsilon)=\tilde{y} \end{matrix} \right\}\cdot P^\varepsilon(x^\varepsilon(t),y^\varepsilon(t),\tau-\tau_\varepsilon,t,d\tilde{x}d\tilde{y})\,d\tau \tag{50}$$

The second and fourth equalities in Eq. (50) are definitions, and the third equality is the Chapman-Kolmogorov equation. Before continuing with Eq. (50), we need to prove

$$\lim_{\substack{\varepsilon\to 0\\ \varepsilon\in J_o}} \mathrm{E}\left\{ \left| G_1(x^\varepsilon(\tau),y^\varepsilon(\tau)) - G_1(x^\varepsilon(\tau-\tau_\varepsilon),y^\varepsilon(\tau))\right| \,\middle|\, \begin{matrix} x^\varepsilon(\tau-\tau_\varepsilon)=\tilde{x} \\ y^\varepsilon(\tau-\tau_\varepsilon)=\tilde{y} \end{matrix} \right\} = 0 \tag{51}$$

Applying the triangle inequality, we have

$$\left| G_1(x^\varepsilon(\tau),y^\varepsilon(\tau)) - G_1(x^\varepsilon(\tau-\tau_\varepsilon),y^\varepsilon(\tau))\right|$$

$$\leq \left| G_1(x^{\varepsilon}(\tau),y^{\varepsilon}(\tau)) - G_1(x^{o}(\tau),y^{\varepsilon}(\tau)) \right| + \left| G_1(x^{o}(\tau),y^{\varepsilon}(\tau)) - G_1(x^{o}(\tau-\tau_{\varepsilon}),y^{\varepsilon}(\tau)) \right| + \left| G_1(x^{o}(\tau-\tau_{\varepsilon}),y^{\varepsilon}(\tau)) - G_1(x^{\varepsilon}(\tau-\tau_{\varepsilon}),y^{\varepsilon}(\tau)) \right| \tag{52}$$

By (A2), $G_1(x,y)$ is continuous in the first variable uniformly on compact (x,y)-sets. We also know that $x^{\varepsilon}(\cdot)$ converges uniformly wp1 to $x^{o}(\cdot)$ on compact intervals. Finally, $\{y^{\varepsilon}(s) : s \in [t,t+\delta_{\varepsilon}],\ \varepsilon \in (0,1]\}$ is tight by (A5) and we have the uniform integrability of (A3b). These four conditions imply the conditional expectation of first and third terms on the rhs of (52) go to zero uniformly on compact τ-sets. To be more explicit consider the conditional expectation of the first term on the rhs of Eq. (52).

$$E\left\{ \left| G_1(x^{\varepsilon}(\tau),y^{\varepsilon}(\tau)) - G_1(x^{o}(\tau),y^{\varepsilon}(\tau)) \right| \,\middle|\, \begin{matrix} x^{\varepsilon}(\tau-\tau_{\varepsilon}) = \tilde{x} \\ y^{\varepsilon}(\tau-\tau_{\varepsilon}) = \tilde{y} \end{matrix} \right\}$$

$$= E\left\{ \Delta G \,\middle|\, \begin{matrix} x^{\varepsilon}(\tau-\tau_{\varepsilon}) = \tilde{x} \\ y^{\varepsilon}(\tau-\tau_{\varepsilon}) = \tilde{y} \\ \Delta G \leq N_1 \\ y^{\varepsilon}(\tau) \in B(0,N_2) \end{matrix} \right\} P\left\{ \{\Delta G \leq N_1\} \cap \{y^{\varepsilon}(\tau) \in B(0,N_2)\} \right\}$$

$$+ E\left\{ \Delta G \,\middle|\, \begin{matrix} x^{\varepsilon}(\tau-\tau_{\varepsilon}) = \tilde{x} \\ y^{\varepsilon}(\tau-\tau_{\varepsilon}) = \tilde{y} \\ \Delta G \leq N_1 \\ y^{\varepsilon}(\tau) \notin B(0,N_2) \end{matrix} \right\} P\left\{ \{\Delta G \leq N_1\} \cap \{y^{\varepsilon}(\tau) \notin B(0,N_2)\} \right\}$$

$$+ E\left\{ \Delta G \,\middle|\, \begin{matrix} x^{\varepsilon}(\tau-\tau_{\varepsilon}) = \tilde{x} \\ y^{\varepsilon}(\tau-\tau_{\varepsilon}) = \tilde{y} \\ \Delta G > N_1 \end{matrix} \right\} P\{\Delta G > N_1\} \tag{53}$$

The wp1 convergence of $x^{\varepsilon}(\tau)$ to $x^{o}(\tau)$ plus the uniform continuity of assumption (A2) implies that the first term on the rhs of Eq. (53) goes to zero for each fixed N_1 and N_2. The tightness in assumption (A5b) implies the second term goes to zero as N_2 goes to ∞. The uniform integrability in (A3b) implies the third term goes to zero as $N_1 \to \infty$. (The proper order is to choose N_1 , N_2, then ε.) Similar reasoning shows that the third term on the rhs of Eq. (52) goes to zero. Using the continuity of $x^{o}(\cdot)$ rather than the convergence of $x^{\varepsilon}(\cdot)$ will show the expectation of the second term on the rhs of Eq. (52) converges uniformly to zero. Thus Eq. (51) is true. Applying Eq. (51) to Eq. (50) we get

$$\lim_{\substack{\varepsilon \to 0 \\ \varepsilon \in J_o}} A_d^{\varepsilon} f(x^{\varepsilon}(t))$$

$$= \lim_{\substack{\varepsilon \to 0 \\ \varepsilon \in J_o}} \frac{1}{\delta_\varepsilon} \int_t^{t+\delta_\varepsilon} \int_{(\tilde{x},\tilde{y})} \mathrm{E}\left\{ G_1(x^\varepsilon(\tau), y^\varepsilon(\tau)) \,\middle|\, \begin{matrix} x^\varepsilon(\tau - \tau_\varepsilon) = \tilde{x} \\ y^\varepsilon(\tau - \tau_\varepsilon) = \tilde{y} \end{matrix} \right\} \cdot P^\varepsilon(x^\varepsilon(t), y^\varepsilon(t), \tau - \tau_\varepsilon, t, d\tilde{x}d\tilde{y})\, d\tau$$

$$= \lim_{\substack{\varepsilon \to 0 \\ \varepsilon \in J_o}} \frac{1}{\delta_\varepsilon} \int_t^{t+\delta_\varepsilon} \int_{(\tilde{x},\tilde{y})} \mathrm{E}\left\{ G_1(x^\varepsilon(\tau - \tau_\varepsilon), y^\varepsilon(\tau)) \,\middle|\, \begin{matrix} x^\varepsilon(\tau - \tau_\varepsilon) = \tilde{x} \\ y^\varepsilon(\tau - \tau_\varepsilon) = \tilde{y} \end{matrix} \right\} \cdot P^\varepsilon(x^\varepsilon(t), y^\varepsilon(t), \tau - \tau_\varepsilon, t, d\tilde{x}d\tilde{y})\, d\tau$$

$$= \lim_{\substack{\varepsilon \to 0 \\ \varepsilon \in J_o}} \frac{1}{\delta_\varepsilon} \int_t^{t+\delta_\varepsilon} \int_{(\tilde{x},\tilde{y})} \int_{(x,y)} G_1(\tilde{x}, y) P^\varepsilon(\tilde{x}, \tilde{y}, \tau, \tau - \tau_\varepsilon, dxdy) \cdot P^\varepsilon(x^\varepsilon(t), y^\varepsilon(t), \tau - \tau_\varepsilon, t, d\tilde{x}d\tilde{y})\, d\tau$$

$$= \lim_{\substack{\varepsilon \to 0 \\ \varepsilon \in J_o}} \frac{1}{\delta_\varepsilon} \int_t^{t+\delta_\varepsilon} \int_{(\tilde{x},\tilde{y})} \int_{y} G_1(\tilde{x}, y) \tilde{P}_{z^\varepsilon}(\tilde{y}, u, 0, dy) \cdot P^\varepsilon(x^\varepsilon(t), y^\varepsilon(t), \tau - \tau_\varepsilon, t, d\tilde{x}d\tilde{y})\, d\tau$$

$$= \lim_{\substack{\varepsilon \to 0 \\ \varepsilon \in J_o}} \frac{1}{\delta_\varepsilon} \int_t^{t+\delta_\varepsilon} \int_{(\tilde{x},\tilde{y})} \int_{y} G_1(\tilde{x}, y) \tilde{P}(\tilde{y}, u, 0, dy|\, \tilde{x}) \cdot P^\varepsilon(x^\varepsilon(t), y^\varepsilon(t), \tau - \tau_\varepsilon, t, d\tilde{x}d\tilde{y})\, d\tau$$

$$= \lim_{\substack{\varepsilon \to 0 \\ \varepsilon \in J_o}} \frac{1}{\delta_\varepsilon} \int_t^{t+\delta_\varepsilon} \int_{(\tilde{x},\tilde{y})} G_a(\tilde{x}, \tilde{y}) P^\varepsilon(x^\varepsilon(t), y^\varepsilon(t), \tau - \tau_\varepsilon, t, d\tilde{x}d\tilde{y})\, d\tau$$

$$= \lim_{\substack{\varepsilon \to 0 \\ \varepsilon \in J_o}} \frac{1}{\delta_\varepsilon} \int_t^{t+\delta_\varepsilon} \mathrm{E}_{e^{t+t_\varepsilon}}\left\{ G_a(x^\varepsilon(\tau), y^\varepsilon(\tau)) \right\} d\tau \tag{54}$$

Where $G_a(\cdot,\cdot)$ is defined as

$$G_a(\tilde{x}, \tilde{y}) = \int G_1(\tilde{x}, y) \tilde{P}(\tilde{y}, u, 0, dy|\, \tilde{x}) \tag{55}$$

By (A10a), $G_a(\cdot,\cdot)$ is uniformly continuous (in *both* variables). Using a proof similar to that for Eq. (51) we have

$$\lim_{\substack{\varepsilon \to 0 \\ \varepsilon \in J_o}} E_{e^{t+t_\varepsilon}}\left\{ \left| G_a(x^\varepsilon(\tau), y^\varepsilon(\tau)) - G_a(x^\varepsilon(t), y^\varepsilon(\tau)) \right| \right\} = 0 \tag{56}$$

holds uniformly on compact (τ, x, y)-sets. Continuing on from Eq. (54), we have

$$\lim_{\substack{\varepsilon \to 0 \\ \varepsilon \in J_o}} A_a^\varepsilon f(x^\varepsilon(t)) = \lim_{\substack{\varepsilon \to 0 \\ \varepsilon \in J_o}} \frac{1}{\delta_\varepsilon} \int_t^{t+\delta_\varepsilon} \int_{(\tilde{x},\tilde{y})} G_a(\tilde{x}, \tilde{y}) P^\varepsilon(x^\varepsilon(t), y^\varepsilon(t), \tau - \tau_\varepsilon, t, d\tilde{x}d\tilde{y})\, d\tau$$

$$= \lim_{\substack{\varepsilon\to 0\\ \varepsilon\in J_o}} \frac{1}{\delta_\varepsilon} \int_t^{t+\delta_\varepsilon} \int_{(\tilde{x},\tilde{y})} G_a(x^\varepsilon(t),\tilde{y}) P^\varepsilon(x^\varepsilon(t),y^\varepsilon(t),\tau-\tau_\varepsilon,t,d\tilde{x}d\tilde{y})\,d\tau$$

$$= \lim_{\substack{\varepsilon\to 0\\ \varepsilon\in J_o}} \int_{\tilde{y}} G_a(x^\varepsilon(t),\tilde{y}) \left(\frac{1}{\delta_\varepsilon} \int_t^{t+\delta_\varepsilon} P^\varepsilon(x^\varepsilon(t),y^\varepsilon(t),\tau-\tau_\varepsilon,t,R_x\times d\tilde{y})\,d\tau \right)$$

$$= \lim_{\substack{\varepsilon\to 0\\ \varepsilon\in J_o}} \int G_a(x^\varepsilon(t),y) Q(t,\varepsilon,dy)$$

$$= \int G_a(x^o(t),\tilde{y}) P^{x_o(t)}(d\tilde{y})$$

$$= \int_{\tilde{y}} \int_y G_1(x^o(t),y)\tilde{P}(\tilde{y},u,0,dy|\, x^o(t))\, P^{x_o(t)}(d\tilde{y})$$

$$= \int G_1(x^o(t),\tilde{y})\, P^{x_o(t)}(d\tilde{y}) \tag{57}$$

The second equality follows from Eq. (56). The third equality is interchanging the order of integration and the fourth is the definition of $Q(\cdot,\cdot,\cdot)$. The fifth is the weak convergence of Lemma 4. The sixth is notation. The last inequality is the invariance of the stationary measures with respect to the transition measures of the fixed x process. The definition of $G_1(\cdot,\cdot)$ and Eq. (57) gives us

$$\lim_{\substack{\varepsilon\to 0\\ \varepsilon\in J_o}} A_a^\varepsilon f(x^\varepsilon(t)) = \int f_x^T(x^o(t)) F_1(x^o(t),\tilde{y}) P^{x^o(t)}(d\tilde{y}) = f_x^T(x^o(t)) \underline{F}(x^o(t)) \tag{58}$$

♦♦ (LEMMA 5)

We have identified the limit of the convergent subsequence $\{\Gamma^\varepsilon(\cdot)\}_{\varepsilon\in J_o}$ and now we have to show $x^o(\cdot)$ satisfies the correct ODE.

LEMMA 6:

$$f(x^o(t+s)) - f(x^o(t)) = \int_t^{t+s} f_x^T(x^o(\tau)) \underline{F}(x^o(\tau))\,d\tau = \int_t^{t+s} Af(x^o(\tau))\,d\tau \tag{59}$$

The Ito differentiation rule applied to $f(x(\cdot))$ ($x(\cdot)$ is the process described by Eq. II.1) gives

$$f(x(t+s)) - f(x(t)) = \int_t^{t+s} f_x^T(x(\tau)) F_1(x(\tau),y(\tau)) \frac{d\tau}{\tau} + \int_t^{t+s} f_x^T(x(\tau)) F_2(x(\tau),y(\tau)) \frac{dw(\tau)}{\tau}$$

$$+\frac{1}{2}\int_{t}^{t+s} F_2^T(x(\tau),y(\tau))\, f_{xx}(x(\tau))\, F_2(x(\tau),y(\tau))\, \frac{d\tau}{\tau^2}$$

(60)

Using the change of variables $x^{\varepsilon}(t)=x(e^{t+t_{\varepsilon}})$ in the first and third integrals, we would have

$$f(x^{\varepsilon}(t+s))-f(x^{\varepsilon}(t))=\int_{t}^{t+s} f_x^T(x^{\varepsilon}(\tau))\,F_1(x^{\varepsilon}(\tau),y^{\varepsilon}(\tau))\,d\tau$$
$$+\int_{e^{t+t_{\varepsilon}}}^{e^{t+s+t_{\varepsilon}}} f_x^T(x^{\varepsilon}(\log(\tau)-t_{\varepsilon}))F_2(x^{\varepsilon}(\cdots),y^{\varepsilon}(\cdots))\frac{dw(\tau)}{\tau}$$
$$+\frac{e^{-t_{\varepsilon}}}{2}\int_{t}^{t+s} F_2^T(x^{\varepsilon}(\tau),y^{\varepsilon}(\tau))\, f_{xx}(x^{\varepsilon}(\tau))\, F_2(x^{\varepsilon}(\tau),y^{\varepsilon}(\tau))\,\frac{d\tau}{e^{\tau}}$$

(61)

We are not assuming an underlying $x(.)$ process, but Eq. (61) still holds for processes defined by Eq. II.2. A proof similar to that of Lemma 2 shows the second term on the rhs of Eq. (61) converges to zero in probability uniformly on compact t-sets. (A4) implies the third term converges to zero in probability uniformly on compact t-sets. The weak convergence as $\varepsilon \to 0$ in J_o implies

$$\mathrm{E}_{e^{t+t_{\varepsilon}}} f(x^{\varepsilon}(t+s))-f(x^{\varepsilon}(t)) \Rightarrow f(x^{o}(t+s))-f(x^{o}(t)) \tag{62}$$

Thus, by Eq. (61) and Theorem A.5 , it suffices show that

$$\mathrm{E}_{e^{t+t_{\varepsilon}}} \int_{t}^{t+s} f_x^T(x^{\varepsilon}(\tau))F_1(x^{\varepsilon}(\tau),y^{\varepsilon}(\tau))\,d\tau \Rightarrow \int_{t}^{t+s} f_x^T(x^{o}(\tau))\underline{F}(x^{o}(\tau))\,d\tau \tag{63}$$

From Lemma 5 we know

$$\int_{t}^{t+s} \mathrm{E}_{e^{\tau+t_{\varepsilon}}}\left\{\frac{1}{\delta_{\varepsilon}}\int_{\tau}^{\tau+\delta_{\varepsilon}} f_x^T(x^{\varepsilon}(\sigma))F_1(x^{\varepsilon}(\sigma),y^{\varepsilon}(\sigma))\,d\sigma\right\}d\tau \Rightarrow \int_{t}^{t+s} f_x^T(x^{o}(\tau))\underline{F}(x^{o}(\tau))\,d\tau \tag{64}$$

By Theorem A.5 and the weak convergence of Eq. (64), it suffices show that the difference between the left hand sides of Eqs. (63) and (64) go to zero in probability uniformly on compact intervals. Since the Riemann sums converge uniformly in probability to their respective integrals, it suffices to show that the lhs of Eq. (63) is a Riemann sum for the lhs of Eq. (64). Readopt the $G_1(\cdot,\cdot)$ notation from Eq. (49) and start with Eq. (63):

$$\mathrm{E}_{e^{t+t_{\varepsilon}}}\left\{\int_{t}^{t+s} G_1(x^{\varepsilon}(\tau),y^{\varepsilon}(\tau))\,d\tau\right\} \approx \mathrm{E}_{e^{t+t_{\varepsilon}}}\left\{\sum_{k=0}^{N_{\varepsilon}}\left\{\int_{t+k\delta_{\varepsilon}}^{t+(k+1)\delta_{\varepsilon}} G_1(x^{\varepsilon}(\tau),y^{\varepsilon}(\tau))\,d\tau\right\}\right\}$$

$$= \mathrm{E}_{e^{t+t_\varepsilon}}\left\{\sum_{k=0}^{N_\varepsilon} \mathrm{E}_{e^{k\delta_\varepsilon+t+t_\varepsilon}}\left\{\int_{t+k\delta_\varepsilon}^{t+(k+1)\delta_\varepsilon} G_1(x^\varepsilon(\tau),y^\varepsilon(\tau))\,d\tau\right\}\right\}$$

$$= \mathrm{E}_{e^{t+t_\varepsilon}}\left\{\sum_{k=0}^{N_\varepsilon}\left[\frac{1}{\delta_\varepsilon}\mathrm{E}_{e^{k\delta_\varepsilon+t+t_\varepsilon}}\left\{\int_{t+k\delta_\varepsilon}^{t+(k+1)\delta_\varepsilon} G_1(x^\varepsilon(\sigma),y^\varepsilon(\sigma))\,d\sigma\right\}\right]\delta_\varepsilon\right\} \tag{65}$$

N_ε is greatest integer less than s/δ_ε - 1. The only approximation in Eq. (65) is that the final term in the summation is missing when s is not exactly divisible by δ_ε. (The final term was left off for notational convenience only.) The final lhs formula in Eq. (65) is a Riemann sum for the function inside its square brackets. So the last formula in Eq. (65) is a Riemann sum for

$$\mathrm{E}_{e^{t+t_\varepsilon}}\left\{\int_t^{t+s} \mathrm{E}_{e^{\tau+t_\varepsilon}}\left\{\frac{1}{\delta_\varepsilon}\int_\tau^{\tau+\delta_\varepsilon} G_1(x^\varepsilon(\sigma),y^\varepsilon(\sigma))\,d\sigma\right\}d\tau\right\} \tag{66}$$

Thus the weak convergence in Eq. (64) implies the weak convergence in Eq. (63) and hence the lemma.

♦♦(LEMMA 6)

Equation (59) holds for any bounded continuous function $f(\cdot)$. Since the integrand in Eq. (59) is continuous, we can divide both sides of Eq. (59) by s take limits as $s \to 0$ to get

$$\frac{d}{dt}\Big(f(x^o(t))\Big) = f_x^T(x^o(t))\underline{F}(x^o(t)) \tag{67}$$

By taking $f(\cdot)$ to be equal to coordinate functions inside compact sets and zero outside larger compact sets we can show

$$\frac{d}{dt}\Big(x^o(t)\Big) = \underline{F}(x^o(t)) \tag{68}$$

on arbitrarily large compact sets. Thus $x^o(\cdot)$ is differentiable and satisfies the appropriate ODE. By hypothesis the solutions to Eq. (68) are unique. Thus we have shown that every subsequence of $\{x^\varepsilon(\cdot)\}_{\varepsilon\in(0,1]}$ has a weakly convergent (sub)subsequence which converges, as $\varepsilon \to 0$, to the same limit $x^o(\cdot)$, the *unique* solution of Eq. (68). This implies that original sequence, $\{x^\varepsilon(\cdot)\}_{\varepsilon\in(0,1]}$, converges weakly to $x^o(\cdot)$.

♦♦♦(THEOREM II.1)

Proof of THEOREM II.2:

$x^\varepsilon(\cdot)$ and $y^\varepsilon(\cdot)$ are defined by equations (II.1), (II.2):

$$dx = \frac{1}{t}[F_1(x,y)dt + F_2(x,y)dw] \qquad x(t_s) = x_o$$
$$dy = H_1(x,y)dt + H_2(x,y)dw \qquad y(t_s) = y_o$$

$$x^{\varepsilon}(t+s) - x^{\varepsilon}(t) = \int_t^{t+s} F_1(x^{\varepsilon}(\tau), y^{\varepsilon}(\tau))d\tau$$
$$+ \int_{e^{t+t_\varepsilon}}^{e^{t+s+t_\varepsilon}} F_2(x^{\varepsilon}(\log(\tau) - t_\varepsilon), y^{\varepsilon}(\log(\tau) - t_\varepsilon))\frac{dw(\tau)}{\tau}$$

$$d\tilde{y}^{\varepsilon}(t) = H_1(x^{\varepsilon}(\log(t) - t_\varepsilon), \tilde{y}^{\varepsilon}(t))dt + H_2(x^{\varepsilon}(\log(t) - t_\varepsilon), \tilde{y}^{\varepsilon}(t))dw(t)$$
$$y^{\varepsilon}(t) = \tilde{y}^{\varepsilon}(e^{t+t_\varepsilon})$$
$$x^{\varepsilon}(0) = x_\varepsilon \qquad y^{\varepsilon}(0) = y(e^{t_\varepsilon}) \qquad (\therefore \tilde{y}^{\varepsilon}(e^{t_\varepsilon}) = y(e^{t_\varepsilon})) \tag{69}$$

We need to verify (A1)-(A10). The tightness of $\{x^{\varepsilon}(t) : t \geq 0, \varepsilon > 0\}$ follows from (b). (A2) and (A6) are explicit hypotheses of Theorem II.2.

(A1) :

By Theorem A.12, it suffices to show

$$x^T F_1(x, y_1, y_2) + y_1^T(A_1(x, y_2)y_1 + B_1(x)u(t)) + y_2^T A_2(x, y_2)$$
$$\leq C\left(1 + \|x\|^2 + \|y_1\|^2 + \|y_2\|^2\right)$$
$$\sup_{\substack{\|z_1\|=1 \\ \|z_2\|=1}} \left\{\left\|S_1^T(x, y_2)z_1\right\|^2 + \left\|F_2^T(x, y_2)z_2\right\|^2\right\} \leq C\left(1 + \|x\|^2 + \|y_2\|^2\right) \tag{70}$$

Assumptions (a) and (c) imply that Eq. (70) is satisfied.

(A5):

By definition, $y^{\varepsilon}(t) = y(e^{t+t_\varepsilon})$so the tightness of $\left\{y^{\varepsilon}(t) : \varepsilon > 0, t \geq 0\right\}$ is equivalent to the tightness of $\left\{y(t) : t \geq t_s\right\}$. The boundedness of $y_2(\cdot)$ from (f) implies that it suffices to show that $\left\{y_1(t) ; t \geq 0\right\}$ is tight. Define

$$A_1^o(t) = A_1(x(t), y_2(t))$$
$$S_1^o(t) = S_1(x(t), y_2(t)) \tag{71}$$

For each sample path let $\Phi_1(\cdot,\cdot)$ be the state transition matrix corresponding to $A_1^o(\cdot)$. Since the sample paths are continuous, this defines $\Phi_1(\cdot,\cdot)$ as a random matrix. For any y and every sample path we have

$$\|\Phi_1(t+s, t)y\| \leq Ke^{\lambda_o s}\|y\| \tag{72}$$

We have

$$y_1(t) = \Phi_1(t,0)y_{1,o} + \int_o^t \Phi_1(t,s)B_1(x(s))u(s)ds + \int_o^t \Phi_1(t,s)S_1^o(s)dw_2(s) \tag{73}$$

By Eq. (73) we know

$$P\{\|y_1(t)\| \geq n\} \leq P\left\{\|\Phi_1(t,0)y_1\| \geq \frac{n}{3}\right\} + P\left\{\left\|\int_o^t \Phi_1(t,s)B_1(x(s))u(s)ds\right\| \geq \frac{n}{3}\right\}$$
$$+P\left\{\left\|\int_o^t \Phi_1(t,s)S_1^o(s)dw_2(s)\right\| \geq \frac{n}{3}\right\} \tag{74}$$

Equation (72), Inequality A.2, and Eq. (74) imply

$$P\{\|y_1(t)\| \geq n\} \leq P\left\{Ke^{\lambda_o t}\|y_1\| \geq \frac{n}{3}\right\} + P\left\{\int_o^t Ke^{\lambda_o(t-s)}\|B_1(x(s))u(s)\|ds \geq \frac{n}{3}\right\}$$
$$+\frac{9}{n^2}E\left\{\int_o^t \left\|\Phi_1(t,s)S_1^o(s)\right\|^2 ds\right\} \tag{75}$$

The matrix norm in Eq. (75) is the square root of the sum of the square of the column norms. The single random variable y_1 is tight, hence the first term on the rhs of Eq. (75) goes to zero as n goes to infinity. (c) and the compact x-support of $F_1(\cdot,\cdot)$ and $F_2(\cdot,\cdot)$ essentially make $B_1(x(s))u(s)$ bounded. This implies the second term on the rhs of Eq. (75) is 0 for sufficiently large n. Equation (72) implies

$$\frac{9}{n^2}E\left\{\int_o^t \left\|\Phi_1(t,s)S_1^o(s)\right\|^2 ds\right\} \leq \frac{K'}{n^2}e^{2\lambda_o t}\left(\frac{1}{-2\lambda_o}\right)\left(e^{-2\lambda_o t} - 1\right)$$
$$\leq \frac{K'}{n^2}\left(\frac{1}{-2\lambda_o}\right) \tag{76}$$

Which implies the third term on the rhs of Eq. (75) goes to zero as $n \to \infty$. Thus the family of random variables $\{y_1(t)\,;\, t \geq 0\}$ is tight and (A5) is proven.

(A7):

The tightness in (A7b) is equivalent to the tightness of the collection:

$$\left\{\begin{bmatrix}\tilde{y}_{1,z^\varepsilon}(t;y_1)\\ \tilde{y}_{2,z^\varepsilon}(t;y_2)\end{bmatrix} I_B\left(\begin{bmatrix}y_1\\ y_2\end{bmatrix}\right) \;\middle|\; \begin{matrix}B \text{ compact}\\ e^{t_\varepsilon+\varepsilon\beta_\varepsilon} \leq t \leq e^{t_\varepsilon+T}\\ z^\varepsilon(\cdot) \text{ with } \sup\left\|z^\varepsilon(t)\right\| \leq K\end{matrix}\right\} \tag{77}$$

A development similar to the proof of (A5) shows

$$P\left\{\left\|\tilde{y}_{1,z^{\varepsilon}}(t;y_1)\right\| \geq n\right\} \leq P\left\{Ke^{\lambda_o(t-e^{t_\varepsilon})}\|y_1\| \geq \frac{n}{3}\right\}$$
$$+P\left\{\int_{e^{t_\varepsilon}}^{t} Ke^{\lambda_o(t-s)}\left\|B_1(z^{\varepsilon}(s))u(s)\right\|ds \geq \frac{n}{3}\right\}$$
$$+\frac{9}{n^2}E_{e^{t_\varepsilon}}\left\{\int_{e^{t_\varepsilon}}^{t}\left\|\Phi_1(t,s)S_1^o(s)\right\|^2 ds\right\} \tag{78}$$

If we let $\beta_\varepsilon(B) = \frac{1}{\varepsilon}\log\left(1+\frac{\log\|B\|}{|\lambda_o|}e^{-t_\varepsilon}\right)$ a calculation shows $Ke^{\lambda_o(t-e^{t_\varepsilon})} \leq \frac{1}{\|B\|}$ when $t \geq e^{\varepsilon\beta_\varepsilon - t_\varepsilon}$, which implies the first term on the rhs of Eq. (78) is 0 for $n>2$. The second and third terms on the rhs goes to zero just as in the proof for (A5). Thus the family of random variables in Eq. (77) is tight and (A7) is proven.

(A3b), (A4b):

Calculations similar to those in the proof of (A5) show that $y(t)$ has finite moments of all order, independent of t. This and the quadratic bounds in (c) imply (A3b) and (A4b).

(A8), (A9a), (A10a):

Theorem A.12 implies the fixed x system has a unique (in the sense of distributions) solution and the transition probabilities are measurable and Feller continuous as functions of (t,y). This and the continuity of $F_1(\cdot,\cdot)$ imply (A8a), (A9a) and (A10a). (A8b) is a hypothesis of Theorem II.2.

(A9b),(A10b)

Let $z^{\varepsilon}(\cdot)$ be a bounded nonanticipative process such that

$$\sup_{t_o \leq s \leq u}\left\|z^{\varepsilon}(s)-x\right\| \Rightarrow 0 \quad as \quad \varepsilon \to 0 \tag{79}$$

To show (A9b), we need to show

$$\lim_{\varepsilon\to o}\int f(\tilde{y})\tilde{P}_{z^{\varepsilon}}(y,u,t_o,d\tilde{y}) = \int f(\tilde{y})\tilde{P}(y,u,t_o,d\tilde{y}|x) \tag{80}$$

Eq. (80) can be written

$$\lim_{\varepsilon\to o}E\left\{f(\tilde{y}_{z^{\varepsilon}}(u)) - f(\tilde{y}(u;x))\middle|\tilde{y}_{z^{\varepsilon}}(t_o) = \tilde{y}(t_o;x) = y\right\} = 0 \tag{81}$$

Without loss of generality we can assume $f(\cdot)$ is continuously differentiable with bounded first derivative. Thus to show Eq. (81), it suffices to show

$$\lim_{\varepsilon\to o}E\left\{\left\|\tilde{y}_{z^{\varepsilon}}(u) - \tilde{y}(u;x)\right\| \;\middle|\tilde{y}_{z^{\varepsilon}}(t_o) = \tilde{y}(t_o;x) = y\right\} = 0 \tag{82}$$

Let

$$\tilde{y}_{z^\varepsilon}(t) = \begin{bmatrix} \tilde{y}_{z^\varepsilon,1}(t) \\ \tilde{y}_{z^\varepsilon,2}(t) \end{bmatrix} \quad \text{and} \quad \tilde{y}(t;x) = \begin{bmatrix} \tilde{y}_1(t;x) \\ \tilde{y}_2(t;x) \end{bmatrix} \tag{83}$$

$\tilde{y}_2(u;x)$ and $\tilde{y}_{z^\varepsilon,2}(u)$ satisfy ordinary differential equations, so the usual theory of continuous dependence of solutions on parameters implies

$$E\left\|\tilde{y}_{z^\varepsilon,2}(u) - \tilde{y}_2(u;x)\right\| \to 0 \tag{84}$$

uniformly on compact intervals. Since the initial conditions are the same we have

$$\begin{aligned} E\left\|\tilde{y}_{z^\varepsilon,1}(u) - \tilde{y}_1(u;x)\right\| \le\ & E\left\|\int_{t_o}^{u}\begin{bmatrix} A_1(z^\varepsilon(t),\tilde{y}_{z^\varepsilon,2}(t))\tilde{y}_{z^\varepsilon,1}(t) \\ -A_1(x,\tilde{y}_2(t;x))\tilde{y}_1(t;x) \end{bmatrix} dt\right\| \\ &+ E\left\|\int_{t_o}^{u}\left[B_1(z^\varepsilon(t)) - B_1(x)\right]u(t)dt\right\| \\ &+ E\left\|\int_{t_o}^{u}\begin{bmatrix} S_1(z^\varepsilon(t),\tilde{y}_{z^\varepsilon,2}(t)) \\ -S_1(x,\tilde{y}_2(t;x)) \end{bmatrix} dw(t)\right\| \end{aligned} \tag{85}$$

Equations (79) and (84) imply the second and third terms on the rhs of Eq. (85) $\to 0$ wp1 as $\varepsilon \to 0$. Defining $\eta^\varepsilon(u) = E\left\|\tilde{y}_{z^\varepsilon,1}(u) - \tilde{y}_1(u;x)\right\|$, we have

$$\begin{aligned} \eta^\varepsilon(u) &\le \delta_1^\varepsilon + \int_{t_o}^{u}\left\|\begin{matrix} A_1(z^\varepsilon(t),\tilde{y}_{z^\varepsilon,2}(t))\tilde{y}_{z^\varepsilon,1}(t) \\ -A_1(x,\tilde{y}_2(t;x))\tilde{y}_1(t;x) \end{matrix}\right\| dt \\ &\le \delta_1^\varepsilon + \int_{t_o}^{u}\left\|\begin{matrix} A_1(z^\varepsilon(t),\tilde{y}_{z^\varepsilon,2}(t))\tilde{y}_{z^\varepsilon,1}(t) \\ -A_1(x,\tilde{y}_2(t;x))\tilde{y}_{z^\varepsilon,1}(t) \end{matrix}\right\| dt \\ &\qquad + \int_{t_o}^{u}\left\|\begin{matrix} A_1(x,\tilde{y}_{z^\varepsilon,2}(t))\tilde{y}_{z^\varepsilon,1}(t) \\ -A_1(x,\tilde{y}_2(t;x))\tilde{y}_1(t;x) \end{matrix}\right\| dt \\ &\le (\delta_1^\varepsilon + \delta_2^\varepsilon) + \int_{t_o}^{u}\left\|A_1(x,\tilde{y}_{z^\varepsilon,2}(t))\right\| \eta^\varepsilon(t)dt \end{aligned} \tag{86}$$

where $(\delta_1^\varepsilon + \delta_2^\varepsilon) \to 0$ wp1 as $\varepsilon \to 0$. The Gronwall-Bellman Lemma says

$$\eta^{\varepsilon}(u) \leq (\delta_1^{\varepsilon} + \delta_2^{\varepsilon}) \exp\left(\int_{t_o}^{u} \| A_1(x, \bar{y}_2(t;x)) \| dt \right) \tag{87}$$

The exponential in Eq. (87) is bounded by hypothesis (a), hence the rhs of Eq. (87) $\rightarrow 0$ wp1. This proves Eq. (82) and therefore (A9b). (A10b) is proved in a similar fashion.

♦♦♦(THEOREM II.2)

Proof of Theorem II.3a:

For the convergence in first part of Theorem II.3a we would like to show

$$\exists\ \varepsilon_o, \text{ such that if } \varepsilon \leq \varepsilon_o \text{ then } \qquad P\left\{ \lim_{t \to \infty} d(x^{\varepsilon}(t), R_c) = 0 \right\} = 1 \tag{88}$$

We will proceed by contradiction, so assume

$$\exists\ \delta,\ \varepsilon_k \to 0,\ t_k \to \infty, \text{ such that },\ P\{ d(x^{\varepsilon_k}(t_k), \mathrm{R_c}) \geq \delta\} \geq \pi_o > 0. \tag{89}$$

For $\tau \geq 0$, (C3) says the family of random variables

$$\{ x^{\varepsilon_k}(t_k - \min(\tau, t_k)) : k \geq 1\} \tag{90}$$

is tight, hence for each τ there is a weakly convergent subsequence. Denote $t_k - \min(\tau, t_k)$ by $\gamma_k(\tau)$. To ease the notation, denote the subsequential indices by k and the limit $x_{o,\tau}$. Thus we have

$$x^{\varepsilon_k}(\gamma_k(\tau)) \Rightarrow x_{o,\tau} \tag{91}$$

By (C3) and the convergence in Eq. (91), the family $\{ x_{o,\tau} : \tau \geq 0 \}$ is tight, hence

$$\exists\ M,\ \tau_o \quad P\{ x_{o,\tau} \notin \mathrm{B}_M \} \leq \pi_o/4 \ \text{ for } \tau \geq \tau_o \tag{92}$$

(B_M is the ball of radius M in $\mathbf{R}^{n_x}$). For any t and $\tau \geq \tau_o$ we have

$$\begin{aligned} P\Big\{ d(x^{\varepsilon_k}(\gamma_k(\tau)+t), \mathrm{R_c}) \geq \delta \Big\} & \\ \leq P\Big\{ x_{o,\tau} \notin \mathrm{B}_M \Big\} &+ P\Big\{ d(x^{\varepsilon_k}(\gamma_k(\tau)+t), \mathrm{R_c}) \geq \delta \ \Big|\ x_{o,\tau} \in \mathrm{B}_M \Big\} \\ \leq \frac{\pi_o}{4} &+ P\left\{ d(x^{\varepsilon_k}(\gamma_k(\tau)+t), x_\tau^o(t)) \geq \frac{\delta}{2} \ \Big|\ x_{o,\tau} \in \mathrm{B}_M \right\} \\ &+ P\left\{ d(x_\tau^o(t), R_c) \geq \frac{\delta}{2} \ \Big|\ x_{o,\tau} \in \mathrm{B}_M \right\} \end{aligned} \tag{93}$$

where $x^o{}_\tau(\cdot)$ is the ODE solution with initial condition $x_{o,\tau}$. By (B2) we can choose $\tau_1 \geq \tau_o$ such that $d(x^o(t), \mathrm{R_c}) < \delta/2$ for any $x^o(0) \in \mathrm{B}_M$ and $t \geq \tau_1$. (C2) tells us that

$$x^{\varepsilon_k}(\gamma_k(\tau) + \cdot) \Rightarrow x_\tau^o(\cdot) \tag{94}$$

The Skorohod embedding says we can assume the convergence in Eq. (94) is uniform wp1 on compact t-intervals, and in particular, the interval $[0, 2\tau_1]$. Applying these last two observations to Eq. (93) with $t = \tau = \tau_1$ gives us

$$\pi_o \leq P\Big\{ d(x^{\varepsilon_k}(\gamma_k(\tau)+t), R_c) \geq \delta \Big\} \leq \frac{\pi_o}{4} + \frac{\pi_o}{4} + 0 = \frac{\pi_o}{2} \tag{95}$$

which is a contradiction. This proves the first part of Theorem II.3a.

For the convergence in second part of Theorem II.3a we would like to show

$$\forall\, T\ \exists\ \varepsilon_o, \text{ such that if } \varepsilon \leq \varepsilon_o \text{ then} P\left\{\lim_{t\to\infty} d(x^{\varepsilon}(\lambda(t+T)), x^{\varepsilon}(\lambda(t))) = 0\right\} = 1 \quad (96)$$

Proceed by contradiction and assume

$\exists\ \delta, T,\ \varepsilon_k \to 0,\ t_k \to \infty$, such that for each k,

$$P\{\ d(x^{\varepsilon_k}(x^{\varepsilon}(\lambda(t_k+T)),\ x^{\varepsilon}(\lambda(t_k))\ ) \geq \delta\} \geq \pi_o > 0. \quad (97)$$

Arguments similar to those for the first part of the theorem lead to the same contradiction.

♦ ♦ ♦ (Theorem II.3a)

Before proceeding with Theorem II.3b a lemma on ODE's will be proved.

Lemma 1: *Assume* $0 < \eta_o \leq \|\underline{F}(x)\| \leq \eta_1$ *and* $\|\underline{F}_x(x)\| \leq \eta_2$ *for each* x *in* $B(x_c, \rho_o)$. *Let* $x^o(\cdot;x)$ *be the solution to the differential equation* $\dot{x}^o = \underline{F}(x^o)$ *with initial condition* x. *Then there are* ρ_1, $0<\rho_1<\rho_o$, *and* $T_1 > 0$ *such that if* $x \in B(x_c, \rho_1/2)$ *then* $x^o(\cdot;x)$ *exits* $B(x_c, \rho_1)$ *within time* T_1.

Proof of Lemma 1:

Taylor's theorem tells us $\left\|x^o(t;x) - x\right\| = \|\underline{F}(x)t + f(t)\|$ where $\|f(t)\| \leq K\eta_1\eta_2 t^2$ and K depends only on dim(x). So we have $\left\|x^o(t;x) - x\right\| \geq \left|\eta_o t - K\eta_1\eta_2 t^2\right|$. If we set $T_1 = \frac{1}{2}\frac{\eta_o}{K\eta_1\eta_2}$, $\rho_1 = \frac{1}{6}\frac{\eta_o^2}{K\eta_1\eta_2}$, and if $x \in B(x_c, \rho_1/2)$ then $x^o(\cdot;x)$ exits $B(x_c, \rho_1)$ within time T_1.

♦ ♦ (Lemma 1)

Proof of Theorem II.3b:

We will proceed by contradiction. First assume $\underline{F}(x_o) \neq 0$. Then there are ρ_o, η_o, η_1, η_2, such that $0 < \eta_o \leq \|\underline{F}(x)\| \leq \eta_1$ and $\|\underline{F}_x(x)\| \leq \eta_2$ for each x in $B(x_o, \rho_o)$. Lemma 1 says there are ρ_1, $0<\rho_1<\rho_o$, and $T_1 > 0$ such that if $x \in B(x_o, \rho_1/2)$ then $x^o(\cdot;x)$ exits $B(x_o, \rho_1)$ within time T_1.

By the hypothesis of the theorem we can choose T_2 such that

$$P\left\{\sup_{t \geq T_2}\left\|x^{\varepsilon}(t) - x_o\right\| < \frac{\rho_1}{4}\right\} \geq \pi_o \quad (98)$$

By (C3), $\{x^{\varepsilon}(t) : t \geq 0,\ \varepsilon \geq 0\}$ is tight. So there exists a random variable z and $s_{\varepsilon_k} \to 2T_2$ as $\varepsilon_k \to 0$ such that $T_2 \leq s_{\varepsilon_k} \leq 2T_2$ and $x^{\varepsilon_k}(s_{\varepsilon_k}) \Rightarrow z$. Since $s_{\varepsilon_k} \geq T_2$ we have

$$P\left\{\sup_{k\geq 0}\left\|x^{\varepsilon_k}(s_{\varepsilon_k}) - x_o\right\| < \frac{\rho_o}{4}\right\} \geq \pi_o \tag{99}$$

The weak convergence and (C2) imply $x^{\varepsilon_k}(s_{\varepsilon_k} + \cdot) \Rightarrow x^{o}(\cdot)$, where $x^{o}(0)=z$. The triangle inequality inequality says

$$\sup_{t\in[0,T_1]}\left\|x^o(t) - x_o\right\| \leq \sup_{t\in[0,T_1]}\left\|x^o(t) - x^{\varepsilon_k}(s_{\varepsilon_k} + t)\right\| + \sup_{t\in[0,T_1]}\left\|x^{\varepsilon_k}(s_{\varepsilon_k} + t) - x_o\right\| \tag{100}$$

Refer to the three terms in Eq. (100) as A, B_k, and C_k. Equation (99) implies

$$P\left\{A \geq \frac{\rho_o}{2}\right\} \leq P\left\{B_k \geq \frac{\rho_o}{4}\right\} + P\left\{C_k \geq \frac{\rho_o}{4}\right\} \tag{101}$$

Either $z \in B(x_c, \rho_1/2)$ and $x^o(\cdot; z)$ exits $B(x_c, \rho_1)$ within time T_1, or $z \notin B(x_c, \rho_1/2)$. Either of these imply $A \geq \rho_1/2$ and the "A" term in Eq. (101) is 1. The weak convergence, $x^{\varepsilon_k}(s_{\varepsilon_k} + \cdot) \Rightarrow x^{o}(\cdot)$, and the Skorohod embedding imply the "B_k." term in Eq. (41) goes to zero as k goes to infinity. Equation (99) implies the "C_k." term in Eq. (99) is less than 1-π_o. Thus Eq. (101) implies $1 \leq 1 - \pi_o$ which is a contradiction. Thus $\underline{F}(x_o)=0$ and the first assertion is proved.

Now assume that $\underline{F}_x(x_o)$ has an eigenvalue in the (open) right half plane and denote the eigenpair by (λ, υ). Define the processes

$$\begin{aligned}\zeta^o(t) &= x^o(t) - x_o\\ \zeta^{\varepsilon}(t) &= x^{\varepsilon}(t) - x_o\end{aligned} \tag{102}$$

and linearize to get

$$\dot{\zeta}^o = \underline{F}_x(x_o)\zeta^o + f(t) \tag{103}$$

where

$$\|f(t)\| \leq K\left\|\zeta^o(t)\right\|^2 \tag{104}$$

Define the scalar processes $r^o(\cdot)$ and $r^{\varepsilon}(\cdot)$ by $r^o(\cdot)=\upsilon^T\zeta^o(\cdot)$ and $r^{\varepsilon}(\cdot)=\upsilon^T\zeta^{\varepsilon}(\cdot)$. The various processes satisfy

$$\begin{aligned}&\lim_{T\to\infty} P\left\{\sup_{\varepsilon<\varepsilon_o}\sup_{t\leq T}\left\|\zeta^{\varepsilon}(t)\right\| < \rho\right\} \geq \pi_o > 0\\ &\lim_{T\to\infty} P\left\{\sup_{\varepsilon<\varepsilon_o}\sup_{t\leq T}\left\|r^{\varepsilon}(t)\right\| < \rho\right\} \geq \pi_o > 0\\ &\left|r^o(t)\right| \geq e^{(\mathrm{Re}(\lambda)-K\rho^2)t}\left|r^o(0)\right| \qquad \text{if } \zeta^o(t) \in B(x_o, \rho)\end{aligned} \tag{105}$$

The triangle inequality and Eq. (44) give

$$e^{(\mathrm{Re}(\lambda)-K\rho^2)t}\left|r^o(0)\right| \leq \left|r^o(t)\right| \leq \left|r^o(t) - r^{\varepsilon}(t)\right| + \left|r^{\varepsilon}(t)\right| \tag{106}$$

Reasoning virtually identical to the first part of this proof (choosing s_{ε_k} etc.) and Eq. (106) will lead to the same contradiction, $1 \leq 1-\pi_o$. The are only two differences One is to choose ρ_1 small enough so that $\text{Re}(\lambda) - K\rho_1^2 > 0$, which is possible only under the assumption that λ is in the (open) right half plane. The other is to insure that $r^o(0) \neq 0$ wp1, but this is true since the covariance of $x^\varepsilon(t)$ is positive definite for each $t \geq 0$ and $\varepsilon > 0$.

♦♦♦(Theorem II.3b)

Proof of Theorem II.4:

(a) let $t_\varepsilon \to \infty$ and define the accelerated processes for $x(\cdot)$ via Eq. IV.1. Theorem II.2 implies the accelerated process satisfies the hypothesis of Theorem II.1 and (C3). Theorem II.3 applies so there exists an ε_o such that $x^{\varepsilon_o}(t) \to R_c$ wp1 for any choice of $x^{\varepsilon_o}(0)$. Choosing $x^{\varepsilon_o}(0) = x(e^{t_{\varepsilon_o}})$ we have $x(e^{t+t_{\varepsilon_o}}) = x^{\varepsilon_o}(t)$ and therefore $x(t) \to R_c$ wp1.

(b),(c) The proof for II.4b,c is similar to II.4a

♦♦♦(Theorem II.4)

APPENDIX C: NUMERICAL TECHNIQUES FOR SIMULATING SDE'S

Several method have been proposed for numerical simulation of stochastic differential equations (e.g. [22],[23],[24]). Runge-Kutta techniques have the advantage that no derivatives of the drift and diffusion functions are required (except for the Wong-Zakai corrections). The three Runge-Kutta techniques to be examined are Euler's method, the modified Huen method, and a 4th order Runge-Kutta with $\lambda=.5$. Euler's method is a first order Runge-Kutta and the modified Huen technique is a second order Runge-Kutta technique with $\lambda=.5$. For the rest of this appendix the 4th order Runge-Kutta with $\lambda=.5$ will be referred to as RK4. Several theorems of Rumelin, Fahrmeir, and Kushner concerning convergence as a function of step size are gathered here. Simulation results on a particular second order SDE are used to illustrate the theorems. In addition, the stepsize required for simulating the RPEM example in Section IV is derived here. This appendix is primarily background, and no new theory is developed.

If the SDE to be simulated is

$$dx = a(x)dt + \sigma(x)dw(t) \tag{1}$$

then an Euler step of size h is

$$x(t+h) = x(t) + a(x(t))h + \sigma(x(t))\Delta w_t \tag{2}$$

$\Delta w_t = w(t+h)-w(t)$ is a zero mean Gaussian random vector. Each component of Δw_t has variance h. A modified Huen step is

$$\begin{aligned}
x_1 &= x(t) + a(x(t))h + \sigma(x(t))\Delta w_t \\
x_2 &= x(t) + a(x_1)h + \sigma(x_1)\Delta w_t \\
x(t+h) &= x(t) + \left(\frac{a(x(t)) + a(x_1)}{2}\right)h + \left(\frac{\sigma(x(t)) + \sigma(x_1)}{2}\right)\Delta w_t \\
&= \frac{1}{2}(x_1 + x_2)
\end{aligned} \tag{3}$$

The RK4 step is

$$\begin{aligned}
x_1 &= x(t) + a(x(t))h + \sigma(x(t))\Delta w_t \\
x_2 &= x(t) + a(x_1)\frac{h}{2} + \sigma(x_1)\frac{\Delta w_t}{2} \\
x_3 &= x(t) + a(x_2)h + \sigma(x_2)\Delta w_t \\
x_4 &= x(t) + a(x_3)\frac{h}{2} + \sigma(x_3)\frac{\Delta w_t}{2} \\
x(t+h) &= x(t) + \left(\frac{a(x(t))}{6} + \frac{a(x_1)}{3} + \frac{a(x_2)}{3} + \frac{a(x_3)}{6}\right)h + \\
&\qquad \left(\frac{\sigma(x(t))}{6} + \frac{\sigma(x_1)}{3} + \frac{\sigma(x_2)}{3} + \frac{\sigma(x_3)}{6}\right)\Delta w_t \\
&= \frac{1}{3}(x_1 + 2x_2 + x_3 + x_4 - 2x(t))
\end{aligned} \tag{4}$$

Equations (2)-(4) show that these stochastic SDE solvers are the same as the deterministic ODE solvers applied to the drift and diffusions terms separately, with Δw_t used instead of h for the diffusion term.

Equations (2)-(4) can be considered as a method for generating a random process that approximates the Ito solution of Eq. (1) or as a method of generating a realization of the Ito solution of Eq. (1) (or a related equation). If $\{\Delta w_t : t=kh,\ k=0,1,2,3,\ldots\}$ is considered a sequence of random vectors, then the sequence $\{x_h(kh) : k=0,1,2,3,\ldots\}$ is another sequence of random vectors. Define a piecewise continuous process $x_h(t)$ by

$$x_h(t) = x_h(kh) \quad \text{for} \quad t \in [kh\ ,\ (k+1)h). \tag{5}$$

In many cases the process $x_h(\cdot)$ will converge weakly or in the mean square to a solution of Eq. (1) or a related equation. A SDE can be simulated by generating a realization of Wiener process increments, $\{\Delta w_t : t=kh,\ k=0,1,2,3,\ldots\}$, using a Gaussian random number generator and applying the one step formulas to get the sequence of numbers, $\{x_h(kh) : k=0,1,2,3,\ldots\}$. Several realizations can be averaged (or otherwise processed) to generate sample statistics.

Before stating the convergence theorems some additional notation is needed. Consider the SDE

$$dx = \left(a(x) - \lambda \sum_{r=1}^{m} \left(\nabla_x \sigma^r(x) \right) \sigma^r(x) \right) dt + \sigma(x) dw(t)$$
$$\sigma(x) = \left[\sigma^1(x) \,|\, \ldots \,|\, \sigma^m(x) \right], \quad \sigma^r \text{ is the r}^{\text{th}} \text{column of } \sigma$$
$$\nabla_x \sigma^r = \left[\frac{\partial \sigma_i^r}{\partial x_j} \right]_{1 \le i, j \le n} \tag{6}$$

The second part of the drift term in Eq. (6) will be referred to as the Wong-Zakai correction. Different SDE simulation techniques lead to different values of λ for the correction term. Define the quantity $S(r,s)$ by

$$S(r,s) = (\nabla_x \sigma^r(\cdot))\sigma^s(\cdot) - (\nabla_x \sigma^s(\cdot))\sigma^r(\cdot) \tag{7}$$

$S(r,s)$ is the Lie Bracket of the vector fields σ^r and σ^s. Two vector fields whose Lie Bracket is zero are said to commute.

THEOREM 1: *Assume the drift and diffusion coefficients of Eq. (1) and their first and second partial derivatives are bounded and continuous. The processes, $x_h(\cdot)$, corresponding to the Euler, modified Huen, and RK4 methods applied to Eq. (6), with λ = 0,.5, and .5 respectively, converge uniformly in the mean square to the Ito solution of Eq. (1) .*

THEOREM 2: *Suppose $a(\cdot)$ is continuous, $\sigma(\cdot)$ is twice continuously differentiable and for each initial condition Eq. (6) has a unique (in the sense of distributions) solution. The processes, $x_h(\cdot)$, corresponding to the Euler and modified Huen methods applied to Eq. (6), with λ = 0, and .5 respectively, converge weakly to the Ito solution of Eq. (1).*

CONJECTURE 3: *Under the hypothesis of Theorem 2, the process, $x_h(\cdot)$, corresponding to the RK4 method applied to Eq. (6), with λ = .5, converges weakly to the Ito solution of Eq. (1).*

THEOREM 4: *Let $x(\cdot)$ be the Ito solution of Eq. (1) and $x_h(\cdot)$ the best approximation to $x(\cdot)$ which is conditioned upon the discrete set of random variables $\{w(kh+h) - w(kh)$: $k=0,1,2,3,\ldots$; $kh+h \le T\}$. If the coefficients of Eq. (1) have continuous 4th order partial derivatives, then*

(a) *if $S(r,s) \equiv 0$ for all $r \neq s$ then the one step mean square error is at best $O(h^3)$*
$O(h^3)$ is achieved by the modified Huen method applied to Eq. (6) with λ = .5

(b) *if $S(r,s) \neq 0$ for some $r \neq s$ then the one step mean square error is* $O(h^2)$
$O(h^2)$ *is achieved by Euler's method applied to Eq.* (6) *with* $\lambda = 0$ (*i.e. Eq.* (1))

THEOREM 5: *If the one step mean square error is* $O(h^{n+1})$ *then the mean square error over a fixed interval* $[0,T]$ *is* $\leq T \bullet O(h^n)$.

Theorem 1 is a special case of theorem 1 in [24]. Theorem 4 is essentially Theorem 3 in [24]. Theorem 5, the relation between one step mean square error and the mean square error over a finite interval, is proved in [23]. This bound on the finite interval error is often pessimistic since it assumes the errors add constructively over the finite interval. Fahrmeir showed the weak convergence of Euler's method [24], while Kushner showed the weak convergence of the modified Huen method [5].

Theorem 4 shows a fundamental difference between stochastic and deterministic differential equations. Deterministic equations with sufficiently smooth coefficients can be solved with errors of arbitrarily high order in h by using high order Runge-Kutta methods. In the stochastic case, the one step mean square error is limited to $O(h^2)$ or $O(h^3)$ depending upon whether the column vector fields of $\sigma(\cdot)$ commute. It should be noted that the one step order of convergence is one order higher for scalar SDE's and vector SDE's driven by a scalar Wiener process and in this case RK4 is one order better than Huen's method [24].

It is very important to apply the SDE solving methods to Eq. (6) with the appropriate λ to solve Eq. (1). Ignoring the Wong-Zakai correction can lead to large biases in the numerical solutions.

The three algorithms were tested with a family of two dimensional SDE's. The family is given by

$$\begin{bmatrix} dx_1 \\ dx_2 \end{bmatrix} = \begin{bmatrix} 0 & -\omega \\ \omega & 0 \end{bmatrix} \begin{bmatrix} x_1 \\ x_2 \end{bmatrix} dt + \begin{bmatrix} \dfrac{x_1^2}{r} & \pm\dfrac{x_1 x_2}{r} \\ \pm\dfrac{x_1 x_2}{r} & \dfrac{x_2^2}{r} \end{bmatrix} \begin{bmatrix} dw_1 \\ dw_2 \end{bmatrix}$$
$$r = 1 + x_1^2 + x_2^2 \tag{8}$$

The Wong-Zakai corrections are

$$\frac{1}{r^3} \begin{bmatrix} 2rx_1(x_1^2 + \delta_+ x_2^2) - 2x_1(r-1)(x_1^2 \pm x_2^2) \\ 2rx_2(\delta_+ x_1^2 + x_2^2) - 2x_2(r-1)(\pm x_1^2 + x_2^2) \end{bmatrix} \quad \delta_+ = \begin{cases} 1 & for\ '+' \\ 0 & for\ '-' \end{cases} \tag{9}$$

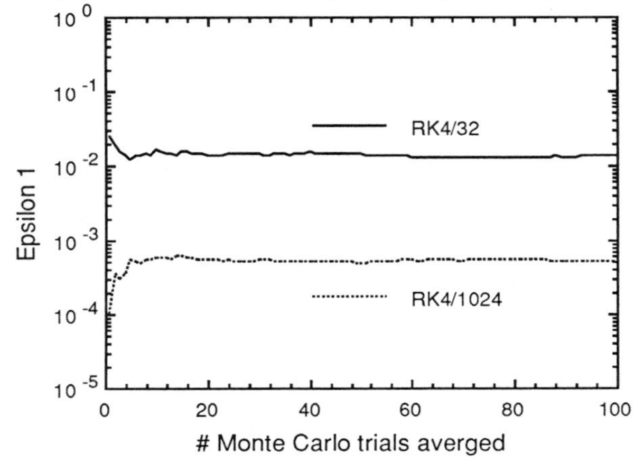

Fig. 3: ε_1 plotted vs the number of Monte Carlo trials averaged. 40 trials seems sufficient for evaluating the SDE solvers on Eqn. 8.

The Lie Bracket, $S(1,2)$, is given by

$$S(1,2) = \begin{cases} 0 & \textit{for } '+' \\ -\dfrac{2x_1x_2}{r^2}\begin{bmatrix} x_1 \\ x_2 \end{bmatrix} & \textit{for } '-' \end{cases} \tag{10}$$

In the '+' case, the system is commutative, while in the '-' case it is not. In the commutative case we expect RK4 and Huen's method to have a faster rate of convergence than Euler's method. In the noncommutative case we expect similar rates of convergence for all three methods.

Equation (8) was simulated with ω=4 over the time interval [0,1]. The initial conditions were $N(0,1)$ for each component. The total number of increments was varied from 32 to 32K (32,768=32x1024). For each Monte Carlo trial, 32K Wiener process increments were generated using a pseudo random Gaussian number generator. The generator is essentially 'RAN1' of [25]. Each algorithm was used to calculate $x(1)$ based upon these increments. The 32K increments were then summed pairwise to give 16K Wiener process increments and the algorithms were again used to calculate $x(1)$. This process was repeated until only 32 increments were used to find $x(1)$. By summing the increments in this way, each case uses increments from the same Wiener process realization and the corresponding $x(1)$'s are directly comparable. 100 different Wiener process realizations were generated since the accuracy is dependent on the realization.

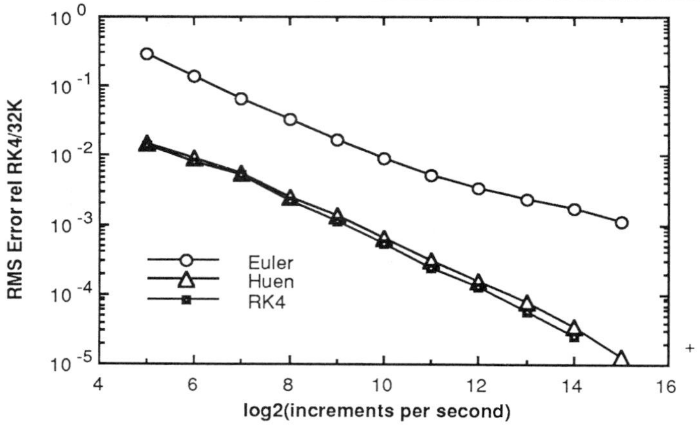

Fig. 4: The figure of merit ε_1 as a function of the number of increments per second for the commutative '+' case.

Two figures of merit were calculated based upon the simulation results. Errors relative to RK4/32K were calculated as well the the relative convergence of each method to the others. Denote the approximated solution of Eq. (8) at t=1 by x(*method, # increments, Monte Carlo run #*). The figures of merit calculated were:

$$\varepsilon_1(mthd,N,incr)=\sqrt{\frac{1}{100}\sum_{mc=1}^{100}\frac{\|x(1;mthd,N,mc)-x(1;\mathrm{RK}4,32K,mc)\|^2}{\|x(1;\mathrm{RK}4,32K,mc)\|^2}}$$

$$\delta_1(m1,m2,N,incr)=\sqrt{\frac{1}{100}\sum_{mc=1}^{100}\frac{\|x(1;m1,N,mc)-x(1;m2,N,mc)\|^2}{\|x(1;m1,N,mc)\|\,\|x(1;m2,N,mc)\|}} \quad (11)$$

Equation (11) was formulated with square roots and norms so the exponents on the vertical axis of the figures would correspond to number of digits to which the methods agree. Relative errors are used to account for the varying magnitude of $x(1)$. The use of 100 Monte Carlo trials was based more on computer time constraints than on a priori statistical analysis. Figure 3 shows ε_1(RK4, 32) and ε_1(RK4, 1024) as functions of the number of Monte Carlo trials (i.e. in Eq. (11), "100" is replaced by the number of trials). ε_1 settles down after ~40 trials which indicates 100 Monte Carlo trials is sufficient for evaluating the SDE solving algorithms when used on Eq. (8).

Figure 4 shows performance relative to RK4/32K (i.e. ε_1) for the commutative '+' case. Curve fits to the data in Fig 4 show that the average mean square error ($=\varepsilon_1{}^2$) is $O(h^2)$ for RK4 and the modified Huen methods and $O(h^{1.6})$ for Euler's method. Figure 5 shows performance relative to RK4/32K

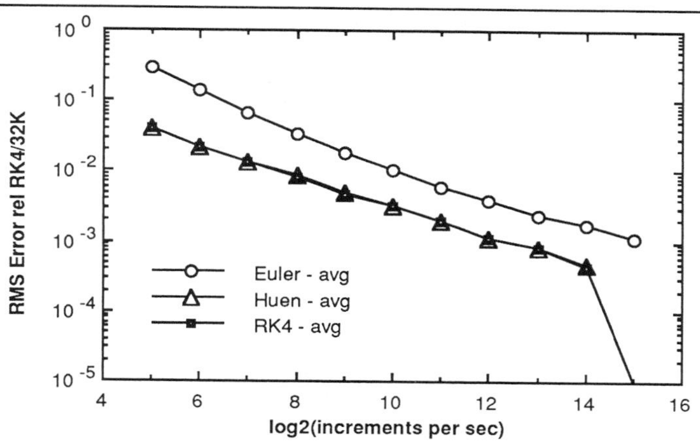

Fig. 5: The figure of merit ε_1 as a function of the number of increments per second for the noncommutative '-' case.

(i.e. ε_1) for the noncommutative '-' case The error over the unit interval is ~$O(h^{1.5})$ all methods.

The commutative case finite interval error of $O(h^2)$ is compatible with $O(h^3)$ one step error and is the worst case predicted by Theorem 5. The ~$O(h^{1.5})$ finite interval error in the noncommutative case is consistent with the $O(h^2)$ theoretical one step errors and is somewhat better than predicted. The order of convergence is not the only factor to consider. If we accept RK4/32K as the 'true' solution to Eqn (8), RK4 and Huen require from 4 to 64 times fewer partition points for the same 'accuracy'.

Another way to compare the SDE solvers uses δ_1 and δ_2 to judge the rate of convergence of each technique to to a common (unknown) solution. Figures 6 and 7 plot these performance measures against the number of partition points. The δ_1 between Euler and either of the two other methods is $O(h^{1.7})$ in the commutative case and $O(h^{1.6})$ in the noncommutative case. Huen and RK4 are converging (in the δ_1 sense) towards a common solution at $O(h^{1.9})$ in the commutative case and $O(h^{1.8})$ in the noncommutative case. At any stepsize Huen and RK4 are much closer to each other than either is to the Euler's method solution.

The above simulations were for a specific family of two dimensional SDE's designed to isolate certain theoretical convergence properties. Parameter estimators such as RPEM have much higher orders, but are stable rather than only marginally stable. 32k intervals per second would require vast amounts of computer time and storage for simulation. The RPEM SDE's for estimating 'a' in the scalar system

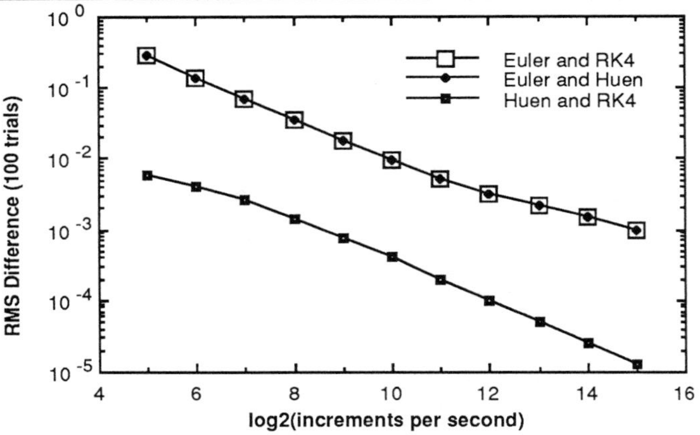

Fig. 6: The figure of merit δ_1 as a function of the number of increments per second for the commutative '+' case.

$$dx = ax\,dt + bu(t)\,dt + \sigma\,dw_1$$
$$dy = x\,dt + \rho\,dw_2 \tag{12}$$

were simulated using the "$S=(tV)^{-1}$" formulation from [6]. This is a 7th order system driven by a two dimensional Wiener process. This system is commutative so there should be some difference between Euler's method and Huen's method. The initial conditions and parameters were

$$\begin{array}{ll} x_o \sim N(0,1) & \hat{x}_o = 0 \\ a = -1 & \hat{a}_o \sim N(0,.3) \\ P_o = 1 & S_o = 1 \\ \sigma = 1 & \rho = 1 \\ bu(t) = 4\cos(2\pi(.2)t + \phi) & \phi \sim \mathrm{U}(0,2\pi) \end{array} \tag{13}$$

The error criterion is

$$\varepsilon(t{:}mthd, N, incr) = \sqrt{\frac{1}{40}\sum_{mc=1}^{40} \max_{1\le i\le 7} \frac{|x_i(t;mthd,N,mc) - x_i(t;\mathrm{RK}4,32K,mc)|^2}{|x_i(t;\mathrm{RK}4,32K,mc)|^2}} \tag{14}$$

This measures the error in the worst component of 'x'=$[x,\hat{x},\hat{a},P,W,S,\Pi]$. The convergence rates for this norm should be slower than predicted by Theorems 4 and 5. The error criteria of Eq. (11) are not appropriate since the components of the filter have very different dynamic ranges. The l_2 norm ignores errors in the smaller components compared to the l_∞ norm. This is inappropriate for a parameter estimator since the most interesting components (the parameter estimation errors) are going to zero

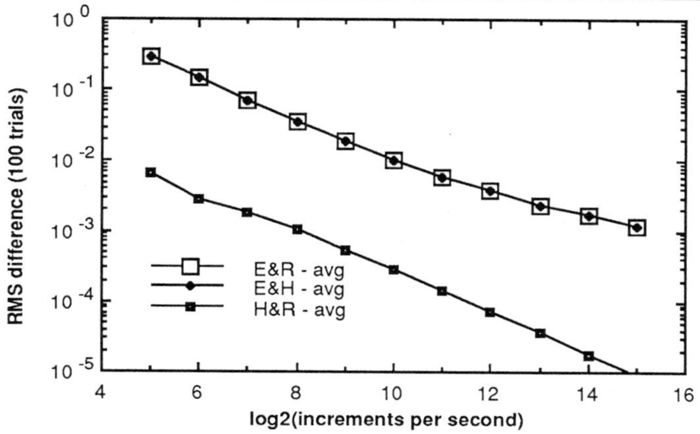

Fig. 7: The figure of merit δ_1 as a function of the number of increments per second for the noncommutative '-' case.

Figures 8-10 are plots of ε for each method as a function of time and stepsize relative to RK4/2048. The (worst component) RMS error was recorded each second for 16 seconds and for 64 through 2048 increments per second. As expected Huen's method and RK4 performed about the same and both were better than Euler's method. For Huen's method and RK4 (Figs. 8 and 10), the RMS error curves dropped by a factor of ~2 when the number of increments per second doubled, while the curves for Euler's method (Fig. 9) only dropped by a factor of ~1.4. This shows the finite time interval mean square error was ~$O(h^2)$ for Huen and RK4, and ~$O(h)$ for Euler. One interesting phenomenon was that the mean square error did not increase with time, contrary to the bound in Theorem 5. Theorem 5 upper bounds the finite interval error by assuming the incremental errors always add constructively over the finite interval, a very pessimistic assumption.

Several issues have been addressed in this appendix. Theoretical results on convergence rates of three SDE solvers were presented. The modified Huen method is most efficient for the vector SDE's simulated, even when Euler's method has the same theoretical order of convergence. For the second order system, it was clear that all the methods were converging to the same solution at rates consistent with the theory, but ~32K increments per second were needed before the three methods agreed to more than 3-5 significant digits after one second. The two dimensional example highlighted the difference in convergence rates when commutative and noncommutative vector fields are integrated. When a simple RPEM case was simulated with Huen's method, 512 increments per second was required to match the RK4/2048 sample paths

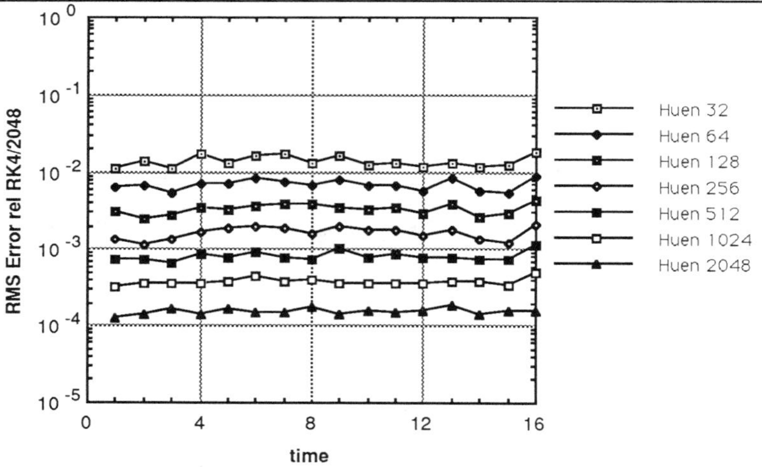

Fig. 8: RMS error for Huen's method (applied to the RPEM equations) relative RK4 w/ 2048 increments per second. Mean square error is ~ $O(h^2)$, h= 1/(# increments per sec).

to within 3 significant digits and this accuracy was stable over the 16 seconds tested. Euler's method could not reach this accuracy with 2048 increments per second. For RPEM, the choice of proper stepsize depends upon the particular system parameters. For examining particular systems, experimentation plus a conservative stepsize choice is adequate. For a general purpose SDE solver an adaptive stepsize algorithm is required.

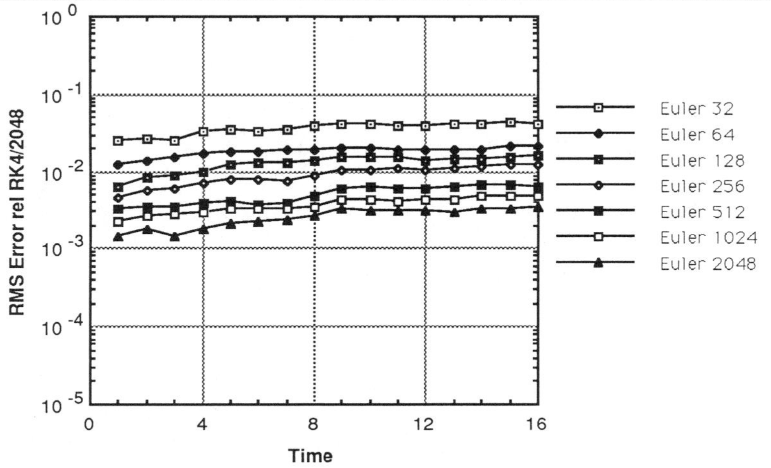

Fig. 9: RMS error for Euler's method (applied to the RPEM equations) relative RK4 w/ 2048 increments per second. Mean square error is ~ O(h), h= 1/(# increments per sec)

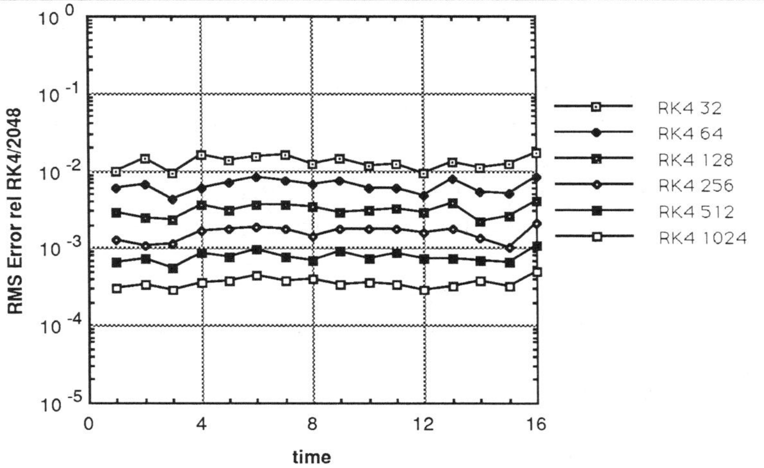

Fig. 10: RMS error for RK4 (applied to the RPEM equations) relative RK4 w/ 2048 increments per second. Mean square error is ~ O(h^2) and virtually the same as Huen's method.
h= 1/(# increments per sec).

REFERENCES:

1, L. Ljung, "Analysis of Recursive Stochastic Algorithms," *IEEE Trans. Automatic Control*, vol. AC-22, no. 4, pp. 551-575, Aug. 1977.

2. D.M. Wiberg, "Towards a Globally Convergent Approximation of Optimal On-Line Parameter Estimation," *Proc. 8th IFAC/IFORS Symp. on Identification and System Parameter Estimation*, Beijing, vol.2, session S.10.1 .

3. D. G. DeWolf, An Ordinary Differential Equation Technique for the Analysis of Continuous Time Parameter Estimators, Ph.D Thesis, School of Engineering and Applied Science, University of California, Los Angeles, 1989.

4. L. Ljung, "Asymptotic Behavior of the Extended Kalman Filter as a Parameter Estimator for Linear Systems," *IEEE Trans. Automatic Control*, vol. AC-24, no. 1, pp. 36-50, Feb. 1979.

5. H.J. Kushner, *Approximation and Weak Convergence Methods for Random Processes, with Applications to Stochastic Systems Theory*, Cambridge, Mass.: MIT Press, 1984.

6. H.J. Kushner and D.S. Clark, *Stochastic Approximation for Constrained and Unconstrained Systems*, Berlin: Springer-Verlag, 1978.

7. A. Shwartz, Convergence of Stochastic Approximations: An Invariant Measure Approach, Ph.D Thesis, Division of Engineering, Brown University, 1982.

8. H.J. Kushner and H. Huang, "Rates of Convergence for Stochastic Approximation Type Algorithms," *SIAM Journal of Control and Optimization*, vol. 17, pp. 607-617, 1979.

9. G.C. Papanicolaou, D. Stroock, and S.R.S. Varadhan, "Martingale Approach to Some Limit Theorems," *Proceedings of the 1976 Duke University Conference on Turbulence*, April 1986.

10. J.H. van Schuppen, "Convergence Results for Continuous Time Adaptive Filtering Algorithms," *Stochastic Processes and their Application*, vol. 96, pp. 209-225, 1983.

11. H.F. Chen, *Recursive Estimation and Control for Stochastic Systems*,New York: John Wiley and Sons, 1985.

12. J.B. Moore, "Convergence of Continuous Time Stochastic ELS Parameter Estimation," *Stochastic Processes and their Application*, vol. 27, pp. 195-215, 1988.

13. M. Gevers, G.C. Goodwin, and V. Wertz, "A Parameter Estimator for Continuous Time Adaptive Control," in *IEEE Conference on Decision and Control*, p. 1922, 1988.

14. M. Gevers, G.C. Goodwin, and V. Wertz, "Continuous Time Stochastic Adaptive Control," Technical Report EE8743, University of Newcastle, October 1988.

15. P. Baldi, "Limit Set of Inhomogeneous Orenstein-Uhlenbeck Processes, Destabilization and Annealing," *Stochastic Processes and their Application*, vol. 23, pp. 153-167, 1986.

16. L. Ljung and T. Soderstrom, *Theory and Practice of Recursive Identification*, Cambridge, Mass: MIT Press, 1983.

17. C. Kenney and A.J. Laub, "Controllability and Stability Radii for Companion Form Systems," *Mathematics of Control, Signals, and Systems*, vol. 1, No. 3, pp. 239-256, 1988.

18. A.C.M. Ran and L. Rodman, "On Parameter Dependence of Solutions of Algebraic Riccati Equations," *Mathematics of Control, Signals, and Systems*, vol. 1, No. 3, pp. 269-284, 1988.

19. L. Ljung, *System Identification: Theory for the User*, Englewood Cliffs, New Jersey: Prentice Hall, 1987.

20. P. Billingsley, *Convergence of Probability Measures:*, New York: John

Wiley and Sons, 1968.

21. D. Stroock and S.R.S. Varadhan, *Multidimensional Diffusion Processes*, New York: Springer-Verlag, 1979.

22. G.N. Milshtein, "Approximate Integration of Stochastic Differential Equations," *Theory Prob. Appl.* vol. 19, pp. 557-562, 1974.

23. G.N. Milshtein, "Approximate Integration of Stochastic Differential Equations," *Theory Prob. Appl.* vol. 23, pp. 396-401, 1978.

24. W. Rumelin, "Numerical Treatment of Stochastic Differential Equations," *SIAM Journal of Numerical Analysis*, vol. 19, No. 3, pp. 604-613, June 1982.

25. W.H. Press, B.P. Flannery, S.A. Teukolsky, W.T. Vetterling, *Numerical Recipes: The Art of Scientific Computing*, Cambridge, UK: Cambridge University Press, 1986.

IN - FLIGHT ALIGNMENT OF INERTIAL NAVIGATION SYSTEMS

Itzhack Y. Bar-Itzhack

Aerospace Engineering Department
Technion - Israel Institute of Technology
Haifa, 32000 Israel

I. INTRODUCTION

Inertial Navigation Systems (INS) are dead-reckoning systems which provide position, velocity and attitude information. Being dead-reckoning systems, INS have to be initialized before the beginning of their normal operation; that is, before the start of the navigation phase. The initialization is a crucial phase in the operation of the system since an INS can be only as good as its initial condition. Initialization implies setting the system initial position and velocity, and determining the current INS attitude and setting it to the desired orientation. The hardest of these tasks is the determination of the INS attitude which is called ***alignment***. Therefore a considerable amount of effort has been invested in devising accurate alignment processes. In the early days of INS, alignment was performed when the system was at rest, but later, when INS was installed in modern air and sea launched vehicles, it became necessary to execute the alignment in-flight. When the INS alignment process is performed while the INS is airborne, the alignment operation is known as ***in-flight alignment***.

The need to perform the alignment in-flight is based on the following rationale. Consider a carrier vehicle which launches a disposable weapon. Obviously, a disposable weapon is characterized by the quality that even when it completes its mission successfully, it ends with the destruction of the weapon and its INS, which is being used to guide the weapon through its entire trajectory or a part of it. Therefore a cost effective design of the integrated weapon system calls for an accurate, and thus, expensive frequently - updated carrier navigation system, and an inexpensive, an thus, inaccurate weapon INS. Since the latter is inaccurate, a too early initialization will result in an accumulation of an enormous error already at lunch. In addition, due to the low quality of the INS, its life cycle is short, hence every aborted mission will bring the INS closer to its MTBF. These problems can be minimized by in-flight alignment (IFA) of the weapon INS just before its launch.

II. ALIGNMENT AT REST (AAR)

In order to better understand the process of IFA, let us first discuss the alignment process while the INS is at rest. Alignment is basically an attitude determination problem. Therefore a necessary condition for alignment is the existence of two non-collinear vectors which are measured in the INS coordinates and are either known or measured in the reference coordinates. When the alignment is performed while the INS is at rest, such two vectors do indeed exist. They are the Earth gravity vector and the Earth angular velocity which is also known as Earth rate. Gravity constitutes a strong signal with respect to the noise generated by the accelerometers which measure its components. On the other hand, Earth rate is of a very small magnitude with respect to the noise of the gyros which measure its components. Therefore to extract the signal from the noisy gyro measurements, we have to accumulate the gyro outputs over some length of time. We note that, usually, the INS coordinates rotate with respect to the reference coordinates during the alignment process and the longer the alignment process lasts, the larger is the rotation. We encounter, therefore, a dynamic problem, whose nature is described by differential equations. Put in state space representation,

$$\dot{\underline{x}}' = A'\underline{x}' + \underline{e} + \underline{n}' \qquad (1.a)$$

the differential equations which adequately describe the undamped INS error behavior during initial alignment at rest are [1]:

$$\frac{d}{dt}\begin{bmatrix} V_N \\ V_E \\ V_D \\ \psi_N \\ \psi_E \\ \psi_D \end{bmatrix} = \left[\begin{array}{ccc|ccc} 0 & 2W_D & 0 & 0 & g & 0 \\ -2W_D & 0 & 2W_N & -g & 0 & 0 \\ 0 & -2W_N & 0 & 0 & 0 & 0 \\ \hline & & & 0 & W_D & 0 \\ & 0 & & -W_D & 0 & W_N \\ & & & 0 & -W_N & 0 \end{array}\right] \begin{bmatrix} V_N \\ V_E \\ V_D \\ \psi_N \\ \psi_E \\ \psi_D \end{bmatrix} + \begin{bmatrix} B_N \\ B_E \\ B_D \\ D_N \\ D_E \\ D_D \end{bmatrix} + \begin{bmatrix} n_{BN} \\ n_{BE} \\ n_{BD} \\ n_{DN} \\ n_{DE} \\ n_{DD} \end{bmatrix} \qquad \ldots (1.b)$$

where W denotes Earth rate, the subscripts N, E and D denote, respectively, north, east and down components, and g denotes the magnitude of the Earth gravity vector. The state vector $\underline{x}'$ consists of the following INS error variables:

V_N - north component of the velocity error
V_E - east component of the velocity error
V_D - down component of the velocity error
ψ_N - north component of the misalignment
ψ_E - east component of the misalignment
ψ_D - down component of the misalignment

$\underline{e}$ consists of the following sensor errors:

B_N - north accelerometer bias
B_E - east accelerometer bias
B_D - down accelerometer bias
D_N - north gyro constant drift
D_E - east gyro constant drift
D_D - down gyro constant drift

and $\underline{n}'$ consists of the following sensor noise elements which are adequately assumed to be zero mean white noise.

n_{BN} - north accelerometer white noise component
n_{BE} - east accelerometer white noise component
n_{BD} - down accelerometer white noise component
n_{DN} - north gyro white noise component
n_{DE} - east gyro white noise component
n_{DD} - down gyro white noise component

(Note that at AAR, the vertical velocity error is not influenced by the misalignment angles, and its model is decoupled from the rest of the error model. Moreover, in accurate INS which are designed for a lengthy operation, the vertical channel is damped. Therefore the model which describes the error behavior during AAR does not include the 3rd raw and column of (1.b). In the accurate INS, the error model for AAR usually contains more sensor error terms).

Coarse alignment is a necessary first step in the alignment process. However, coarse alignment is easy to achieve, therefore we shall concentrate on the fine alignment stage which is really the problematic part of the alignment. It is easy to see that at the fine alignment stage the INS velocity error components are small. Neglecting in (1.a) products of small magnitude variables, the equations which involve the misalignment angles are:

$$\dot{v}_N = g.\psi_E + B_N + n_{BN} \tag{2.a}$$

$$\dot{v}_E = -g.\psi_N + B_E + n_{BE} \tag{2.b}$$

$$\dot{\psi}_N = W_D.\psi_E + D_N + n_{DN} \tag{2.c}$$

$$\dot{\psi}_E = -W_D.\psi_N + W_N.\psi_D + D_E + n_{DE} \tag{2.d}$$

$$\dot{\psi}_D = -W_N.\psi_E + D_D + n_{DD} \tag{2.e}$$

The signals which are indicative of the INS misalignment are the velocity error components v_N and v_E. For INS at rest, they are the readings of the INS-indicated components of the horizontal velocity. This is so since for an INS at rest the correct readings are zero thus any reading other than zero is an error value. By inspection of (2.a) and (2.b) it is evident that in the ideal case where the accelerometer errors are negligible, v_N and v_E are indeed indicative of the misalignments about the local east and north axes respectively. Moreover, it seems that a proper rotation of the INS coordinates about their north and east axes which will zero v_N and v_E, will also zero ψ_N and ψ_E. This process is known as ***leveling***. Note from (2.d) that even if ψ_N is zero, as long as ψ_D is not zero, ψ_E will keep changing even though ψ_E may have been temporarily eliminated as a result of leveling. Therefore, to achieve full alignment, the INS coordinate system has to be rotated about all three of its axes until v_N and v_E are zero. In the early days of inertial navigation, feedback loops utilized v_N and v_E generated by the INS to torque the INS coordinate system about its three axes in order to zero v_N and v_E [2]. In the ideal case where the accelerometer and gyro errors are zero, zero v_N and v_E imply zero ψ_N, ψ_E and ψ_D; that is, perfect alignment. However, since the accelerometers and the gyros are not ideal, the misalignment errors reach a non-zero steady state. From (2) it is evident that the lower bounds on the misalignment angles are:

$$\psi_{N,min} = \frac{-B_E}{g} \qquad (3.a)$$

$$\psi_{E,min} = \frac{B_N}{g} \qquad (3.b)$$

$$\psi_{D,min} = \frac{-D_E}{W_N} \qquad (3.c)$$

The method of using v_N and v_E to torque the INS coordinates through feedback loops is known as ***self alignment***.

With the introduction of the Kalman filter (KF) in the mid sixties, it was realized that it could be successfully used for INS alignment [3,4,5]. The place of the KF in the INS alignment is depicted in Fig. 1. When a KF is used in the alignment process, the process can be divided into two tasks [5]. The first task is the ***estimation*** of the misalignment angles based on the readings v_N and v_E and the second task is the ***torquing*** of the INS coordinates to eliminate the misalignment angles. (Note that in the classical self alignment process described above there is no estimation task but rather v_N and v_E are fed directly as torquing commands). Normally, the INS coordinates are torqued continuously in order to maintain a nearly constant attitude with respect to the desired orientation during the estimation process. The estimated misalignment angle could be used to either generate an additional torquing command to continuously eliminate the misalignment angles or be used only once at the end of the estimation process to align the INS coordinate system in a single shot. The torquing process, whether performed continuously during the estimation process or whether performed only once at the end of the estimation stage, is

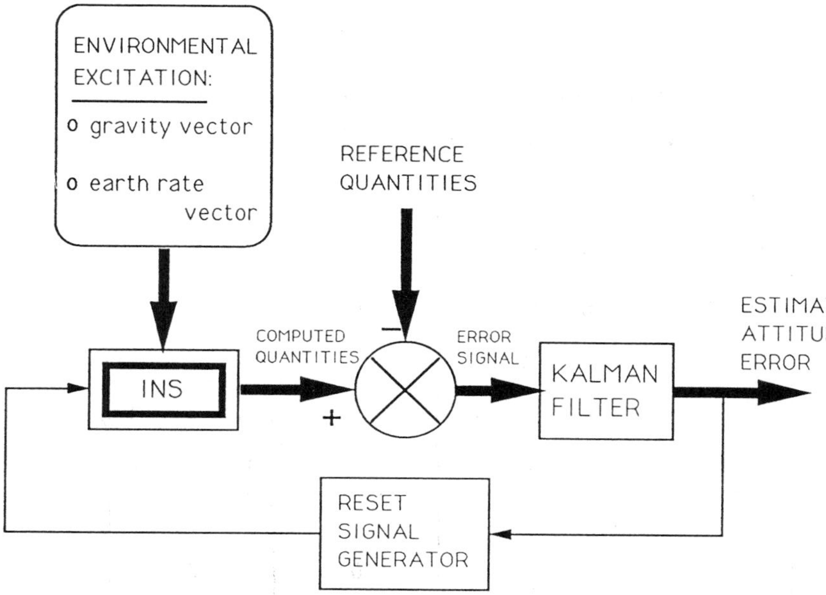

Fig. 1: The role of the KF in INS initial alignment.

a straight forward procedure. It is the estimation process which is the crucial task for its accuracy directly influences the alignment accuracy. Therefore, from now on we will only consider the estimation part of the alignment process.

The model of (1) is inadequate for use in the KF since $\underline{e}$ is not a white noise vector as required by the filter. For the relatively short time span of the alignment process, it can be adequately assumed that $\underline{e}$ is constant, thus we may write:

$$\dot{\underline{e}} = 0 \tag{4}$$

When (1) and (4) are augmented, we obtain the augmented INS error model:

$$\dot{\underline{x}} = A\underline{x} + \underline{n} \tag{5}$$

Where the dynamics matrix A, which contains both the system and the sensor dynamics models, is as follows:

$$A = \left[\begin{array}{ccc|ccc|ccc|c}
0 & 2W_D & 0 & 0 & g & 0 & 1 & 0 & 0 & \\
-2W_D & 0 & 2W_N & -g & 0 & 0 & 0 & 1 & 0 & 0 \\
0 & -2W_N & 0 & 0 & 0 & 0 & 0 & 0 & 1 & \\
\hline
 & & & 0 & W_D & 0 & & & & \begin{array}{ccc} 1 & 0 & 0 \end{array} \\
 & 0 & & -W_D & 0 & W_N & & 0 & & \begin{array}{ccc} 0 & 1 & 0 \end{array} \\
 & & & 0 & -W_N & 0 & & & & \begin{array}{ccc} 0 & 0 & 1 \end{array} \\
\hline
 & 0 & & & 0 & & & 0 & & 0 \\
\hline
 & 0 & & & 0 & & & 0 & & 0
\end{array}\right] \tag{6}$$

The components of the augmented state vector, $\underline{x}$, are:

$$\underline{x}^T = [V_N, V_E, V_D, \psi_N, \psi_E, \psi_D, B_N, B_E, B_D, D_N, D_E, D_D]$$

where T denotes the transpose. The augmented noise vector $\underline{n}$ is given as follows:

$$\underline{n}^T = [n_{BN}, n_{BE}, n_{BD}, n_{DN}, n_{DE}, n_{DD}, 0, 0, 0, 0, 0, 0]$$

As the KF is set to estimate the whole state vector, $\underline{x}$, it has the potential of estimating not only the misalignment angles but also the INS velocity errors, the accelerometer biases and the gyro constant drifts. The estimation of accelerometer or gyro error sources and their removal is known as ***calibration***. We realize that the fact that the KF cannot accept dynamics models whose driving forces are not white, is really a blessing in disguise since it forces us to include the sensor error sources in the state where they may be estimated and removed. The ability to estimate the sensor error sources is not assured for it depends on their observability condition. This issue will be discussed at a later point.

It is interesting to note that the same lower bounds on the misalignment angles which exist in the classical alignment process (3), also exist when a KF is used in the alignment. To see this, let us re-write, for example, equation (2.a):

$$\dot{v}_N = g.\psi_E + B_N + n_{BN} \qquad (2.a)$$

Since we are dealing with a lower bound, let us assume that all misalignment errors other than ψ_E, as well as D_E are zero. The KF processes v_N and obtains an estimate of ψ_E. As long as

$$g.\psi_E \neq - B_N \qquad (7)$$

v_N will not be constantly zero and the KF will continue to estimate ψ_E. But when the inequality of (7) turns into an equality, the added part to v_N will be just an integral of a zero mean white noise, the remaining part of ψ_E will bear no signature on the observation used by the KF and the estimation process of ψ_E will come to a halt. Consequently, the lower bound on the estimate of ψ_E will be again that specified in (3.b).

Figure 2 describes the influence of the misalignment errors on the INS horizontal velocity error components when the INS is aligned at rest and torquing of the INS coordinates is performed continuously. From (2) it is realized that the misalignment angles which have an immediate and noticeable effect on the growth of v_N and v_E are ψ_E and ψ_N respectively. This is so because they are multiplied by the strong signal g. Therefore their better part is estimated right away and eliminated through torquing. The influence of ψ_D on v_N is felt only at a later stage and after integration (see 2.d). Because of this sequence of events, the block diagram of Fig. 2 is a good approximate description of the model expressed in (2). Note that while g which multiplies ψ_E and ψ_E in (2.a) and (2.b) amplifies the misalignments, W_N which

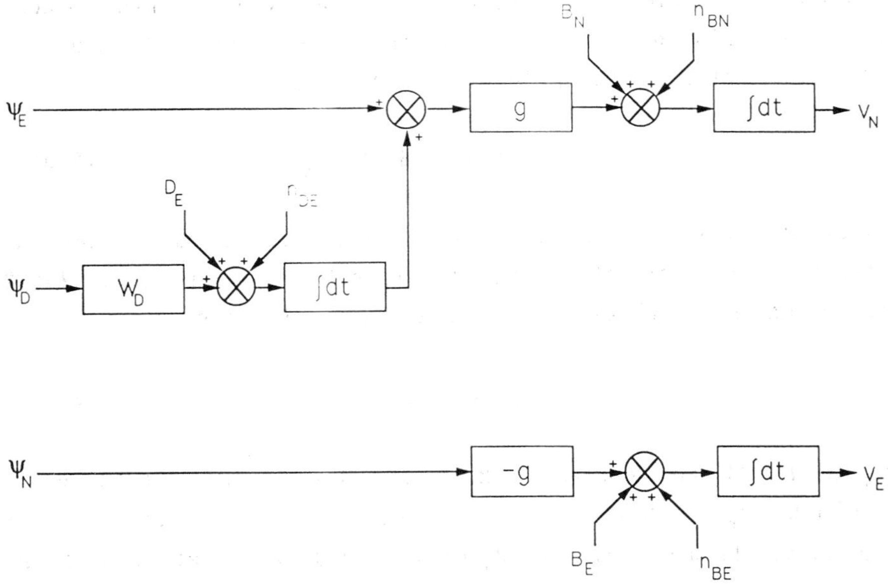

Fig. 2: A simplified block diagram description of the influence of the misalignment angles on the INS horizontal velocity error components during AAR.

multiplies ψ_D in (2.d) actually attenuates it. The process described above of estimating ψ_D through v_N is known as ***gyrocompassing***. Gyrocompassing is the limiting factor in the speed at which AAR can be performed.

III. IN - FLIGHT ALIGNMENT (IFA)

In IFA the correct velocity is not zero anymore, therefore, in order to know the INS velocity errors (which, as in AAR, are indicative of the misalignment angles) the correct velocity has to be furnished by a system outside of the INS being aligned. This system may be a velocity measuring device like a doppler radar or it may be an accurate INS, in which case the initial alignment is called ***transfer alignment***.

It seems that IFA is more complicated than AAR, but the truth is that IFA opens new opportunities for faster and more accurate alignment. The reason for that is the ability to pull accelerations which excite modes whose observability in AAR is low. This will be explained next.

The model which adequately describes the behavior of the undamped INS error during IFA is as follows:

$$\frac{d}{dt}\begin{bmatrix} V_N \\ V_E \\ V_D \\ \psi_N \\ \psi_E \\ \psi_D \end{bmatrix} = \left[\begin{array}{ccc|ccc} 0 & 2W_D & 0 & 0 & -f_D & f_E \\ -2W_D & 0 & 2W_N & f_D & 0 & -f_N \\ 0 & -2W_N & 0 & -f_E & f_N & 0 \\ \hline & & & 0 & W_D & 0 \\ & 0 & & -W_D & 0 & W_N \\ & & & 0 & -W_N & 0 \end{array}\right] \begin{bmatrix} V_N \\ V_E \\ V_D \\ \psi_N \\ \psi_E \\ \psi_D \end{bmatrix} + \begin{bmatrix} B_N \\ B_E \\ B_D \\ D_N \\ D_E \\ D_D \end{bmatrix} + \begin{bmatrix} n_{BN} \\ n_{BE} \\ n_{BD} \\ n_{DN} \\ n_{DE} \\ n_{DD} \end{bmatrix} \quad \ldots (8)$$

where f_N, f_E and f_D are, respectively the north, east and down components of the specific force experienced by the INS. Note that the corresponding model for AAR given in (1.b) is a special case of this model. Equation (8) comprises the following six scalar equations:

$$\dot{v}_N = -f_D \cdot \psi_E + f_E \cdot \psi_D + B_N + n_{BN} \qquad (9.a)$$

$$\dot{v}_E = f_D \cdot \psi_N - f_N \cdot \psi_D + B_E + n_{BE} \qquad (9.b)$$

$$\dot{v}_D = -f_E \cdot \psi_N + f_N \cdot \psi_E + B_D + n_{BD} \qquad (9.c)$$

$$\dot{\psi}_N = W_D \cdot \psi_E + D_N + n_{DN} \qquad (9.d)$$

$$\dot{\psi}_E = -W_D \cdot \psi_N + W_N \cdot \psi_D + D_E + n_{DE} \qquad (9.e)$$

$$\dot{\psi}_D = -W_N \cdot \psi_E + D_D + n_{DD} \qquad (9.f)$$

When comparing (9) to (2) we note a main difference between AAR and IFA. As described in the preceding section, the process of estimating ψ_D in AAR is that of gyrocompassing; that is, while the north and east misalignments influence the change of the observables v_N and v_E strongly and directly, ψ_D is attenuated and integrated before it influences the change of v_N. In IFA, though, the azimuth misalignment angle, ψ_D, influences ***both*** v_N and v_E directly. The strength of that influence is determined by the strength of the specific force components f_N and f_E. We see, then, that the lateral components of the specific force excite ψ_D such that its influence on v_N and v_E can be made (depending on f_N and f_E) equal or even larger than that of ψ_N and ψ_E. Another way to look at this difference is the following. It was mentioned in the beginning of the preceding section that the two vectors which are used to align the INS during AAR are Earth gravity vector and Earth angular rate vector and that the magnitude of the latter is much smaller than the former. In fact, when using the MKS units, the ratio between them is $7.4 * 10^{-6}$. The gravity vector is the signal used for leveling (i.e. for ψ_N and ψ_E determination) and Earth rate is used for azimuth determination; i.e. for ψ_D determination, and that is why in AAR, leveling is so much superior to azimuth determination. On the other hand, in IFA, Earth rate is replaced by the horizontal part of the specific force vector operating on the INS and this may be made quite large which brings azimuth determination to the same level of speed and accuracy as leveling. Indeed, at AAR, leveling typically achieves steady state in a few tens of seconds while azimuth determination reaches steady state in about ten minutes. On the other hand, at IFA, both leveling and azimuth determination reach typically their steady state values in several tens of seconds.

Using the arguments that led to the construction of the block diagram shown in Fig. 2, we can use the pertinent part of (9) to construct in Fig. 3 a block diagram which illustrates the influence the misalignment angles have on the development of v_N and v_E during IFA. A comparison between this block diagram and the one presented in Fig. 2 further demonstrates the difference between AAR and IFA discussed in the preceding paragraph.

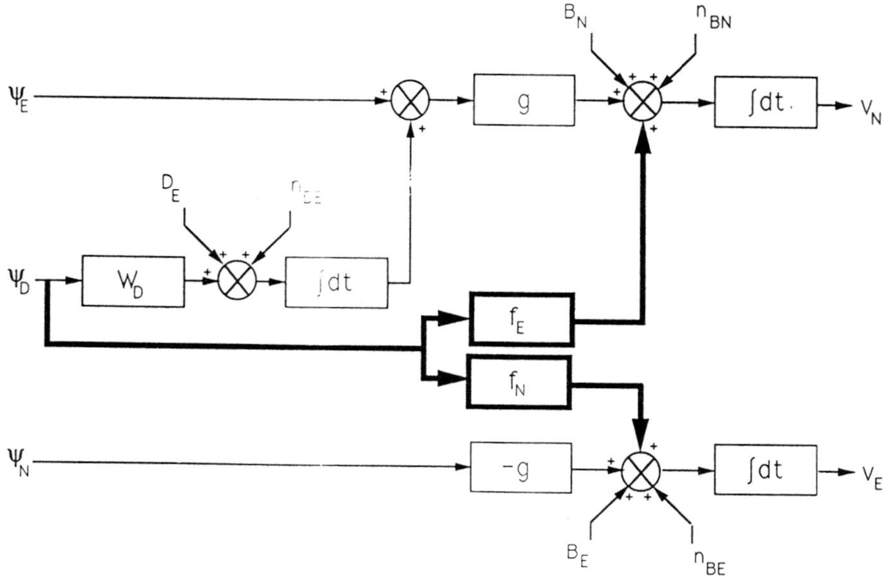

Fig. 3: A simplified block diagram description of the influence of the misalignment angles on the INS horizontal velocity error components during IFA.

When comparing (9) to (2) we find yet another difference between IFA and AAR. The difference is the existence of an extra connection between the misalignment angles and the velocity error of the INS. This relationship is expressed by (9.c) and although it looks redundant to (9.a) and (9.b), it adds a very important relations which breaks the lower bound on the misalignment estimation error [6]; thus, theoretically, the misalignment estimates can reach the correct values of the misalignment angles. This point will be further discussed in the section which deals with observability.

As in the case of AAR, here too, the basic misalignment angles dynamic model given by (8) is inadequate for use in the KF since $\underline{e}$ is not a white noise vector as required by the filter. Similarly to the procedure applied to AAR, we augment (4) with (8) to arrive at a dynamics model of the form of (5), only that here the dynamics matrix is as follows:

$$A = \left[\begin{array}{c|c|c|c}
\begin{matrix} 0 & 2W_D & 0 \\ -2W_D & 0 & 2W_N \\ 0 & -2W_N & 0 \end{matrix} & \begin{matrix} 0 & -f_D & f_E \\ f_D & 0 & -f_N \\ -f_E & f_N & 0 \end{matrix} & \begin{matrix} 1 & 0 & 0 \\ 0 & 1 & 0 \\ 0 & 0 & 1 \end{matrix} & 0 \\ \hline
0 & \begin{matrix} 0 & W_D & 0 \\ -W_D & 0 & W_N \\ 0 & -W_N & 0 \end{matrix} & 0 & \begin{matrix} 1 & 0 & 0 \\ 0 & 1 & 0 \\ 0 & 0 & 1 \end{matrix} \\ \hline
0 & 0 & 0 & 0 \\ \hline
0 & 0 & 0 & 0
\end{array}\right] \qquad (10)$$

V. TRANSFER ALIGNMENT (TA)

As mentioned earlier, when the reference velocity which is used to determine the INS velocity error is another INS, the IFA is called ***transfer alignment*** (TA). In order to understand the peculiar characteristics of TA, let us consider the computation of the INS velocity error needed for the KF which is used in the IFA.

The device which measures the velocity of the parent vehicle on which the INS is installed, yields the velocity of a certain point of the vehicle. Let us denote this point by the subscript 1 and the true velocity of this point by $\underline{V}_1$. The measuring device actually yields the vector $\underline{V}_1 + \underline{v}_m$ where $\underline{v}_m$ denotes the error generated by the measuring device. Normally, the INS being aligned is not installed at point 1. To obtain the velocity measurement at the INS, we have to add a lever arm compensation to the velocity of point 1. Therefore the velocity measured at the INS is:

$$\underline{V}_m = \underline{V}_1 + \underline{w}x\underline{r} + \underline{v}_m \tag{11}$$

where $\underline{w}$ is the angular velocity vector of the vehicle with respect to the reference coordinates and $\underline{r}$ is the lever arm vector of the INS location relative to point 1. A block diagram representation of the TA process is presented in Fig.4

Usually the INS is installed in a weapon which is attached to the wing of the parent vehicle. In this case due to random structural vibrations, both $\underline{w}$ and $\underline{r}$ vary randomly in an oscillatory fashion [7]. Let $d\underline{w}$ and $d\underline{r}$ be the random variations in $\underline{w}$ and $\underline{r}$ then the actual added velocity due to the lever arm is:

$$\underline{V}_L = (\underline{w} + d\underline{w})x(\underline{r} + d\underline{r}) = \underline{w}x\underline{r} + \underline{w}xd\underline{r} + d\underline{w}x\underline{r} \tag{12}$$

The correct velocity at the INS point is $\underline{V}$ where:

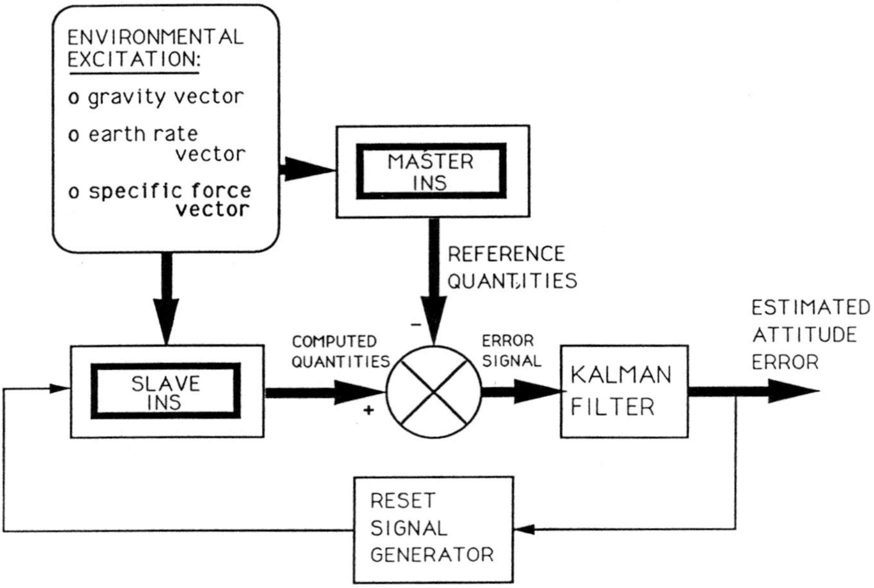

Fig. 4: Block diagram representation of Transfer Alignment (TA)

$$\underline{V} = \underline{V}_1 + \underline{V}_L \tag{13}$$

The INS provides:

$$\underline{V}_{INS} = \underline{V} + \underline{v} = \underline{V}_1 + \underline{V}_L + \underline{v} \tag{14}$$

The ***effective*** measurement, $\underline{z}$, processed by the KF is:

$$\underline{z} = \underline{V}_{INS} - [I - \underline{\Psi}\times]\underline{V}_m \tag{15}$$

where I is the 3rd order identity matrix and $\underline{\Psi}\times$ is the cross product matrix of the misalignment angles [see Eq. (8) in 8]. The inclusion of the latter matrix is explained as follows [2]. In this discussion we assume that $\underline{V}_m$, the measured velocity of point 1, is obtained in terms of components on vehicle fixed axes. Since the relative orientation of the so called "platform" axes [2] with respect to the vehicle axes is known, $\underline{V}_m$ is transformed into the "platform" axes and then these components are subtracted from from the components of $\underline{V}_{INS}$ to form $\underline{z}$. However, the components of $\underline{V}_{INS}$ are in the "computer" coordinate system [2]. Therefore, in effect, the components of $\underline{V}_m$ which are given in the "platform" axes are taken to be the components on the computer axes. The outcome of this mismatch between the coordinate systems amounts to a rotation of the measured velocity by the rotation matrix $[I - \underline{\Psi}x]$. (It should noted that not always $\underline{V}_m$ is given in vehicle fixed axes and not always do we know the exact transformation between the axes in which $\underline{V}_m$ is given and the body fixed axes. Therefore each case has to be examined and its error model developed individually).

Now when substituting (12) into (14) and the result together with (11) into (15) the following is obtained:

$$\underline{z} = \underline{v} - (\underline{V}_1 + \underline{w}x\underline{r}) \times \underline{\Psi} - \underline{v}_m + \underline{w}xd\underline{r} - \underline{r}xd\underline{w} \qquad (16)$$

In order to use the effective measurement, $\underline{z}$, in the IFA KF, we have to express $\underline{z}$ as a combination of a linear function of the states of the dynamic system and a white noise vector. The vectors $\underline{v}$ and $\underline{\Psi}$ are already included in the state vector, $\underline{x}$. The noise vectors $d\underline{r}$ and $d\underline{w}$ which appear in the term $\underline{w}xd\underline{r} - \underline{r}xd\underline{w}$ can be modeled as the output of a linear system whose inputs are white noise signals, hence $d\underline{r}$ and $d\underline{w}$ can be augmented with the state vector. Finally we have to consider $\underline{v}_m$ in the device which measures the velocity of point 1. The nature of $\underline{v}_m$ depends on the type of the measuring device. All standard devices have acceptable error models [9] which allow the augmentation of $\underline{v}_m$ with the system state vector. This results in the required expression of $\underline{z}$ as a sum of a linear function of the state vector and a white noise vector.

In TA, the independent source which supplies the measured velocity vector of point 1 is another (master) INS. The dynamic behavior of the error term, $\underline{v}_m$, of an INS is described by the classical INS error model: that is, $\underline{v}_m$ is included in the state $\underline{x}''$ whose dynamics model is:

$$\dot{\underline{x}}'' = A'' \underline{x}'' + \underline{n}''$$

where A" is an augmented INS error dynamics matrix which, for an accurate INS, may be quite elaborate. It may include complicated sensor error models and models of measurement systems involved in the updating of the master INS. Accordingly, $\underline{x}''$ includes the basic INS error states, sensor states and states of the systems involved in the updating of the master INS. It should be noted though that quite often the master INS is so accurate with respect to the required slave INS accuracy that the former is considered a perfect system. Consequently $\underline{v}_m$ is modeled as white noise only. In other words, the master computed velocity is considered as a generic velocity measurement. A main characteristic of TA is that the alignment of the slave INS can only as good as that of the master [9]. In other words, if the alignment is perfect, the slave is perfectly aligned with the master. This is why this mode of alignment is known as transfer alignment.

IV. OPTIMAL IFA MANEUVERS

As mentioned earlier, the advantage of IFA over AAR stems from the ability to generate, during IFA, specific forces that do not exist at AAR. The specific force is generated when the parent vehicle, on which the INS is flown, performs maneuvers. A question that comes immediately to mind is, is there a certain maneuver which is optimal in some sense. We can ask ourselves the question, what kind of maneuvers will yield the smallest sum of the variance of the alignment errors at a given final time. That is, what maneuver will minimize the following cost function:

$$J = \text{trace}\ [WP(T_f)]$$

where W is a diagonal matrix whose main diagonal consists of zeros and ones which pull the misalignment error variances form P, the KF covariance matrix. T_f is the given final time of the alignment. We realize that this optimality problem is that of constrained optimization since, obviously, the magnitude of the specific force which the parent vehicle can pull is limited. It can be shown [10] that for a constant velocity of the parent vehicle, this problem has a bang-bang solution. That is, the vehicle switches between maximum turn rate to one side and a maximum turn rate to the other side. On the other hand, a constant specific force results in an azimuth misalignment estimation time which is smaller than that needed to achieve the same accuracy when an alternating specific force is generated during IFA maneuvers [11]. Another question which may be asked is, what kind of specific force is superior, a lateral or an axial one? There is no clear cut answer to this question. The answer depends on the initial conditions yielding in certain cases a more accurate and a shorter azimuth misalignment estimation time when axial specific force is generated, and, for other conditions, lateral acceleration yields better results [12].

It should be noted that while the questions discussed above have an academic merit, in practice though, all it takes to achieve excellent results, is a benign s-shaped maneuver.

In the foregoing discussion, only planar IFA maneuvers were considered. This was so because in the vertical direction we have the constant specific force, -g, supplied by earth gravity. In a stable platform INS this condition is sufficient for leveling. However, it is impossible to distinguish between the down accelerometer bias and its scale factor. (Note that in the simple INS error models used here we didn't include scale factor errors. To see the effect of the down accelerometer scale factor, one has to add on the right hand side of (9.c) the term $k_D \cdot f_D$ where k_D is the scale factor error). A time varying f_D is needed to separate the bias from the scale factor. In a strapdown INS, though, when the parent vehicle rolls, the INS vertical rotates with respect to the local vertical and the down accelerometer measures a changing specific force. Consequently the down accelerometer scale factor error is separable from the bias.

VI. OBSERVABILITY CONSIDERATIONS

The issue of AAR and IFA, and the difference between them can be examined from a purely theoretical point of view as an observability problem as defined in linear control theory [13], [10], [14], [6]. Adopting this approach, we first investigate the observability qualities of INS at AAR.

As mentioned in Section II, the relevant model for AAR is that of (5) where the 3rd and 9th rows in (5) and the 3rd and 5th columns in (6) are eliminated since, as seen in (6), the vertical velocity error is not influences by the misalignment angles, and its model is decoupled from the rest of the error model altogether. Then, if we denote the AAR error model as follows:

$$\dot{\underline{x}}^* = A^*\underline{x}^* + \underline{n}^* \qquad (17)$$

The components of the AAR state vector, $\underline{x}^*$, are:

$$\underline{x}^{*T} = [V_N, V_E, \psi_N, \psi_E, \psi_D, B_N, B_E, D_N, D_E, D_D] \qquad (18)$$

and the noise vector $\underline{n}^*$ is:

$$\underline{n}^{*T} = [n_{BN}, n_{BE}, n_{DN}, n_{DE}, n_{DD}, 0, 0, 0, 0, 0]$$

the dynamics matrix, A^*, is:

$$A^* = \left[\begin{array}{cc|ccc|cc|ccc} 0 & 2W_D & 0 & g & 0 & 1 & 0 & & 0 & \\ -2W_D & 0 & -g & 0 & 0 & 0 & 1 & & & \\ \hline & & 0 & W_D & 0 & & & 1 & 0 & 0 \\ & 0 & -W_D & 0 & W_N & & 0 & 0 & 1 & 0 \\ & & 0 & -W_N & 0 & & & 0 & 0 & 1 \\ \hline & 0 & & 0 & & & 0 & & 0 & \\ \hline & 0 & & 0 & & & 0 & & 0 & \end{array}\right] \tag{19}$$

The measurement model for the AAR KF is:

$$\underline{z}^* = H^*\underline{x}^* + \underline{u}^* \tag{20}$$

where $\underline{u}^*$ is a zero mean white measurement error and H^* is the following measurement matrix:

$$H^* = \left[\begin{array}{cc|cccc} 1 & 0 & 0 & 0 & \dots & 0 \\ 0 & 1 & 0 & 0 & \dots & 0 \end{array}\right] \tag{21}$$

The problem of estimating $\underline{x}^*$ is a stochastic one. Therefore, the suitable observability criterion which should be applied in this case is that of stochastic observability. Although the definition of stochastic observability differs from that of the deterministic observability, the algebraic criteria for their existence are identical [15,16]. Consequently the observability study of this linear model of (17) and (20) is carried out using the following steps [17]:

a. Compute the Observability matrix Q:

$$Q^T = [H^{*T} \mid (H^*A^*)^T \mid \ldots\ldots \mid (H^*A^{*\,n-1})^T]$$

where n is the order of the system which in this case is 10.

b. Determine r, the rank of Q.

c. Choose r independent rows in Q and use them to form, M_1, the upper part of a matrix M.

d. Guess p (p=n-r) n-dimensional rows, independent of the rows of M_u and of one another.

e. Use there rows to form, M_2, the lower part of M.

f. Transform the original system to the Observable Canonical form [17, p.74] as follows:

$$\underline{y} = M\underline{x}^*$$

$$\dot{\underline{y}} = L\underline{y}$$

$$\underline{z}^* = H_1\underline{y}$$

where

$$L = MA^*M^{-1}$$

$$H_1 = H^*M^{-1}$$

and $\underline{y}^T = [\underline{y}_1^T \mid \underline{y}_2^T]$ satisfies the equation:

$$\begin{bmatrix} \underline{y}_1 \\ \cdots \\ \underline{y}_2 \end{bmatrix} = \begin{bmatrix} L_{11} & 0 \\ \cdots & \cdots \\ L_{21} & L_{22} \end{bmatrix} \begin{bmatrix} \underline{y}_1 \\ \cdots \\ \underline{y}_2 \end{bmatrix}$$

g. Inspect the relationship between $\underline{y}_1$ (the ***observable*** part of $\underline{y}$) and states of $\underline{x}$, and the relationship between $\underline{y}_2$ (the ***unobservable*** part of $\underline{y}$) and the states of $\underline{x}$.

h. Draw your conclusions on the observability of the system.

When the above eight step procedure is applied to the AAR model we find that r, the rank of the 10th order model, is 7. That is, there are only seven observed ***modes*** in this system. In the $\underline{y}$ domain, one can easily see which are the observed variables and which are not, but, unfortunately, the y-variables are not necessarily physical variables. In the $\underline{x}$ domain, though, the variables are the physical variables listed in (18) but they cannot be divided into observed and unobserved variables. The states are rather a linear combination of the observed and unobserved modes. Therefore the degree of observability of a state is determined by the strength of the presence of the observed modes in the construction of that state.

It should be noted that since the transformation matrix, M, is not unique (see steps c and d in the foregoing procedure), the $\underline{y}$ vector is not unique either, consequently the designation of the observed and the unobserved modes is not unique.

The observability analysis of the INS errors during IFA is more elaborate. During IFA, the parent vehicle maneuvers in order to generate specific forces. These specific forces change in time with the maneuver. From (10) it is obvious that the dynamics matrix of the INS error model is, then, time varying. The observability analysis of time varying systems is well known, but is complicated to perform and is not too informative. Consequently, a direct observability analysis during IFA involves tedious mathematical work with little constructive conclusions. However, it can be shown that instead of examining the continuously varying system describing the true behavior of the INS errors during IFA, it is possible to examine an equivalent piece-wise constant system because its KF yields state estimates which are similar to the estimates obtained by the KF of the time varying INS. The piece-wise constant system is obtained when we assume that a piece-wise constant specific force is generated by the parent vehicle while flying a certain trajectory [14, 6]. Although this trajectory is never performed by the parent vehicle, it is a very good approximation of a typical IFA maneuver and yields similar results. The trajectory consists of the 5 possible segments listed in Table I. Using trajectories which consist of

Table I: Trajectory Definition

Number of Segment	Maneuver Characteristics	F_N	F_E	F_D
1	straight and level flight (SLF)	0	0	-g
2	SLF with north acceleration	F_N	0	-g
3	SLF with east acceleration	0	F_E	-g
4	SLF with horizontal acceleration	F_N	F_E	-g
5	pull up or dive	0	0	F_D

segments defined in the table, and a modified observability matrix which takes in account the switches between segments, the following can be shown [14] for the 2-channel 10-state INS model given in (17) - (21):

- o The observability of the 2-channel INS can be improved by an IFA trajectory which consists of only ***two*** segments in which constant horizontal specific force, which differs from segment to segment, is experienced by the system.

- o The order in which the segments are flown is immaterial.

- o The number of unobservable modes is 2 out of the total 10 modes of the system.

- o The 2 unobservable modes are present in 7 of the system states to a degree which practically render these states unobservable.

- o To estimate the entire state vector, 2 additional states have to be measured. They have to be chosen out of 2 groups of states, each selected from a different group. One group includes the states ψ_N, B_E and D_E. The other group includes the states ψ_E, B_N, D_N and D_D.

The observability analysis of INS at IFA is carried on using the 3-channel 12-state dynamics model presented in Section III. The measurement matrix for this case is:

$$H = \begin{bmatrix} \begin{vmatrix} 1 & 0 & 0 \\ 0 & 1 & 0 \\ 0 & 0 & 1 \end{vmatrix} & 0 \end{bmatrix}$$

The conclusions drawn from the latter analysis are as follows [6]:

- o For any single segment of the IFA trajectory, the rank of the observability matrix is 9. Consequently 9 modes can be estimated at any single segment.

- o An IFA trajectory which consists of 2 segments will render

11 observable modes.

o An IFA trajectory which consists of 3 segments will render a completely observable system, consequently ***all*** the states of the original (untransformed) system are observable.

From the preceding analysis we can see now the advantage of IFA over AAR expressed in terms of observability measures. At AAR the number of unobserved modes is 3 out of the 10 system modes. IFA maneuvers increase the degree of observability of the INS such that the number of unobserved modes is reduced to 2. In addition to exposing the difference between AAR and IFA, the preceding analysis reveals the advantage of aligning in-flight an undamped INS. The latter becomes totally observable at the end of the IFA process. We conclude therefore that when the major sensor error sources are accelerometer bias and gyro constant drift rate, even if the INS vertical channel is damped, it is advantageous to deactivate the damping mechanism during the IFA stage of the INS operation. Then, being, temporarily, a 3-channel INS, all the error states are observable and thus estimable. That is, the undamped system is calibrated during the IFA and then the damping mechanism can be re-activated for the rest of the INS operation time.

References

1. I.Y. Bar-Itzhack and N. Berman, "Control Theoretic Approach to Inertial Navigation Systems," Journal of Guidance Control and Dynamics, 11, pp. 237-245 (1988).

2. J.C. Pinson, "Inertial Guidance of Cruise Vehicles," in "Guidance and Control of Aerospace Vehicles," (C.T. Leondes ed.), McGraw-Hill, New York, 1963.

3. F.D. Jurenka and C.T. Leondes, "Optimum Alignment of an Inertial Autonavigator," IEEE Transactions on Aerospace and Electronic Systems, 3, pp. 880-888 (1967).

4. S.F. Schmidt, J.D. Weinberg and J.S. Lukesh, "Application of Kalman Filtering to the C-5 Guidance and Control System," in "Theory and Applications of Kalman Filtering," AGARD, Paris, France, AD 704 306, 1970.

5. B. Danik, "Gyrocompass Alignment of Inertial Platforms," The Singer Company Kearfott Division, Engineering Technical Report KD-73-9, 1973.

6. D. Goshen-Meskin and I.Y. Bar-Itzhack, "Observability Studies of Inertial Navigation Systems," AIAA Guidance Navigation and Control Conference, Boston, MA, paper no. 89-3580, August 14-16, 1989.

7. J. Baziw and C.T. Leondes, "In - Flight Alignment and Calibration of Inertial Measurement Units - Part I: General Formulation," IEEE Transactions on Aerospace and Electronic Systems, 8, pp. 440-449 (1972).

8. D.O. Benson, Jr., "A Comparison of Two Approaches to Pure-Inertial and Doppler-Inertial Error Analysis, "IEEE Transactions on Aerospace and Electronic Systems, 11, pp. 447-455 (1972).

9. I.Y. Bar-Itzhack and E.F. Malove, "Accurate INS Transfer Alignment Using A Monitor Gyro and External Navigation Measurements," IEEE Transactions on Aerospace and Electronic Systems, 16, pp. 53-65 (1980).

10. A.A. Sutherland Jr., "The Kalman Filter in Transfer Alignment of Inertial Guidance Systems," Journal of Spacecraft and Rockets, 5, pp. 1175-1180 (1968).

11. B. Porat and I.Y. Bar-Itzhack, "Effect of Acceleration Switching During INS In-Flight Alignment," Journal of Guidance and Control, 4, pp. 385-389 (1981).

12. I.Y. Bar-Itzhack and B. Porat, "Azimuth observability Enhancement During Inertial Navigation System In-Flight Alignment," Journal of Guidance and Control, 3, pp. 337-344 (1981).

13. W. Kortum, "Design and Analysis of Low-Order Filters Applied to the Alignment of Inertial Platforms," in "Practical Aspects of Kalman Filtering Implementation," AGARD Neuilly-sur-Saine, France, AD A 024 377, 1976.

14. D. Goshen-Meskin and I.Y. Bar-Itzhack, "Observability Analysis of Inertial Navigation Systems During In-Flight Alignment," AIAA Guidance, Navigation and Control Conference, Minneapolis, MN, paper no. 88-4125, August 15-17, 1988.

15. A.E. Bryson, Jr. and Y.C. Ho, "Applied Optimal Control," Hemisphere Publishing Corp., Washington, D.C., 1975, pp. 369, 370.

16. A. Gelb, ed., "Applied Optimal Estimation," M.I.T. Cambridge, MA, 1986, pp. 131, 132.

17. H. Kwakernaak and R. Sivan, "Linear Optimal Control Systems," Wiley-Interscience, New York, 1972, pp. 67 - 79.

INDEX

A

Abstraction, 44–46
Acceleration
  dynamics, 292
  orientation dynamics, 293
Acceleration state, 287
Adams–Bashforth integrator, two-step, 254–256
  computation cost, 255
Adams–Bashforth method, real-time simulation, 254
Adaptive integration, nonlinear system, 247–248
  block diagram, 248
Adaptive matrix integration
  implementation, 263–265
  integration coefficient computation, 264–265
  Jacobian identification, 263–264
Aerodynamics, nonlinear mathematical model, 7
Aerospace system designer
  end-user psychology theories, 68
  human-related issues, 68
  risk discounting, 68
  user role playing, 68
Aggregation, 44–46
Aiding and adapative interface, 80
Air-to-air interception
  feedback guidance law, 153–202
    accuracy, 190–191
  horizontal plane, 162–176
    approximate solution, 174
    dynamic equations, 162–163
    exact solution, 174
    horizontal guidance law synthesis, 168–172
    medium-range, 174
    modeling considerations, 166–168
    numerical examples, 173–176
    optimal control formulation, 163–166
    problem formulation, 162–166
    reduced order solution, 170
    short range, 176
    singular perturbation analysis, 166–173
  mathematical model, 155–158
  medium-range, 153–202
    defined, 155
  multiple time-scale model, 153–202
  optimal control formulation, 158–160
  phases, 153
  problem formulation, 155–160
  singular perturbation theory, 153–202
    application, 160–162
  three-dimensional, 192–201
    approximate solution, 198
    exact solution, 198
    guidance law synthesis, 193–196
    load factor synthesis, 196
    medium-range, 198
    modeling considerations, 192–193
    numerical examples, 196–201
  vertical plane, 177–191
    accuracy assessment, 190–191
    approximate solution, 190
    basic guidance law synthesis, 182–186
    equations of motion, 177–178
    guidance law improvements, 186–189
    interception time solution, 190
    medium-range, 190
    modeling considerations, 180–182
    optimal control formulation, 178–180
    problem formulation, 177–180
    singular perturbation analysis, 180–191
Aircraft attitude, 23

Aircraft design, nonlinear features, 5
Aircraft mathematical model, vertical takeoff and landing aircraft, 11–13
Aircraft trajectory, singular perturbation theory, 161
Aircrew system design, 58, 59
  model problem, 60
Alignment
  coarse, 372
  inertial navigation system, 369
  Kalman filter
    misalignment angle estimation, 374
    torquing, 374
Alignment at rest, inertial navigation system, 370–378
  earth angular velocity, 370
  earth gravity vector, 370
  reference coordinate, 370
Ames control concept, 9, 10
Angle aspect Kalman filter, 277
Angle of attack, 25, 26
  lift curve slope reversal, 28
  trim map, 7
Angular velocity vector, skew symmetric matrix function, 36–37
Armature current, 244
Augmented extended Kalman filter, 298–299
Augmented lift, 6
Automatic control, 1–38, 5, 62
  nomenclature, 2–5

B

Bell Aircraft UH-1H helicopter, inverse-model-follower control system, 8, 29
Brennan and Leake turbojet engine model
  converged linearized estimated model, 252
  estimated linearized model, eigenvalues plot, 253
  on-line adaptive integration, 252
  three-step Matrix Stability Region Placement simulation, 233–236
Brownian motion process, 274

C

CAD, 97–98
Carrier deck motion, simulation, 7
Closed-loop linear discrete-time system, 225
Closed-loop regulator design, inverse-model-follower control system, 6
Collaborative control, 63
Command generator, 21
  feedback loops, 32
  gains, 32
  parallel channels, 31, 32
  three channel integrators, 32, 33
  three-dimensional flow diagram, 32, 33
Command section, 9–11
  design, 10, 30
  rough trajectory command, 11
  smooth acceleration, 11
  trajectory acceleration, 11
Commanded trajectory acceleration, inverse-model-follower control system, 5
Computer-aided Acquisition and Logistics Support, 84
Constant specific energy, time-optimal zoom interception, 206–210
Continuity equation, 109–110
Continuous forcing term, 274
Continuous time parameter estimation, 307–365
  convergence, 310
Control section, 11–13
Conventional feedback control, inverse-model-follower control system, 6
Converged linearized estimated model, Brennan and Leake
  turbojet engine model, 252
Cost-benefit, information system, 54
Cross-disciplinary information seeking, human-machine system design, 58–59

D

Damped wave equation, 113–114
Data base, human-machine system design, 83–86
  content problems, 85
  growth rate, 83–84
  information retrieval, 85
  size, 83–84
DC motor, impulse response, 245
Deconvolution, 124
  algorithm application, 140–147
  filter tuning, 136–142
  technique, 101–149
Design
  questions/answers metaphor, 43–44

theory, 44–46
Design and evaluation tool, 74, 97–100
application, 74
biomechanical CAD, 97–98
CAD, 97–98
explanation, 81
functionality, 74
general model building package, 97
overhead tasks, 77
problem-independent design support, 75–77
procedure execution, 77
specialized modeling package, 97
support distribution, 75, 76
tutoring, 81
Design information system, 84–85
Design process, distributed nature, 41
Design space
abstraction dimension, 44–46
aggregation dimension, 44–46
task dimension, 45, 46
Design support, 41–91, 97–100
adaptive aiding, 56
architecture, 54–57
cluster support concepts, 51–52
clustering requirements, 50
define tasks, 47
designer verbs, 51, 53
error monitoring, 56
functional analysis, 52
information management, 56
interface, 56
map to general tasks, 47–49
map to limitations, 49
map to support concepts, 50–51
modifying objects, 51, 53
primary objects, 51, 53
psychological limitations, 49
requirement analysis, 49–50
requirements, 47–53
support verbs, 51, 52
task limitation, 49, 50
Designer
decision-maker, 72
decision making, 42
defined, 41
Designers' associate, 42
DHC-6 Twin Otter aircraft, inverse-model-follower control system, 29
Diagonalization, 256–259
Jordan reduction of J, 259
triangular approach, 259
Differential equation, 307–365
deterministic ordinary, asymptotic behavior, 307
limiting ordinary, 307–365
ordinary, 307
stiffness, 211
Differential equation model, stochastic forcing term, 274
Discontinuous acceleration, target trajectory, 284
Discontinuous forcing term, 274
Display construction, human-machine system design, 98
Display/control design, 79
Display/control prototyping tool, human-machine system design, 98
Drag coefficient, trim map, 7

E

Earth angular velocity, 370
Earth gravity vector, 370
Earth rate, 370
Eigenvalue
Jordan co-ordinate, origin, 238
least negative, 212
stiffness, 211
timestep, 211
widespread, 212
Electro-optical sensor, 273, 276
optical transfer function, 281
tracking, 282–284
Engine throttle, trim map, 7
Estimated linearized model, Brennan and Leake turbojet engine model, eigenvalues plot, 253
Estimation algorithm, imager, 278
target orientation, 278
Explicit linear multistep integration technique
real-time simulation, 212–266
vector systems, 212–266
Explicit method, real-time simulation, 212
Extended least squares algorithm, 310

F

Feed-forward controller, 9, 10
control section, 13
Feedback control loop, inverse-model-follower control system, 6

Feedback guidance law, air-to-air interception, 153–202
accuracy, 190–191
Filter tuning, 136–140
Flight alignment, intertial navigation system, 369–370, 379–394
advantages, 386
observability problem, 388–394
optimal maneuvers, 386–388
rationale, 370
Forced singular perturbation technique, 161
application, 161
zeroth-order feedback control approximation, 161–162
Forecasting tool, 79–80
Forward flight, hover, transition run, 24–27
Forward-looking infrared sensor, 276
Function allocation, 79
Function requirements identification and analysis, 80

G

Gain scheduling, 247
Gauss–Markov model, random target path, 284
Gauss–Markov process, 273
General modeling package, 69
Guidance, image interpretation, 273–303
Guidance law synthesis, vertical plane, 182–189
Gyro measurement, 370
Gyrocompassing, 378

H

Horizontal guidance law synthesis, 168–172
Horizontal plane, air-to-air interception, 162–176
approximate solution, 174
dynamic equations, 162–163
exact solution, 174
horizontal guidance law synthesis, 168–172
medium-range, 174
modeling considerations, 166–168
numerical examples, 173–176
optimal control formulation, 163–166
problem formulation, 162–166
reduced order solution, 170
short range, 176
singular perturbation analysis, 166–173
Hover, forward flight, transition run, 24–27
Human engineering
failure to utilize, 67
organizational barriers, 66
quality, 66
Human engineering standards electronic retrieval, human machine system design, 99
Human-machine interaction paradigm, 62
Human-machine system design, 57–70
adapative aiding, 73
analysis method, 59
cross-disciplinary information seeking, 58–59
data base, 83–86
content problems, 85
growth rate, 83–84
information retrieval, 85
size, 83–84
design contributors, 72
display construction, 98
display/control prototyping tool, 98
error monitoring, 73
execute function, 87
explain function, 87
human engineering standards electronic retrieval, 99
human information, 71
human workload evaluation tool, 100
indicate function, 87
information management, 73
information transfer, 58
inputs, 63–65
designer limitations, 65–70
interface function, 87–88
location, 72
macro properties, 60
manpower cost spreadsheet, 99
micro attributes, 60
multi-disciplinary, 58
nature, 60–63
outputs, 63–65
designer limitations, 65–70
parameters, 72
personnel cost spreadsheet, 99
problems, 63–65
designer limitations, 65–70
proprietary information, 88

quality characteristics, 70
responsibility, 71
search function, 86
selection motivation, 57–60
support system objectives, 70–73
supporting, 70–82
assessment, 80–82
human-machine problem areas, 77–80
problem-independent, 75–77
state of the art review, 74
tasks, 63–65
designer limitations, 65–70
technological advances requiring, 57
technology transition, 88–90
timeline construction and analysis, 98–99
training cost spreadsheet, 99
transform function, 87
tutor function, 87
Human-machine system design tool, 74, 97–100
application, 74
biomechanical CAD, 97–98
CAD, 97–98
explanation, 81
functionality, 74
general model building package, 97
overhead tasks, 77
problem-independent decision support, 75–77
procedure execution, 77
specialized modeling packages, 97–98
support distribution, 75, 76
tutoring, 81
Human performance model, system designer, 69
Human pilot, rough trajectory command input, 29
Human-system evaluation, 77–78
Human workload evaluation tool, human-machine system design, 100

I

Image interpretation
guidance, 273–303
tracking, 273–303
Image interpretation algorithm, 285
Imager, estimation algorithm, 278
target orientation, 278
Imaging sensor, 279
Implicit integration technique, real-time simulation, 212
Impulse response, DC motor, 245
In-flight alignment, inertial navigation system, undamped error, 379
In situ transducer
operational difficulties, 102
survivability, 102
transducer calibration scale factor, 102
Inertial navigation system
alignment, 369
alignment at rest, 370–378
earth angular velocity, 370
earth gravity vector, 370
reference coordinate, 370
in-flight alignment, undamped error, 379
initialization, 369
Kalman filter, initial alignment, 374–375
misalignment, 373
self alignment, 374
transfer alignment, 383–386
Information
categorized, 43
generated, 43
input, 43
output, 43
usability, 69
usefulness, 69
value, 68, 69
Information environment, 54, 55
Information seeking and utilization, 43
Information system, 41–91, 97–100
cost-benefit, 54
design process support, 41–91, 97–100
Information transfer, human-machine system design, 58
Initialization, inertial navigation system, 369
Input Jacobian G, 246
Integrated services digital network, 84
Integration, 256–259
Jordan reduction of J, 259
triangular approach, 259
Integration coefficient computation, adaptive matrix
integration, 264–265
Integration operator, 215
Integrator character, 215
Intelligent systems technology, 62
Interactive multiple model, 285
Interception geometry, three-dimensional, 156

Intertial navigation system, flight alignment, 369–370, 379–394
advantages, 386
observability problem, 388–394
optimal maneuvers, 386–388
rationale, 370
Inverse-model-follower control system, 1–38
Bell Aircraft UH-1H helicopter, 8, 29
closed-loop regulator design, 6
commanded trajectory acceleration, 5
conventional feedback control, 6
decoupled dynamics, 29
DHC-6 Twin Otter aircraft, 29
feedback control loop, 6
historical background, 6–8
linear feedback control theory, 6
linearized dynamics, 29
Newton–Raphson inversion scheme
angular velocity vector skew symmetric matrix function, 36–37
command section, 9–11
control section, 11–13
coordinate systems, 35–36
direction cosine matrix derivative, 37
feed-forward controller, 9, 10
forward flight-hover transition, 24–27
model inversion process, 13–20
ongoing research, 27–28
rotational dynamic equation matrix form, 37–38
simulation results, 20–27
system concept, 8–11
trim process failure, 27
turning, climbing, accelerating trajectory, 27
vector cross product, 37
vertical attitude maneuver, 21–24
open-loop feed-forward control, 6
Quiet Short Haul Research Aircraft, 29
short takeoff and landing research, 6, 29
nonlinear aerodynamic effects, 6
vertical takeoff and landing aircraft, 7
Newton–Raphson approach, 7, 8–38

J

Jacobian identification, adaptive matrix integration, 263–264
Jacobian matrix J, 246
Jordan co-ordinate, 228
eigenvalue, origin, 238
Jordan reduction of J, 259

K

Kalman filter
alignment
misalignment angle estimation, 374
torquing, 374
augmented, 298–299
inertial navigation system, initial alignment, 374–375
parallel, 284
Kalman filter paradigm, 284
Kalman filter problem, 308

L

Lift coefficient, trim map, 7
Lift-curve slope reversal, 24
angle of attack, 28
Linear control theory, observability problem, 388
Linear feedback control theory, inverse-model-follower control system, 6
Linear-Gauss-Markov estimation/prediction paradigm, 275
Linear system
multistep matrix integrator, J singular, 236–246
p-step matrix integrator, 225–236
Local-area network, 84

M

Maneuver probability vector, 289
Maneuvering target, tracking, 284–298
acceleration estimators, 285–290
encounter model, 285–290
Manpower cost spreadsheet, human-machine system design, 99
Many-to-many map, 277
Many-to-one map, 277
Matrix Stability Region Placement, 213
nonlinear systems, 213
three-step method, Brennan and Leake turbojet engine, 233–236
two-step, 254–256
computation cost, 255
off-line computational cost, 262

on-line computational cost, 262
two-step integrator, implementation, 256–263
two-step matrix integrator, 214–224
block diagram implementation, 219
Matrix system, upper triangular form, 232
Measurement model, point target tracking, 280
Military system design, human user, 66
Minimum-time optimal control problem, 153
Minimum variance deconvolution, 125–147
assumed model form, 125–126
Minimum variance deconvolution algorithm, 134–135
Minimum variance deconvolution technique, 101–149
Misalignment error, 373–378
Model building package, 97
Model estimation algorithm, 289
Model inversion, 1–38
nomenclature, 2–5
Model inversion process, 13–20
Newton–Raphson, 13–20
trim procedure, 15
perturbation procedure, 16
rotation matrix, 15
trim variables, 16
Momentum equation, 111–112
Motion model
deterministic drift, 274
random perturbation, 274
Multiple model adapative filter, 284–285
Multiple time-scale model, air-to-air interception, 153–202
Multistep matrix integrator
linear system, J singular, 236–246
nonlinear system, 246–253
input Jacobian G, 246
Jacobian matrix J, 246
recursive identification of J, 246–247
repeated linearization, 246–247
single linearization, 246–247
system matrix J, 246
real-time simulation, 211–266
computational aspects, 254
implementation, 256

N

Navier–Stokes equation of continuity, 109–110
Navier–Stokes momentum equation, 111–112
Newton–Raphson inversion scheme, inverse-model-follower control system
angular velocity vector skew symmetric matrix function, 36–37
command section, 9–11
control section, 11–13
coordinate systems, 35–36
direction cosine matrix derivative, 37
feed-forward controller, 9, 10
forward flight-hover transition, 24–27
model inversion process, 13–20
ongoing research, 27–28
rotational dynamic equation matrix form, 37–38
simulation results, 20–27
system concept, 8–11
trim process failure, 27
turning, climbing, accelerating trajectory, 27
vector cross product, 37
vertical attitude maneuver, 21–24
Nonlinear model, orthogonal projection algorithm, 251
eigenvalues, 251
stepsize, 251
Nonlinear system
adaptive integration, 247–248
block diagram, 248
multistep matrix integrator, 246–253
input Jacobian G, 246
Jacobian matrix J, 246
recursive identification of J, 246–247
repeated linearization, 246–247
single linearization, 246–247
system matrix J, 246
Nonmaneuvering target, tracking, 279–284
Numerical integration, 215

O

On-line adaptive integration, Brennan and Leake turbojet engine model, 252
Open-loop feed-forward control, inverse-model-follower control system, 6
Optical transfer function, electro-optical sensor, 281
Optimal control solution, 154
feedback approximation, 154
Ordinary differential equation, 307
advantages, 324

Ordinary differential equation (*continued*)
continuous time, 309
discrete time, 309
invariant measure approach, 309
numerical examples, 324–328
recursive prediction error method, 307, 319
calculations, 324
mismodeling, 326
true parameter value, 326
Wiberg Estimator, 307
Orientation dynamics
acceleration, 293
radial acceleration, 286
Orthogonal projection algorithm, 249
nonlinear model, 251
eigenvalues, 251
stepsize, 251

P

P-step matrix integrator
implementation, 230
linear system, 225–236
P-step Matrix Stability Region Placement integrator
block diagram implementation, 226
difference equation, 225
P-step method, real-time simulation
additional roots, 213
principal roots, 213
spurious roots, 213
Parallel observation path, 277
Perpendicular axis system, 25, 26
Perpendicular inertial-axis reference system, 28
Perpendicular system, 17
Personnel cost spreadsheet, human-machine system design, 99
Perturbation procedure, 16
Pitch angle, 25, 26
Pitch channel signal, 22, 23
Pneumatic distortion, pressure sensing device, 101–149
causes, 101
Pneumatic distortion model, 102–103, 107–123
compensation algorithm derivation, 124–135
continuity equation, 109–110
damped wave equation, 113–114
deconvolution, 124
deconvolution algorithm application, 140–147
deconvolution analysis results, 135–147
discrete state variable model expression, 122–124
downstream boundary condition, 115
equation of state, 112
filter tuning, 136–140
flight maneuver time history comparison, 144–147
frequency response comparisons, 120–121, 121–123
governing equations reduced to wave model, 113–114
idealized configuration geometry, 108–109
initial condition, 114
mathematical model derivation, 107–124
minimum variance deconvolution, 125–147
assumed model form, 125–126
minimum variance deconvolution algorithm, 134–135
minimum variance estimation fundamental lemma, 126–127
momentum equation, 111–112
Navier–Stokes equation of continuity, 109–110
Navier–Stokes momentum equation, 111–112
numerical solutions, 115–119
overdamped sensor configuration, 120–121
post-flight smoothing algorithm, 131–132
real time filtering algorithm, 132–134
residual error recursive equation derivation, 127–131
underdamped sensor configuration, 121–123
upstream boundary condition, 114
wave model reduced to second order filter, 119–122
Pneumatic tubing, 101–102, 107
Point-target sensor, 279
Point target tracking, 279–282
measurement model, 280
observation model, 280–282
Pole placement constraint, 227
Polytropic process, 112
Prediction
target location, 300, 301

tracking, 298–302
Pressure response
acoustic resonance, 107
frictional attenuation, 101, 107
magnitude attenuation, 107
phase lag, 107
pneumatic resonance, 101, 107
spectral attenuation, 107
zero volume configuration, 107
Pressure sensing device
pneumatic distortion, 101–149
causes, 101
schematic, 110
in situ mounting, 102
Pressure sensing technology
laser/Doppler, 102
piezo-film, 102
Pressure transducer, 101–102, 107
Procedures design, 79
Projection error, 291–292, 297
Proprietary information, human-machine system design, 88
Propulsion, nonlinear mathematical model, 7
Protection and life support system design, 80

Q

Quantization error, 297
Quiet Short Haul Research Aircraft, inverse-model-follower control system, 29

R

Radar, 273
Radial acceleration, 292
orientation dynamics, 286
Random target path, Gauss-Markov model, 284
Real-time simulation
Adams–Bashforth method, 254
explicit linear multistep integration technique, 212–266
explicit method, 212
hardware constraints, 211
implicit integration technique, 212
multistep matrix integrator, 211–266
computational aspects, 254
implementation, 256
p-step method
additional roots, 213
principal roots, 213
spurious roots, 213
stiff systems, 211–266
timestep, size, 211
Recursive identification of J, 246–247
Recursive least squares algorithm
covariance resetting, 249
selective data weighting, 249
Recursive prediction error method
continuous time, convergence, 319–324
continuous time analogues, 319
convergence, 307
discrete time, 310
ordinary differential equation, 307, 319
calculations, 324
mismodeling, 326
true parameter value, 326
state space model, 319
state estimation error, 319
true parameter values, 319
Reduced-order trajectory, vertical plane, 184
Reference flight trajectory, 201–202
Regulator section, design, 10, 30
Repeated linearization, 246–247
Risk discounting, 68
Role playing, 68
Rotational change, 287
Rotational command generator, trim procedure, 18
Rotor angular displacement, 244
Rotor angular velocity, 244
Rough horizontal acceleration command, 21
Rough trajectory command input, human pilot, 29
Rough trajectory input section, 30–31
Runway reference-axis, 25, 26

S

Self alignment, inertial navigation system, 374
Semiautomatic control, 5
Sensor-image processor block, 287
Sensor-image processor error, 288
Sensor-preprocessor, 274
Sensor-processor error, 295–298
Short takeoff and landing research, inverse-model-follower control system, 6, 29
nonlinear aerodynamic effects, 6

Signal interpretation, 273
Simulation
  carrier deck motion, 7
  wind turbulence, 7
Single linearization, 246–247
Singular perturbation theory
  air-to-air interception, 153–202
    application, 160–162
  aircraft trajectory, 161
Skew symmetric matrix function, angular velocity vector, 36–37
Smart tracker, 273
Smooth command, 21
Software engineering
  common sense psychology, 67
  designers' beliefs about users, 67
  user knowledge theories, 67
Stability region placement, 213
State space model, recursive prediction error method, 319
  state estimation error, 319
  true parameter values, 319
Steady-state cruise, 153
Step deceleration command, 24
Stochastic differential equation, 307
  Euler's method, 355
  modified Huen method, 355
  numerical simulation, 355–365
  Runge-Kutta techniques, 355
  4th order Runge-Kutta, 355
Stochastic parameter estimator, asymptotic behavior, 307
System designer, human performance model, 69
System matrix J, 246
System transient response, 215

T

Target holofeature, 277, 278
Target location, prediction, 300, 301
Target motion, 275
Target motion model, 273
Target trajectory, discontinuous acceleration, 284
Target velocity, 300, 301
Task limitation, 49, 50
Technology transition, human-machine system design, 88–90
Three-dimensional interception geometry, 156
Time-optimal zoom interception, constant specific energy, 206–210
Timeline construction and analysis, human-machine system design, 98–99
Timestep
  eigenvalue, 211
  real-time simulation, size, 211
Toeplitz structure, 241
Tracker architecture, 273, 298
  hybrid, 298
Tracking
  electro-optical sensor, 282–284
  image interpretation, 273–303
  long range, 279
  maneuvering target, 284–298
    acceleration estimators, 285–290
    encounter model, 285–290
  nonmaneuvering target, 279–284
  prediction, 298–302
Tracking algorithm, 273
Training cost spreadsheet, human-machine system design, 99
Trajectory control system, 5
Trajectory input section, 30
Transducer calibration scale factor, 102
Transfer alignment, 379
  block diagram representation, 384
  inertial navigation system, 383–386
Transition rate matrix, 288
Translational command generator, 19, 31–34
  trim procedure, 18
Translational regulator, 19, 34
Trim
  four-degree-of-freedom, 18
  six-degree-of-freedom, 18
Trim map
  angle of attack, 7
  drag coefficient, 7
  engine throttle, 7
  lift coefficient, 7
Trim procedure, 16–20
  control section, 18, 19
  error equations, 16–17
  model inversion process, 15
    perturbation procedure, 16
    rotation matrix, 15
    trim variables, 16
  perpendicular system, 17
  rotational command generator, 18
  translational command generator, 18
  vertical attitude, 17
Trim process failure, 27

Trim variable, 16, 18
Turn-climb-acceleration phase, 153
Turning, climbing, accelerating trajectory, 27
Two-step matrix integrator, vector system, 215–224
aircraft mathematical model, 11–13
design, 8–9
inverse-model-follower control system, 7
Newton–Raphson approach, 7, 8–38
Vertical velocity error, 372

V

Vector system
explicit linear multistep integration technique, 212–266
two-step matrix integrator, 215–224
Velocity error, 300, 301, 302, 303
Velocity error component, 373
Velocity estimate, 300
Vertical plane
air-to-air interception, 177–191
accuracy assessment, 190–191
approximate solution, 190
basic guidance law synthesis, 182–186
equations of motion, 177–178
guidance law improvements, 186–189
interception time solution, 190
medium-range, 190
modeling considerations, 180–182
optimal control formulation, 178–180
problem formulation, 177–180
singular perturbation analysis, 180–191
reduced-order trajectory, 184
Vertical takeoff and landing aircraft

W

Weak convergence theorem
mathematical background, 330–332
proof, 333–355
statement, 310–318
White-noise acceleration model, 275
Wiberg Estimator, ordinary differential equation, 307
Wide-area network, 84
Wind gust, 28
Wind turbulence, simulation, 7
Workspace layout, 77–78

Z

Zero mean white noise, 372
Zero-order-hold method, generalization, 225
Zeroth-order feedback control approximation, forced singular perturbation technique, 161–162
Zoom interception, 206–210